*Friedrich Siebert and
Peter Hildebrandt*
**Vibrational Spectroscopy
in Life Science**

Related Titles

Heimburg, T.

Thermal Biophysics of Membranes

approx. 330 pages
2007
Hardcover
ISBN: 978-3-527-40471-1

Wartewig, S.

IR and Raman Spectroscopy

Fundamental Processing

191 pages with 177 figures and 4 tables
2003
Hardcover
ISBN: 978-3-527-30245-1

Chalmers, J., Griffiths, P. (Eds.)

Handbook of Vibrational Spectroscopy

3880 pages in 5 volumes
2002
Hardcover
ISBN: 978-0-471-98847-2

Friedrich Siebert and Peter Hildebrandt

Vibrational Spectroscopy in Life Science

WILEY-VCH Verlag GmbH & Co. KGaA

The Authors

Friedrich Siebert
Institut für Molekulare Medizin und
Zellforschung
Albert-Ludwigs-Universität Freiburg
e-mail: frisi@biophysik.uni-freiburg.de

Peter Hildebrandt
Institut für Chemie
Technische Universität Berlin
e-mail: hildebrandt@chem.tu-berlin.de

Cover Picture
Vibrational spectroscopy, i.e. Raman (bottom) and infrared (top) spectroscopy, has considerably contributed to the understanding of the function of proteins, here of the light-driven proton pump bacteriorhodopsin.

■ All books published by Wiley-VCH are carefully produced. Nevertheless, authors, editors, and publisher do not warrant the information contained in these books, including this book, to be free of errors. Readers are advised to keep in mind that statements, data, illustrations, procedural details or other items may inadvertently be inaccurate.

Library of Congress Card No.: applied for

British Library Cataloguing-in-Publication Data
A catalogue record for this book is available from the British Library.

Bibliographic information published by the Deutsche Nationalbibliothek
Die Deutsche Nationalbibliothek lists this publication in the Deutsche Nationalbibliografie; detailed bibliographic data are available in the Internet at ⟨http://dnb.d-nb.de⟩.

© 2008 WILEY-VCH Verlag GmbH & Co. KGaA, Weinheim

All rights reserved (including those of translation into other languages). No part of this book may be reproduced in any form – by photoprinting, microfilm, or any other means – nor transmitted or translated into a machine language without written permission from the publishers. Registered names, trademarks, etc. used in this book, even when not specifically marked as such, are not to be considered unprotected by law.

Printed in the Federal Republic of Germany
Printed on acid-free paper

Typesetting   Asco Typesetter, North Point, Hong Kong
Printing   betz-druck GmbH, Darmstadt
Binding   Litges & Dopf GmbH, Heppenheim
Wiley Bicentennial Logo   Richard J. Pacifico

ISBN: 978-3-527-40506-0

Contents

Preface *IX*

1 **Introduction** *1*
1.1 Aims of Vibrational Spectroscopy in Life Sciences *2*
1.2 Vibrational Spectroscopy – An Atomic-scale Analytical Tool *3*
1.3 Biological Systems *4*
1.4 Scope of the Book *7*
1.5 Further Reading *9*
References *10*

2 **Theory of Infrared Absorption and Raman Spectroscopy** *11*
2.1 Molecular Vibrations *12*
2.1.1 Normal Modes *15*
2.1.2 Internal Coordinates *18*
2.1.3 The *FG*-Matrix *19*
2.1.4 Quantum Chemical Calculations of the *FG*-Matrix *23*
2.2 Intensities of Vibrational Bands *25*
2.2.1 Infrared Absorption *25*
2.2.2 Raman Scattering *28*
2.2.3 Resonance Raman Effect *32*
2.3 Surface Enhanced Vibrational Spectroscopy *38*
2.3.1 Surface Enhanced Raman Effect *39*
2.3.2 Surface Enhanced Infrared Absorption *43*
References *60*

3 **Instrumentation** *63*
3.1 Infrared Spectroscopy *63*
3.1.1 Fourier Transform Spectroscopy *64*
3.1.1.1 Interferometer *67*
3.1.1.2 Infrared Detectors *69*
3.1.2 Advantages of Fourier Transform Infrared Spectroscopy *70*
3.1.3 Optical Devices: Mirrors or Lenses? *71*
3.1.4 Instrumentation for Time-resolved Infrared Studies *72*

Vibrational Spectroscopy in Life Science. Friedrich Siebert and Peter Hildebrandt
Copyright © 2008 WILEY-VCH Verlag GmbH & Co. KGaA, Weinheim
ISBN: 978-3-527-40506-0

3.1.4.1 Time-resolved Rapid-scan Fourier Transform Infrared Spectroscopy   72
3.1.4.2 Time-resolved Studies Using Tunable Monochromatic Infrared Sources   74
3.1.4.3 Time-resolved Fourier Transform Infrared Spectroscopy Using the Step-scan Method   74
3.1.5 Time-resolved Pump-probe Studies with Sub-nanosecond Time-resolution   76
3.2 Raman Spectroscopy   79
3.2.1 Laser   80
3.2.1.1 Laser Beam Properties   81
3.2.1.2 Optical Set-up   83
3.2.2 Spectrometer and Detection Systems   84
3.2.2.1 Monochromators   84
3.2.2.2 Spectrographs   86
3.2.2.3 Confocal Spectrometers   87
3.2.2.4 Fourier Transform Raman Interferometers   89
References   97

4 Experimental Techniques   99
4.1 Inherent Problems of Infrared and Raman Spectroscopy in Life Sciences   99
4.1.1 The "Water" Problem in Infrared Spectroscopy   99
4.1.2 Unwanted Photophysical and Photochemical Processes in Raman Spectroscopy   101
4.1.2.1 Fluorescence and Raman Scattering   102
4.1.2.2 Photoinduced Processes   104
4.2 Sample Arrangements   105
4.2.1 Infrared Spectroscopy   106
4.2.1.1 Sandwich Cuvettes for Solution Studies   106
4.2.1.2 The Attenuated Total Reflection (ATR) Method   108
4.2.1.3 Electrochemical Cell for Infrared Spectroscopy   113
4.2.2 Raman and Resonance Raman Spectroscopy   116
4.2.2.1 Measurements in Solutions   116
4.2.2.2 Solid State and Low-temperature Measurements   117
4.3 Surface Enhanced Vibrational Spectroscopy   118
4.3.1 Colloidal Suspensions   119
4.3.2 Massive Electrodes in Electrochemical Cells   120
4.3.3 Metal Films Deposited on ATR Elements   122
4.3.4 Metal/Electrolyte Interfaces   123
4.3.5 Adsorption-induced Structural Changes of Biopolymers   127
4.3.6 Biocompatible Surface Coatings   128
4.3.7 Tip-enhanced Raman Scattering   130
4.4 Time-resolved Vibrational Spectroscopic Techniques   131
4.4.1 Pump–Probe Resonance Raman Experiments   132
4.4.1.1 Continuous-wave Excitation   133

4.4.1.2	Pulsed-laser Excitation	138
4.4.1.3	Photoinduced Processes with Caged Compounds	141
4.4.2	Rapid Mixing Techniques	141
4.4.2.1	Rapid Flow	144
4.4.2.2	Rapid Freeze–Quench	145
4.4.3	Relaxation Methods	146
4.4.4	Spatially Resolved Vibrational Spectroscopy	148
4.5	Analysis of Spectra	149
	References	151

5 Structural Studies 155

5.1	Basic Considerations	155
5.2	Practical Approaches	158
5.3	Studies on the Origin of the Sensitivity of Amide I Bands to Secondary Structure	161
5.4	Direct Measurement of the Interaction of the Amide I Oscillators	167
5.5	UV-resonance Raman Studies Using the Amide III Mode	169
5.6	Protein Folding and Unfolding Studies Using Vibrational Spectroscopy	171
	References	178

6 Retinal Proteins and Photoinduced Processes 181

6.1	Rhodopsin	183
6.1.1	Resonance Raman Studies of Rhodopsin	185
6.1.2	Resonance Raman Spectra of Bathorhodopsin	188
6.1.3	Fourier Transform Infrared Studies of the Activation Mechanism of Rhodopsin	195
6.1.3.1	Low-temperature Photoproducts	197
6.1.3.2	The Active State Metarhodopsin II (MII)	201
6.2	Infrared Studies of the Light-driven Proton Pump Bacteriorhodopsin	206
6.3	Study of the Anion Uptake by the Retinal Protein Halorhodopsin Using ATR Infrared Spectroscopy	214
6.4	Infrared Studies Using Caged Compounds as the Trigger Source	217
	References	222

7 Heme Proteins 227

7.1	Vibrational Spectroscopy of Metalloporphyrins	228
7.1.1	Metalloporphyrins Under D_{4h} Symmetry	228
7.1.2	Symmetry Lowering	231
7.1.3	Axial Ligation	232
7.1.4	Normal Mode Analyses	233
7.1.5	Empirical Structure–Spectra Relationships	234
7.2	Hemoglobin and Myoglobin	236
7.2.1	Vibrational Analysis of the Heme Cofactor	237

7.2.2	Iron–Ligand and Internal Ligand Modes	239
7.2.3	Probing Quaternary Structure Changes	240
7.3	Cytochrome c – a Soluble Electron-transferring Protein	244
7.3.1	Vibrational Assignments	245
7.3.2	Redox Equilibria in Solution	246
7.3.3	Conformational Equilibria and Dynamics	248
7.3.4	Redox and Conformational Equilibria in the Immobilised State	253
7.3.5	Electron Transfer Dynamics and Mechanism	260
7.3.6	The Relevance of Surface-enhanced Vibrational Spectroscopic Studies for Elucidating Biological Functions	267
7.4	Cytochrome c Oxidase	268
7.4.1	Resonance Raman Spectroscopy	268
7.4.2	Redox Transitions	271
7.4.3	Catalytic Cycle	274
7.4.4	Oxidases from Extremophiles and Archaea	277
	References	278
8	**Non-heme Metalloproteins**	**283**
8.1	Copper Proteins	284
8.2	Iron–Sulfur Proteins	290
8.3	Di-iron Proteins	296
8.4	Hydrogenases	300
	References	302
	Index	**305**

Preface

Vibrational spectroscopy and life sciences, how do they fit together? For more than 30 years vibrational spectroscopy was the classical tool used for the study of small molecules and an analytical tool to characterise unknown chemical compounds, and therefore, it is not obvious that these two subjects would indeed fit together. Nevertheless, the fact that K. P. Hofmann asked us to write a book on the application of vibrational spectroscopy in life sciences, within the newly created series *Tutorials in Biophysics*, clearly demonstrates that this subject has reached a mature stage.

The success of vibrational spectroscopy in life sciences is certainly due, largely, to technical developments leading, for instance, to the commercial availability of lasers for Raman spectroscopy and rapid-scan interferometric detection systems for Fourier transform infrared (IR) spectroscopy. In this way, the sensitivity of vibrational spectroscopy increased considerably, allowing experiments that were hitherto unimaginable to be carried out. However, it is still not clear how these developments made it possible for the basic questions on protein function to be addressed, considering that proteins are very complex systems consisting of thousands of atoms.

Thus, the main goal of this tutorial is to provide arguments as to why vibrational spectroscopy is successful in biophysics research. Both of us have had the privilege of taking active roles in these exciting scientific developments right from the beginning. Thus, it should be understood that the material in this book has been influenced by our personal experiences. When we started to devise the content of the book, we soon realised that, when considering the application of vibrational spectroscopy in life sciences, we had to focus on *molecular* biophysics. This meant leaving out the exciting fields in which vibrational spectroscopy is used as a diagnostic tool for the identification of bacteria, cancerous cells and metabolites in living cells. In addition, within the field of molecular biophysics, we had to make compromises, mainly dictated by space limitations. We, therefore, decided to restrict the applications of vibrational spectroscopy to selected classes of proteins and enzymes for the benefit of an instructive illustration of the *principles* of the most important methodologies. The selection of examples was – inevitably – subjective and governed by didactic considerations. Thus, not all colleagues who have made important contributions to this field could be adequately referenced.

Vibrational Spectroscopy in Life Science. Friedrich Siebert and Peter Hildebrandt
Copyright © 2008 WILEY-VCH Verlag GmbH & Co. KGaA, Weinheim
ISBN: 978-3-527-40506-0

As an additional consequence, the vibrational spectroscopy of other classes of proteins, lipids and nucleic acids and of lipid–protein and nucleic-acid–protein interactions, had to be omitted. However, we are convinced that scientists interested in these systems will be able to extract the principle ideas of the various vibrational spectroscopic methods described in the applications to proteins and enzymes.

The present tutorial introduces the *fundamentals* of Raman and infrared spectroscopy, including the concept of molecular vibrations and a basic theoretical treatment of IR absorption and Raman scattering. It further describes, in more detail, instrumental and sampling techniques. The book is intended for students and scientists with backgrounds in life sciences and in physics and chemistry. Hence, in this respect we also had to make compromises to accommodate the interests and backgrounds of a readership coming from very different disciplines.

The book was completed with "a little help from our friends". We would like to thank P. Hamm (Zürich), J. Bredenbeck (Frankfurt) and T. A. Keiderling (Chicago) for their advice on the chapter on structural studies and for providing figures for this chapter. Further thanks are due to R. Vogel (Freiburg), for his help in preparing several figures. We thank G. Büldt, (Jülich) for providing figures of the ground and M-state structures of bacteriorhodopsin. Support and assistance in various aspects by M. Böttcher, J. Grochol, A. Kranich, M. A. Mroginski, H. Naumann, D. v. Stetten, N. Wisitruangsakul and I. Zebger (Berlin) are gratefully acknowledged. Special thanks are due to D. H. Murgida (Berlin/Buenos Aires) for continuous critical discussions, providing important stimuli for the book. In particular, we wish to thank I. Geisenheimer (Berlin) for the great work on producing the artwork for the figures. Thanks must also be given to C. Wanka from Wiley for her patience and support. Last but certainly not least, we wish to thank our wives, D. Siebert-Karasek and K. Graf-Hildebrandt, for their steady support and encouragement and specifically for their indulgence when this book occupied our evenings and weekends.

February 2007   *Friedrich Siebert, Peter Hildebrandt*

1
Introduction

Vibrational spectroscopy is a classical technique and one of the oldest spectroscopic methods. Its origins can be traced back two centuries to William Herschel, who discovered infrared (IR) radiation in the electromagnetic spectrum of the sun. At the beginning of last century, IR radiation was being used increasingly to measure interactions with matter, thereby producing the first vibrational spectra. In the 1920s, the discovery of the Raman effect, named according to the Indian scientist Chandrasekhara V. Raman, led to a second area of vibrational spectroscopy. Around that time, the potential of IR and Raman spectroscopy to elucidate molecular structures was soon acknowledged, although technical constraints limited the applications to fairly small molecules. Many of these studies have been considered in the famous textbook by Herzberg (Herzberg 1945), which is still a standard reference for spectroscopists and a rich source of information.

It took a fairly long time until vibrational spectroscopy was introduced into biological studies. This was not only due to the limited sensitivity and poor performance of spectrometers, detectors, and light sources in those early days, but also the state-of-the-art of preparing and purifying biological samples up to a grade that was appropriate for spectroscopic experiments was nowhere near as advanced as it is nowadays. In both spectroscopy and biology, the progress in methodology and technology started to grow exponentially in the 1960s. Important milestones in the exciting development of vibrational spectroscopy were certainly the invention of lasers and their use as light sources in Raman spectroscopy and the development of interferometers for measuring IR spectra. Thus, experiments with large and rather complex molecular systems became possible, and the application of Raman and IR spectroscopy to biomolecules afforded astonishing results, which had not previously been anticipated. The enormous success of the union between vibrational spectroscopy and the life sciences prompted many researchers from very different disciplines to adopt various IR and Raman spectroscopic techniques for the study of biological systems, thereby constituting a highly interdisciplinary research area at the interface between physics, chemistry, and biology.

Vibrational Spectroscopy in Life Science. Friedrich Siebert and Peter Hildebrandt
Copyright © 2008 WILEY-VCH Verlag GmbH & Co. KGaA, Weinheim
ISBN: 978-3-527-40506-0

1.1
Aims of Vibrational Spectroscopy in Life Sciences

The physiological functions of biological macromolecules are determined by the structural organisation at different hierarchical levels, which are the sequence of the individual building blocks in a biopolymeric chain (primary structure), the fold of the chain (secondary structure), and the spatial arrangement of various secondary structural elements within a chain (tertiary structure). Finally, two or more biopolymeric chains may constitute the quaternary structure. In this way, highly complex three-dimensional structures are formed, which have been optimised through evolution to carry out specific biological functions. For example, proteins that possess very similar primary, secondary, and tertiary structures, such as the bacterial retinal proteins bacteriorhodopsin, sensory rhodopsin, or halorhodopsin, can exert very different functions (i.e., signal transduction, proton or anion transport) due to subtle structural differences in critical parts of the proteins. Conversely, the same elementary chemical reaction can be catalysed by structurally different enzymes. A typical example is the reduction of molecular oxygen by the heme-copper enzyme cytochrome c oxidase or by the copper enzyme laccase.

The most challenging task in contemporary molecular biophysics, therefore, is the elucidation of the structure–function relationship of biological macromolecules. However, in view of the powerful techniques used in structural biology, i.e., X-ray crystallography, NMR spectroscopy, and cryogenic electron microscopy, which can provide detailed structures of macromolecules, one might ask what the current and future contributions of vibrational spectroscopy to this field could be.

Of course, knowledge of the three-dimensional structure of a biopolymer is important in the understanding of the functional mechanism as it guides the development of realistic hypotheses. However, a comprehensive elucidation of reaction mechanisms on a molecular level requires structural information usually beyond the resolution of the classical methods used in structural biology. For instance, the positions of hydrogen atoms and protons in the three-dimensional structure and van-der-Waals, hydrogen bonding, or electrostatic intermolecular interactions, which are essential for biochemical and biophysical processes, can only be *assumed* but not determined by X-ray crystallography. NMR spectroscopic techniques could be an alternative, but size limitations impose severe constraints because three-dimensional structures are currently restricted to biopolymers smaller than about 50 kDa.

Biological processes involve a series of structurally different states, such that a full understanding of the reaction mechanism requires knowledge of the initial and final states and of the intermediate species. Identification of intermediate states and the description of their molecular properties are only possible on the basis of techniques that can provide structural data as a function of time. Extending X-ray crystallography to the time-resolved domain is associated with substantial experimental difficulties and, moreover, is restricted to those instances where the crystals are not destroyed during the reaction sequence.

In all these respects, vibrational spectroscopy offers a variety of advantages. Firstly, vibrational spectroscopy can contribute to the elucidation of details in the molecular structures and intermolecular interactions that go far beyond the resolution of even highly resolved crystal structures. Secondly, unlike NMR spectroscopy, vibrational spectroscopy is in principle not restricted by the size of the sample and thus can afford valuable information for small biomolecules in addition to complex biological systems. Thirdly, vibrational spectroscopic methods are applicable regardless of the state of the biomolecule, i.e., they can be used to study biomolecules in solutions, in the solid and crystalline state, or in monolayers. Thus, it is possible to adapt the techniques according to the specific requirements of the sample and the biophysical questions to be addressed. In this sense, vibrational spectroscopy offers the potential to probe molecular events under conditions that are closely related to the physiological reaction environment. Fourthly, this versatility also allows combining vibrational spectroscopy with various time-resolved approaches. Thus, detailed information regarding the dynamics of biological systems can also be obtained, down to the femtosecond time scale.

Thus, it is one of the central objectives of this book to demonstrate that vibrational spectroscopic methods represent powerful tools, which are complementary to the techniques used in structural biology.

1.2
Vibrational Spectroscopy – An Atomic-scale Analytical Tool

Vibrational spectroscopy probes the periodic oscillations of atoms within a molecule. These oscillations do not occur randomly but in a precisely defined manner. This can easily be understood by taking into account that an N-atomic molecule has $3N$ degrees of freedom, of which three refer to translations and three (two) correspond to rotations in the case of a nonlinear (linear) molecule structure. The remaining degrees of freedom represent $3N - 6$ ($3N - 5$) vibrations of a nonlinear (linear) molecule, the so-called normal modes. In each normal mode every atom oscillates in-phase and with the same frequency, albeit with different amplitudes. The frequency, however, the first principle observable in vibrational spectroscopy, has a sensitive dependence on the forces acting on the individual atoms and on the respective masses. These forces do not only result from the chemical bonds connecting the individual atoms but also include contributions from non-bonding interactions within the molecule and with the molecular environment. In this way, the frequencies of the normal modes constitute a characteristic signature of the chemical constitution, the structure, and electron density distribution of the molecule in a given chemical environment, i.e., all of the parameters required for a comprehensive atomic-scale description of a molecule. These parameters also control the second important observable parameter in the vibrational spectrum, the intensities of the bands, which, unlike the frequencies, are not independent of the method by which the vibrational spectrum is probed.

The two main techniques used to obtain vibrational spectra, IR and Raman spectroscopy, are based on different physical mechanisms. In IR spectroscopy, molecules are exposed to a continuum of IR radiation and those photons that have energies corresponding to the frequencies of the normal modes can be absorbed to excite the respective vibrations. The wavelength range of the IR radiation corresponding to the frequencies of molecular vibrations extends typically between 2.5 and 50 µm. In Raman spectroscopy, these so-called vibrational transitions are induced upon inelastic scattering of monochromatic light by the molecule, such that the frequency of the scattered light is shifted by the frequency of the molecular vibration. For a given molecule, absorption- and scattering-induced vibrational transitions are associated with different probabilities, hence IR and Raman spectra may display different vibrational band patterns, which are an additional source of information about the structural and electronic properties of the molecule.

1.3
Biological Systems

The size of the biological systems that are the targets of vibrational spectroscopy in the life sciences can vary substantially. They range from building blocks of biopolymers (e.g., amino acids or lipids) or cofactors of proteins up to protein assemblies, membranes, or DNA–protein complexes. Concomitant with the increasing size of the system, the number of signals in the spectrum, i.e., the vibrational modes, increases with the number of atoms involved. Thus, only for small molecules with less than 50 atoms, corresponding to ca. 150 normal modes, is it usually possible to resolve all the individual vibrational bands in the IR and Raman spectra. For biopolymers such as proteins or nucleic acids, the number of vibrational modes is prohibitively large, resulting in complex spectra with many overlapping bands of slightly different frequencies. This is also true for bands originating from the same modes of the individual building blocks as these entities may be in a slightly different environment. Accordingly, it is not obvious how detailed information, for example, on the interaction of a substrate in the catalytic centre of an enzyme, or on the minute structural changes occurring in the protein during the enzymatic process, can be derived from vibrational spectra of large biological systems. The question should be rephrased: how can vibrational spectroscopy be made to be selective for those molecular groups of the macromolecule that one is interested in?

For proteins, there are two basic principles by which the desired selectivity is accomplished. In Raman spectroscopy, the wavelength of the monochromatic light, which is used for inelastic scattering, is selected to be in resonance with an electronic transition of a chromophoric group of the protein, which may either be a cofactor or a chromophore of the apoprotein. Under these *resonance* conditions, the probability of the scattering-induced transitions, and thus the intensity of the Raman scattered light originating from vibrational modes of the chromophore, is selectively enhanced by several orders of magnitude. Then the resonance Raman spectrum displays the vibrational bands of the chromophore exclusively, whereas the Raman bands of the optically transparent matrix remain largely invisible. This selectivity is associated with an enhanced sensitivity and thus drastically reduces the protein concentration required for high quality spectra.

A more general method uses the "function" of the system, that is, its natural reaction, as a selectivity tool. The underlying idea is simple: the molecular groups involved in the function represent only a small fraction of the total system. As an example, we refer to the membrane protein bacteriorhodopsin, which acts as a light driven proton pump. This function is associated with only relatively small structural changes as shown in Fig. 1.1a and b (Sass et al. 2000). On stabilising two well-defined functional states of protein, in this instance the parent state BR_{570} and an intermediate M_{410}, the difference between the respective spectra only displays contributions from those groups undergoing molecular changes during the $BR_{570} \rightarrow M_{410}$ transition, because all bands that remain unchanged cancel each other out (Fig. 6.19). Correspondingly, the spectra are greatly simplified and, moreover, only reflect the *functionally relevant* structural changes. This method is called reaction-induced difference spectroscopy. The term "reaction" is implied in a very general sense. It can refer to ligand binding, substrate binding and transformation, light-induced reactions, and electron transfer in redox-reactions.

Both methods, i.e., resonance Raman and IR difference spectroscopy, can be extended to time-resolved studies, such that it is possible to probe the dynamics of molecular changes in real time during the reactions and processes of the system.

However, the scope of Raman and IR spectroscopy in the life sciences is broader as it is not restricted to the analyses of minute structural changes. For many proteins and for other biological systems including nucleic acids and membranes, these techniques may provide valuable information about more global structural properties. The individual building blocks of proteins, i.e., the amino acids, are linked via the same chemical entities as are the peptide bonds. Likewise, nucleic acids also form a backbone of repetitive units of sugar–phosphate linkages. As some of the vibrational modes of these units depend on the folding of the biopolymeric chain, vibrational spectra can give insights into the secondary structures of proteins and nucleic acids. Also, bilayer membranes exhibit global structural properties, which result from the periodic arrangement of lipid molecules possessing the same conformation. Characteristic vibrational marker bands for these conformations may be monitored to determine extended structural changes associated, for instance, with phase transitions.

The considerable progress that has been achieved in experimental Raman and IR spectroscopy in recent years is not adequately paralleled by the development of universal strategies for extracting the structural information from the spectra. Still, empirical approaches prevail that are based on the comparison with experimental data for related systems and model compounds. In many instances, isoto-

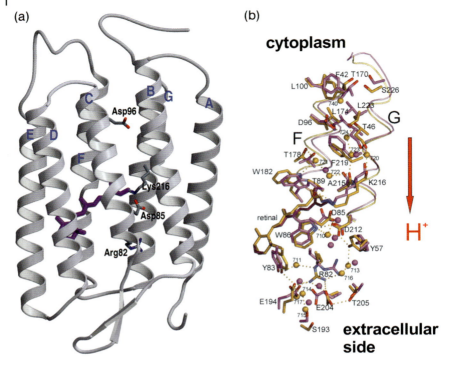

Fig. 1.1 (a) Three-dimensional structure of bacteriorhodopsin in the parent state BR_{570}. The seven transmembrane helices are indicated by the letters A to G. The chromophore is shown in purple. The retinal binding lysine Lys216, the proton acceptor for Schiff base deprotonation, Asp85, and the proton donor for Schiff base reprotonation are indicated. In addition, Arg82 pointing towards the retinal binding site is shown. The C-terminus is up (intracellular side), the N-terminus down (extracellular side), proton pumping is from the intracellular to the extracellular side. Oxygen atoms are coloured in red, nitrogen atoms in blue. Coordinates from crystal structure 1CWK of the protein data bank were used (courtesy of G. Büldt). For details of the mechanism of bacteriorhodopsin see Chapter 6.2. (b) Differences in the crystal structures of the ground and M states of bacteriorhodopsin in the neighbourhood of helices F and G. Ground state is shown in purple, M state in yellow. Oxygen atoms are coloured in red, nitrogen atoms in blue. Resolved water molecules are depicted as purple and yellow balls for the ground and M states, respectively. The direction of proton pumping is indicated. In the M state, the light-induced isomerisation of the chromophore retinal from all-*trans* to 13-*cis* is clearly seen. A distinct molecular change concerns Arg82 (R82), which now points downwards. This is thought to cause proton release from a site close to the extracellular surface and to increase the pK_a of the retinal Schiff base for its reprotonation in the next N state. Several water molecules have been displaced in the M state. However, protonation of Asp85 (D85) and deprotonation of the Schiff base, as deduced from infrared and Raman spectroscopy, cannot be deduced from the M structure as protons cannot be seen directly (adapted from Sass et al. 2000).

pic labelling is an indispensable tool in vibrational spectroscopy for assigning bands to specific modes. This experimental approach is straightforward for molecules including protons that can be exchanged by deuterons in ^2H_2O solutions. For all other isotopic substitutions (e.g., ^{15}N, ^{18}O, ^{13}C, covalently bound ^2H) synthetic work is required either by organic chemists or by microorganisms producing the compounds of interest in isotopically enriched media. These time-demanding and costly procedures are not applicable in each instance, but have been shown to contribute substantially to the vibrational analyses of protein cofactors and building blocks of nucleic acids, proteins, and membranes. Furthermore, it should be emphasised that NMR studies on proteins also require, in most instances, isotopic labelling with ^{13}C, ^{15}N, and ^2H.

The vibrational analyses of proteins are also supported by genetic engineering such that specific bands can be assigned to individual amino acid residues. This approach strongly benefits from the tight interactions of spectroscopists and biologists, inasmuch as the functional consequences of individual mutations have to be assessed as a prerequisite for unambiguous interpretations of the spectra in terms of structure–function relationships.

These empirical approaches typically only focus on small segments of the vibrational spectra and thus the major part of the structural information contained in the spectra remains obscured. More comprehensive methods are based on the classical treatment of the vibrational eigenstate problem. In the past, these normal mode analyses have been the domain of a few specialists, and, in fact, only a small number of biomolecules, i.e., cofactors of proteins such as tetrapyrroles or retinals, have been treated by these tedious methods. The popularisation of quantum chemical programs, the development of efficient program codes, and the increasing availability of powerful personal computers, have all contributed to reducing the exclusivity of theoretical methods and to open up novel possibilities for comprehensive and reliable vibrational analyses. Although a sound application of these methods requires knowledge of theoretical chemistry, they will no doubt develop to become a standard tool to be employed routinely by experimentalists also.

1.4
Scope of the Book

During recent years vibrational spectroscopy has become an important tool in biophysical research, both for structural and functional studies. Whereas in the beginning this research area was the domain of physicists and physico-chemists, who not only had to master the methodological challenges but also to become acquainted with the concepts and emerging problem in the life sciences, more and more biologists have now recognised the high potential of these techniques to elucidate the molecular functioning of biomolecules. Therefore, the main goal of this book is to introduce the basic concepts of vibrational spectroscopy to "newcomers" to this area, and to students specialising in this particular discipline of molecular biophysics, in addition to advanced scientists with a non-spectroscopic background and to spectroscopists who intend to work with biological systems. Specific emphasis is given to the practical aspects of Raman and IR spectroscopy, which, when applied in the life sciences, usually has to be adapted to the specific needs and demands of the systems to be studied. This is reflected by an extensive description of the instrumental and sampling techniques (Chapters 3 and 4) in the first part of the book. Conversely, we restricted the treatise of the theoretical background to the elementary relationships, avoiding lengthy mathematical derivations (Chapter 2). Generally, we will separate more elaborate explanations and derivations from the body of the text. Thus, the main content of the various chapters is easier to follow, and the more specialised or difficult parts can be read later, or even be omitted. For a better understanding of these chapters, a basic knowledge of physics, especially optics and molecular physics, and of general physical chemistry would be helpful.

A basic knowledge of biochemistry, in particular with respect to the structure and processes of proteins, is desired for the second part of the book. Textbooks on biophysics, biophysical chemistry, and biochemistry usually provide an excellent basis. This part includes four chapters (Chapters 5–8) devoted to applications of vibrational spectroscopic methods to the study of biomolecules. Instead of covering the broad range of biological molecules comprehensively, this part is restricted to structural studies of proteins (Chapter 5) and to specific classes of proteins (Chapters 6–8). These chapters are considered to *illustrate* the application of dedicated methods and to point out what type of information they may provide. In this respect, proteins (and among them specific representatives such as rhodopsin or cytochrome *c*) represent the most versatile targets because they have been studied by a large variety of different vibrational spectroscopic techniques and, in some instances, even served as models for methodological developments. We will, therefore, describe not only well-established approaches but also new and emerging techniques that promise to become important analytical tools in the future. The restriction to principle aspects of the applications also implies that only exemplary results are reported. For comprehensive accounts, the reader is referred to original and review articles. According to the concept outlined above, and due to general space restrictions imposed on this book, other important biological systems, such as nucleic acids, lipids, and carbohydrates, will not be covered. However, the methodological approaches usually applied to these systems are equally well covered, on the basis of the specific example proteins.

The applications of vibrational spectroscopy discussed in this book are restricted to problems in *molecular* biophysics. They do not include approaches for characterising bacteria, tissues, and cell cultures, even though Raman and IR spectroscopic analyses of such highly complex systems are of significant importance in microbiological and medical applications. As these studies do not focus on the molecular properties of biomolecules, they are beyond the scope of this book.

For the readers having a background in physics and physical chemistry, we want to demonstrate that, despite the complexity of biological macromolecules, vibrational spectroscopy is a potent tool for the study of their structural properties and their functions at a molecular level. Biologists and biochemists, on the other hand, should be encouraged to utilise the fairly sophisticated IR and Raman spectroscopic techniques and to exploit their specific advantages for studying biological systems. Eventually, we hope that the reader can assess the potential and limitations of vibrational spectroscopy in molecular life sciences and be able to judge whether the system he or she is interested in could be successfully studied using vibrational spectroscopy.

1.5
Further Reading

Vibrational spectroscopy is a method which has developed over many years. Thus a number of excellent books have been published that cover certain aspects, and a few of these monographs should be mentioned here. The book by Colthup provides an excellent introduction to general vibrational spectroscopy (Colthup et al. 1975). It also contains a treatment of the basic theoretical concepts. Lin-Vien et al. have presented a collection of data for organic molecules (Lin-Vien et al. 1991), directed to provide a basis for the identification chemical compounds. However, as the spectral properties of chemical groups are discussed fairly thoroughly, this book also serves as a reference for many more applications. The book by Nakamoto offers similar information on inorganic and coordination compounds (Nakamoto 1986). A compilation of spectra of amino acids serves as a very useful reference (Barth 2000). The effect of isotopic labelling on molecular vibrations is discussed by Pincas and Laulicht (Pinchas and Laulicht 1971). A basic introduction into practical, theoretical, and applied aspects of Raman spectroscopy is given by Smith and Dent (Smith and Dent 2005). A comprehensive treatise of Raman spectroscopy including various applications has been edited by Schrader (Schrader 1995). The technical aspects of Fourier transform spectroscopy, particularly important for IR spectroscopists, are covered in great detail in the book by Griffith and de Haseth (Griffith and de Haseth 1986). An up-to-date account of the theory and practice of surface enhanced Raman spectroscopy is presented in a recent book that includes contributions from various research groups (Kneipp et al. 2006). For the theory of vibrational spectroscopy, we wish to recommend the excellent books by Herzberg (Herzberg 1945) and Wilson et al. (Wilson et al. 1955). Albeit published half a century ago, they are indispensable textbooks and reference books for all vibrational spectroscopists. The book by Long also includes major treatise on the theory of Raman spectroscopy with particular emphasis on polarisation effects (Long 1977). Biological applications of vibrational spectroscopy are described in the textbooks by Carey (Raman) (Carey 1982), Twardowski and Anzenbacher (Raman and IR) (Twardowski and Anzenbacher 1994).

Collections of review articles on specialised topics of vibrational spectroscopy, also on biomolecular applications, can be found in the book series *Advances in Spectroscopy* (edited by Clark and Hester) and in the three-volume edition *Biologi-*

cal Applications of Raman Spectroscopy (Spiro 1987, 1988). A selection of articles on infrared spectroscopy of biomolecules have been published in a book of the same title (Mantsch and Chapman 1996), and more specialised articles on biomolecular infrared and Raman spectroscopy have appeared recently in a book from the series Practical Spectroscopy (Gremlich and Yan 2001).

References

Barth, A., **2000**, "The infrared spectra of amino acid side chains", *Prog. Biophys. Molec. Biol.* **74**, 141–173.

Carey, P. R., **1982**, "*Biochemical Applications of Raman Spectroscopy*", Academic Press, New York.

Colthup, N. B., Daly, L. H., Wiberly, S. E., **1975**, "*Introduction to Infrared and Raman Spectroscopy*", Academic Press, New York.

Gremlich, H. U., Yan, B., **2001**, "*Infrared and Raman Spectroscopy of Biological Materials (Practical Spectroscopy)*", Marcel Dekker, Basel.

Griffith, P. R., de Haseth, J. A., **1986**, "*Fourier Transform Infrared Spectroscopy*", John Wiley & Sons, New York.

Herzberg, G., **1945**, "*Molecular Spectra and Molecular Structure: II, Infrared and Raman Spectra of Polyatomic Molecules*", Van Nostrand Reinhold, New York.

Kneipp, K., Moskovits, M., Kneipp, H. (Eds.), **2006**, "Surface-enhanced Raman scattering: Physics and applications", *Topics Appl. Phys.* **103**, Springer, Berlin.

Lin-Vien, D., Colthup, N. B., Fately, W. G., Grasselli, J. G., **1991**, "*Infrared and Raman Characteristic Frequencies of Organic Molecules*", Academic Press, Boston.

Long, D. A., **1977**, "*Raman Spectroscopy*", McGraw-Hill, New York.

Mantsch, H. H., Chapman, D. (Eds.), **1996**, "*Infrared Spectroscopy of Biomolecules*", Wiley-Liss, New York.

Nakamoto, K., **1986**, "*Infrared and Raman Spectra of Inorganic and Coordination Compounds*", John Wiley & Sons, New York.

Pinchas, S., Laulicht, I., **1971**, "*Infrared Spectra of Labelled Compounds*", Academic Press, London.

Sass, H. J., Büldt, G., Gessenich, R., Hehn, D., Neff, D., Schlesinger, R., Berendzen, J., Ormos, P. **2000**, "Structural alterations for proton translocation in the M state of wild-type bacteriorhodopsin". *Nature* 406, 649–653.

Schrader, B. (Ed.), **1995**, "*Infrared and Raman Spectroscopy*", VCh-Verlag, Weinheim.

Smith, E., Dent, G., **2005**, "*Modern Raman Spectroscopy – A Practical Approach*", Wiley, Chichester.

Spiro, T. G., **1987**, **1988**, "*Biological Applications of Raman Spectroscopy*", Vols. I, II, III, Wiley, Chichester.

Twardowski, J., Anzenbacher, P., **1994**, "*Raman and IR Spectroscopy in Biology and Biochemistry*", Ellis Horwood, New York.

Wilson, E. B., Decius, J. C., Cross, P. C., **1955**, "*Molecular Vibrations: The Theory of Infrared and Raman Vibrational Spectra*", McGraw-Hill, New York.

2
Theory of Infrared Absorption and Raman Spectroscopy

Molecular vibrations can be excited via two physical mechanisms: the absorption of light quanta and the inelastic scattering of photons (Fig. 2.1) (Herzberg 1945). Direct absorption of photons is achieved by irradiation of molecules with polychromatic light that includes photons of energy matching the energy difference $h\nu_k$ between two vibrational energy levels, the initial (i, e.g., ground state) and the final (f, e.g., first excited state) vibrational state.

$$h\nu_k = h\nu_f - h\nu_i \qquad (2.1)$$

As these energy differences are in the order of 0.5 and 0.005 eV, light with wavelengths longer than 2.5 μm, that is infrared (IR) light, is sufficient to induce the vibrational transitions. Thus, vibrational spectroscopy that is based on the direct absorption of light quanta is denoted as IR absorption or IR spectroscopy.

The physical basis of IR light absorption is very similar to light absorption in the ultraviolet (UV)–visible (vis) range, which causes electronic transitions or combined electronic–vibrational (vibronic) transitions. Thus, UV–vis absorption spectroscopy can, in principle, also provide information about molecular vibrations. However, for molecules in the condensed phase at ambient temperature, the vibrational fine structure of the absorption spectra is only poorly resolved , if at all, such that vibrational spectroscopy of biomolecules by light absorption is restricted to the IR range.

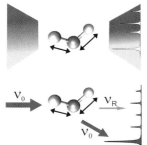

Fig. 2.1 Illustration of the excitation of molecular vibrations in IR (top) and Raman (bottom) spectroscopy. In IR spectroscopy, the vibrational transitions are induced by absorption of light quanta from a continuous light source in the IR spectral region. Vibrational Raman transitions correspond to inelastic scattering (ν_R; thin arrow) of the incident monochromatic light (ν_0) whereas the elastic scattering (ν_0) is represented by the thick arrow.

Vibrational Spectroscopy in Life Science. Friedrich Siebert and Peter Hildebrandt
Copyright © 2008 WILEY-VCH Verlag GmbH & Co. KGaA, Weinheim
ISBN: 978-3-527-40506-0

In contrast to IR spectroscopy, the scattering mechanism for exciting molecular vibrations requires monochromatic irradiation. A portion of the incident photons is scattered inelastically such that the energy of the scattered photons ($h\nu_R$) differs from that of the incident photons ($h\nu_0$). According to the law of conversation of energy, the energy difference corresponds to the energy change of the molecule, which refers to the transition between two vibrational states. Thus, the energy differences

$$h\nu_0 - h\nu_R = h\nu_f - h\nu_i \tag{2.2}$$

lie in the same range as the transitions probed by the direct absorption of mid-IR quanta, although photons of UV, visible, or near-infrared light are used to induce scattering. This inelastic scattering of photons was first discovered by the Indian scientist C. V. Raman in 1928 and is thus denoted as the Raman effect.

Vibrational transitions may be associated with rotational transitions that can only be resolved in high resolution spectra of molecules in the gas phase and is, therefore, not relevant for the vibrational spectroscopy of biomolecules. Thus, vibration–rotation spectra will not be treated in this book.

Depending on the molecule, the same or different vibrational transitions are probed in IR and Raman spectroscopy and both techniques provide complementary information in many instances. Hence, IR and Raman spectra are usually plotted in an analogous way to facilitate comparison. The ordinate refers to the extent of the absorbed (IR) or scattered (Raman) light. In IR absorption spectroscopy, the amount of absorbed light is expressed in units of absorbance or, albeit physically less correct but frequently used, in terms of the optical density. In contrast, Raman intensities are measured in terms of counts per second, i.e., of photons detected per second. As this value depends on many apparatus-specific parameters, in most instances only relative intensities represent physically meaningful quantities. Thus, the Raman intensity scale is typically expressed in terms of arbitrary units or the scale is even omitted. The energy of the vibrational transition, expressed in terms of wavenumbers (cm^{-1}), is given on the abscissa, corresponding to the frequency of the absorbed light ν_{abs} in IR spectroscopy and to the frequency difference between the exciting and scattered light, $\nu_0 - \nu_R$, in Raman spectroscopy.

The principle sources of information in vibrational spectroscopy are the energies of the vibrational transitions and the strength of their interaction with the IR or UV–vis radiation, i.e., the band intensities. Classical mechanics constitutes the basis for describing the relationship between vibrational frequencies and the molecular structure and force fields whereas quantum mechanics is indispensable for understanding the transition probabilities and thus the intensities of vibrational bands in the IR or Raman spectra.

2.1
Molecular Vibrations

As the starting point for introducing the concept of harmonic vibrations, it is instructive to consider molecules as an array of point masses that are connected with each other by mass-less springs representing the intramolecular interactions between the atoms (Wilson et al. 1955). The simplest case is given by two masses, m_A and m_B, corresponding to a diatomic molecule A–B. Upon displacement of the spheres along the x-axis from the equilibrium position by Δx, a restoring force F_x acts on the spheres, which according to Hooke's law, is given by

$$F_x = -f \Delta x \tag{2.3}$$

Here f is the spring or force constant, which is a measure of the rigidity of the spring, that is, the strength of the bond. The potential energy V then depends on the square of the displacement from the equilibrium position

$$V = \frac{1}{2} f \Delta x^2 \tag{2.4}$$

For the kinetic energy T of the oscillating motion one obtains

$$T = \frac{1}{2} \mu (\Delta \dot{x})^2 \tag{2.5}$$

where μ is the reduced mass defined by

$$\mu = \frac{m_A \cdot m_B}{m_A + m_B} \tag{2.6}$$

Because of the conservation of energy, the sum of V and T must be constant, such that the sum of the first derivatives of V and T is equal to zero, as expressed by Eq. (2.7):

$$0 = \frac{dT}{dt} + \frac{dV}{dt} = \frac{1}{2} \mu \frac{d(\Delta \dot{x}^2)}{dt} + \frac{1}{2} f \frac{d(\Delta x^2)}{dt} \tag{2.7}$$

which eventually leads to the Newton equation of motion

$$\frac{d^2 \Delta x}{dt^2} + \frac{f}{\mu} \Delta x = 0 \tag{2.8}$$

Equation (2.8) represents the differential equation for a harmonic motion with the solution given by a sine or cosine function, i.e.,

$$\Delta x = A \cdot \cos(\omega t + \varphi) \tag{2.9}$$

where A, ω, and φ are the amplitude, circular frequency, and phase, respectively. Combining Eq. (2.9) with its second derivative one obtains

$$\frac{d^2 \Delta x}{dt^2} + \omega^2 \Delta x = 0 \tag{2.10}$$

such that comparison with Eq. (2.8) yields

$$\omega = \sqrt{\frac{f}{\mu}} \tag{2.11}$$

Equation (2.11) describes what one intuitively expects: the circular frequency of the harmonic vibration increases when the rigidity of the spring (or the strength of the bond) increases but decreases with increasing masses of the spheres. In order to express the circular frequency in wavenumbers (in cm^{-1}), Eq. (2.11) has to be divided by $2\pi c$ (with c given in cm s^{-1}):

$$\tilde{\nu} = \frac{1}{2\pi c} \sqrt{\frac{f}{\mu}} \tag{2.12}$$

In contrast to the straightforward treatment of a two-body system, including a third sphere corresponding to a triatomic molecule clearly represents a conceptual challenge (Wilson et al. 1955). Let us consider a bent molecule such as H_2O as an example (Fig. 2.2). Following the same strategy as for the diatomic molecule, we analyse the displacements of the individual atoms in terms of the restoring forces. There are two questions to be answered. (a) What are the displacements that lead to vibrations? (b) Are all possible displacements allowed?

In the Cartesian coordinate system, each atom can be displaced in the x-, y-, and z-directions, corresponding to three degrees of freedom. Thus, a molecule of N atoms (α) has in total $3N$ degrees of freedom, but not all of them correspond to vibrational degrees of freedom. If all atoms are displaced in the x-, y-, and z-directions by the same increments, the entire molecule moves in a certain direction, representing one of the three translational degrees of freedom. Furthermore, one can imagine displacements of the atoms that correspond to the rotation of

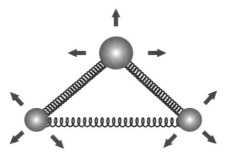

Fig. 2.2 Illustration of the vibrating H_2O molecule represented by spheres that are connected via springs of different strengths. The tighter springs linking the large sphere (oxygen) with each of the small spheres (hydrogen) symbolises the chemical bonds between two atoms, whereas the looser spring refers to weaker interactions between two atoms that are not connected via a chemical bond.

the molecule. It can easily be seen that a nonlinear molecule (i.e., where the atoms are not located along a straight line) has three rotational degrees of freedom, whereas there are only two for a linear molecule. Thus, the remaining $3N-6$ and $3N-5$ degrees of freedom correspond to the vibrations of a nonlinear and a linear molecule, respectively. For the treatment of molecular vibrations in terms of Cartesian coordinates, the rotational and translational degrees of freedom can be separated by choosing a rotating coordinate system with its origin in the centre of mass of the molecule.

As an important implication of these considerations, we note that the vibrational degrees of freedom and thus the number of molecular vibrations are uniquely determined by the number of atoms in the molecule. In our example of a nonlinear three-atomic molecule there are just $3 (= 3 \cdot 3 - 6)$ vibrational degrees of freedom. Thus, molecular vibrations do not represent random motions but well-defined displacements of the individual atoms. Consequently, one may intuitively expect that these vibrations, which are denoted as normal modes, are characteristic of a given molecule. The primary task of the normal mode analysis is to decode the relationships between normal modes, specifically their frequencies, and molecular properties.

2.1.1
Normal Modes

To determine the normal mode frequencies, we begin by expressing the kinetic and potential energy in terms of the displacements of the Cartesian coordinates for each atom α (Wilson et al. 1955). For the kinetic energy one obtains [see Eq. (2.5)]

$$T = \frac{1}{2} \sum_{\alpha=1}^{N} m_\alpha \left[\left(\frac{d\Delta x_\alpha}{dt} \right)^2 + \left(\frac{d\Delta y_\alpha}{dt} \right)^2 + \left(\frac{d\Delta z_\alpha}{dt} \right)^2 \right] \qquad (2.13)$$

At this point it is convenient to introduce so-called mass-weighted Cartesian displacement coordinates, which are defined according to

$$q_1 = \sqrt{m_1}\Delta x_1, \quad q_2 = \sqrt{m_1}\Delta y_1, \quad q_3 = \sqrt{m_1}\Delta z_1 \quad \text{for atom } \alpha = 1$$
$$q_4 = \sqrt{m_2}\Delta x_2, \quad q_5 = \sqrt{m_2}\Delta y_2, \quad q_6 = \sqrt{m_2}\Delta z_2 \quad \text{for atom } \alpha = 2 \qquad (2.14)$$

and correspondingly for all other atoms such that one obtains $3N$ mass-weighted Cartesian displacement coordinates. Substituting Eq. (2.14) in Eq. (2.13) simplifies the expression for the kinetic energy to

$$T = \frac{1}{2} \sum_{i=1}^{3N} \dot{q}_i^2 \qquad (2.15)$$

To derive the appropriate expression for the potential energy, V, is more complicated as it has to take into account all possible interactions between the individual atoms, which primarily include the bonding interactions but also non-bonding (electrostatic, van-der-Waals) interactions. For the three-atomic water molecule in Fig. 2.2 this implies that the displacement of one hydrogen atom depends on the attractive and repulsive forces of both the central oxygen and the second hydrogen atom. Within the framework of the sphere–spring model we therefore also have to connect both hydrogen "spheres" via a spring which, however, is less rigid than those connecting the hydrogen spheres with the oxygen.

It is convenient to expand the potential energy in a Taylor series in terms of the displacement coordinates $\Delta x_i, \Delta y_i, \Delta z_i$, which can be also expressed in terms of the coordinates q_i defined in Eq. (2.14).

$$V = V_0 + \sum_{i=1}^{3N} \left(\frac{\partial V}{\partial q_i}\right)_0 q_i + \frac{1}{2} \sum_{i,j=1}^{3N} \left(\frac{\partial^2 V}{\partial q_i \partial q_j}\right)_0 q_i q_j + \cdots \cdot \tag{2.16}$$

The first term refers to the potential energy at equilibrium, which we can set equal to zero as we are interested in changes to V brought about by displacements of the individual atoms. At equilibrium, infinitesimal changes in q_i do not cause a change in V, such that the second term is also zero. For small displacements q_i within the harmonic approximation, higher order terms can be neglected, such that Eq. (2.16) is simplified to

$$V \cong \frac{1}{2} \sum_{i,j=1}^{3N} \left(\frac{\partial^2 V}{\partial q_i \partial q_j}\right)_0 = \frac{1}{2} \sum_{i,j=1}^{3N} f_{ij} q_i q_j \tag{2.17}$$

where f_{ij} are the force constants.

In books on classical mechanics it is shown that, in the absence of external and non-conservative forces, Newton's equations of motion can be written in the following form:

$$\frac{d}{dt}\frac{\partial T}{\partial \dot{q}_j} + \frac{\partial V}{\partial q_j} = 0 \tag{2.18}$$

which yields

$$\ddot{q}_j + \sum_{i=1}^{3N} f_{ij} q_i = 0 \tag{2.19}$$

Equation (2.19) is equivalent to Eq. (2.10) for the diatomic harmonic oscillator, except that it represents not just one but a set of $3N$ linear second-order differential equations for which we can write the general solution, in analogy to Eq. (2.9),

$$q_i = A_i \cos(\sqrt{\lambda} t + \varphi) \tag{2.20}$$

Inserting Eq. (2.20) into Eq. (2.19) yields

$$-A_j\lambda + \sum_{i=1}^{3N} f_{ij}A_i = 0 \tag{2.21}$$

which corresponds to $3N$ linear equations for A_j. These equations only have a solution different from zero if the $3N \cdot 3N$ determinant vanishes (secular equation):

$$\begin{vmatrix} f_{11} - \lambda & f_{12} & f_{13} & \cdots & f_{1,3N} \\ f_{21} & f_{22} - \lambda & f_{23} & \cdots & f_{2,3N} \\ f_{31} & f_{32} & f_{33} - \lambda & \cdots & f_{3,3N} \\ \cdots & \cdots & \cdots & \cdots & \cdots \\ f_{3N,1} & f_{3N,2} & f_{3N,3} & \cdots & f_{3N,3N} - \lambda \end{vmatrix} = 0 \tag{2.22}$$

There are $3N$ solutions for λ corresponding to $3N$ frequencies $\lambda^{1/2}$. As the summation has been made over all $3N$ degrees of freedom, 6 (5) of these solutions refer to translational and rotational motions of the nonlinear (linear) molecules and, therefore, must be zero. Thus, Eq. (2.22) yields only $3N - 6$ ($3N - 5$) non-zero values for λ. The proof for this is lengthy and is not shown here (Wilson et al. 1955). The non-zero solutions correspond to the so-called normal modes.

Once the individual λ_k values have been determined, the amplitudes A_i for each normal mode have to be determined on the basis of in Eq. (2.21).

$$\begin{aligned} (f_{11} - \lambda_k)A_{1k} + f_{12}A_{2k} + \cdots\cdots + f_{1,3N}A_{3N,k} &= 0 \\ f_{21}A_{1k} + (f_{22} - \lambda_k)A_{2k} + \cdots\cdots + f_{2,3N}A_{3N,k} &= 0 \\ \cdots\cdots\cdots\cdots\cdots\cdots\cdots\cdots\cdots\cdots\cdots\cdots\cdots\cdots & \\ f_{3N,1}A_{1k} + f_{3N,2} + \cdots\cdots + (f_{3N,3N} - \lambda_k)A_{3N,k} &= 0 \end{aligned} \tag{2.23}$$

As Eq. (2.23) represents a set of homogeneous equations, only relative amplitudes can be obtained and a normalisation is required, as will be discussed below. The amplitudes A_{ik} describe the character of a normal mode as they quantify the displacements of each atom i in each normal mode k. Eqs. (2.20 and 2.23) imply that in a given normal mode k all atoms vibrate in-phase and with the same frequency $(\lambda_k)^{1/2}$, but with different amplitudes. Thus, it is always an approximation, albeit a useful one in many instances, to characterise normal modes of polyatomic molecules in terms of specific group vibrations, i.e., if only one coordinate dominates the normal mode.

Although the treatment of normal modes in the Cartesian coordinate system is straightforward, it has the disadvantage of distributing all information for a given normal mode among $3N$ equations. In particular, for describing probabilities of vibrational transitions [see Eq. (2.2)] a more compact presentation is desirable. For this purpose, the mass-weighted Cartesian coordinates q_i are converted into normal coordinates Q_k via an orthogonal transformation according to

$$Q_k = \sum_{i=1}^{3N} l_{ik} q_i \tag{2.24}$$

The transformation coefficients l_{ik} are chosen such that T and V, expressed as a function of Q_k, adopt the same form as Eqs. (2.15 and 2.16) and the potential energy does not depend on cross products $Q_k \cdot Q_{k'}$ (with $k \neq k'$). The solution of Newton's equation of motion thus leads to

$$Q_k = K_k \cos(\sqrt{\lambda_k} t + \varphi_k) \tag{2.25}$$

with arbitrary values of K_k and φ_k. The representation of molecular vibrations in normal coordinates is particularly important for the quantum mechanical treatment of the harmonic oscillator (Box 2A).

2.1.2
Internal Coordinates

The normal coordinate system is, mathematically, a very convenient system and, moreover, is required for the quantum chemical treatment of vibrational transitions. However, it is not a very illustrative system as molecular vibrations are usually imagined in terms of stretching or bending motions of molecules or parts of molecules. Such motions cannot be intuitively deduced from a normal coordinate or the array of mass-weighted Cartesian coordinates (Wilson et al. 1955). It is, therefore, desirable to introduce a coordinate system that is based on "structural elements" of molecules, such as bond lengths and angles, and torsional and out-of-plane angles. These so-called internal coordinates are derived from Cartesian displacement coordinates ($\Delta x_\alpha, \Delta y_\alpha, \Delta z_\alpha$) on the basis of the geometry of the molecule.

The displacement of each atom α is defined by the vector $\vec{p}_\alpha(\Delta x_\alpha, \Delta y_\alpha, \Delta z_\alpha)$, which is related to the internal coordinate S_t according to

$$S_t = \sum_{\alpha=1}^{N} \vec{s}_{t\alpha} \cdot \vec{p}_\alpha \tag{2.26}$$

The vector $\vec{s}_{t\alpha}$ is chosen such that it points in the direction of the largest displacement of $\vec{p}_\alpha$ corresponding to the greatest increase in S_t. This statement is best illustrated on the basis of the most simple internal coordinate, the bond stretching coordinate (Fig. 2.3). A stretching coordinate is defined by two atoms ($\alpha = 1, 2$). Thus, for this coordinate the displacement of all other atoms is zero and the sum in Eq. (2.26) only refers to two terms. The largest displacement from the equilibrium positions occur along the axis of the bond assumed to be the x-axis but in opposite directions for atom 1 and 2. Expressing $\vec{s}_{t1}$ and $\vec{s}_{t2}$ in terms of unit vectors we thus obtain

$$\vec{s}_{t1} = \vec{e}_{21} = -\vec{e}_{12} \quad \text{and} \quad \vec{s}_{t2} = \vec{e}_{12} \tag{2.27}$$

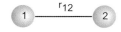
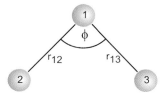

Fig. 2.3 Definition of internal coordinates: top, stretching coordinate; and bottom, bending coordinate.

According to Eq. (2.26), the bond stretching coordinate S_s is then given by

$$S_s = \Delta x_1 - \Delta x_2 \tag{2.28}$$

For the valence angle bending coordinate S_b we have to consider three atoms (Fig. 2.3). To achieve the largest contribution to S_b, the displacements of atoms 1 and 2 and thus $\vec{s}_{t1}$ and $\vec{s}_{t2}$ are perpendicular to the vectors defining the respective bonds between the atoms 1 and 3 and 2 and 3. A unit (infinitesimal) displacement of atom 1 along $\vec{s}_{t1}$ increases ϕ by the amount of $1/r_{31}$. Geometric considerations then yield

$$\vec{s}_{t1} = \frac{\vec{e}_{31} \cdot \cos\phi - \vec{e}_{32}}{r_{31} \cdot \sin\phi} \tag{2.29}$$

and analogous expressions can be derived for $\vec{s}_{t2}$ and $\vec{s}_{t3}$.

$$\vec{s}_{t2} = \frac{\vec{e}_{32} \cdot \cos\phi - \vec{e}_{31}}{r_{32} \cdot \sin\phi} \tag{2.30}$$

$$\vec{s}_{t3} = \frac{[(r_{31} - r_{32}\cos\phi)\vec{e}_{31} + (r_{32} - r_{31}\cos\phi)\vec{e}_{32}]}{r_{31}r_{32}\sin\phi} \tag{2.31}$$

Insertion of the Eqs. (2.29, 2.30, and 2.31) into Eq. (2.26) yields the internal coordinate for the bending motion.

Further internal coordinates are the out-of-plane deformation and torsional coordinates that refer to the angle between a bond and a plane and to a dihedral angle, respectively. In both instances, four atoms are required to define these coordinates (Wilson et al. 1955).

2.1.3
The *FG*-Matrix

For a molecule with $3N - 6$ vibrational degrees of freedom, a complete set of internal coordinates includes $3N - 6$ independent internal coordinates. The definition of this set is straightforward for small molecules such as the triatomic nonlinear molecule in Fig. 2.2, for which two bond stretching and one bond angle deformation coordinates are necessary and sufficient. However, with the increasing size of the molecules, definition of the coordinate set becomes more and more complicated. Thus, it is of particular importance to choose systematic strategies for selection of the internal coordinates. Appropriate protocols have been proposed in the literature, however, sorting out independent and dependent internal coordinates may represent a challenge in many instances (Wilson et al. 1955; Fogarasi et al. 1992). Such redundancies appear in particular in ring systems and have to be removed by appropriate boundary conditions *a posteriori*.

The internal coordinates are independent of the masses of the atoms involved, which are introduced by setting up the so-called G-matrix, which is derived in Wilson et al. (1955). The elements of the G-matrix are given by

$$G_{tt'} = \sum_{\alpha=1}^{N} \mu_\alpha \vec{s}_{t\alpha} \cdot \vec{s}_{t'\alpha} \tag{2.32}$$

where μ_α is the reciprocal mass of atom α.

The G-matrix now contains all the information on the chemical constitution and the structure of the molecule. The elements $G_{tt'}$ represent a $t \cdot t'$ matrix with the number of internal coordinates t being equal to the number of vibrational degrees of freedom. For a nonlinear three-atomic molecule (Fig. 2.2) the G-matrix can be easily calculated using Eq. (2.32) on the basis of the $\vec{s}_{t1}$ vectors defined in Eqs. (2.27–2.31).

$$\begin{vmatrix} \mu_1 + \mu_2 & \mu_3 \cos\phi & \dfrac{-\mu_3 \sin\phi}{r_{32}} \\ \mu_3 \cos\phi & \mu_2 + \mu_3 & \dfrac{-\mu_3 \sin\phi}{r_{31}} \\ \dfrac{-\mu_3 \sin\phi}{r_{32}} & \dfrac{-\mu_3 \sin\phi}{r_{31}} & \dfrac{\mu_1}{r_{31}^2} + \dfrac{\mu_2}{r_{32}^2} - \mu_3\left(\dfrac{1}{r_{31}^2} + \dfrac{1}{r_{32}^2} - \dfrac{2\cos\phi}{r_{31}r_{32}}\right) \end{vmatrix} = G \tag{2.33}$$

In analogy to the treatment in the Cartesian and normal coordinate systems, the next step is to derive expressions for the kinetic and potential energy in terms of the internal coordinates (Wilson et al. 1955). The kinetic energy in terms of internal coordinates is given by

$$T = \frac{1}{2} \sum_{tt'} (G^{-1})_{tt'} \dot{S}_t \dot{S}_{t'} \tag{2.34}$$

where $(G^{-1})_{tt'}$ are the elements of the inverse G-matrix [Eq. (2.33)]. The expression for the potential energy is written as

$$V = \frac{1}{2} \sum_{tt'} F_{tt'} S_t S_{t'} \tag{2.35}$$

The Newton equation of motion then adopts a form similar to Eq. (2.19), with the solution for the differential equation given by

$$s_t = A_t \cos(\sqrt{\lambda} t + \varphi) \tag{2.36}$$

in analogy to Eq. (2.20). Thus, one obtains the secular equation

$$\begin{vmatrix} F_{11} - (G^{-1})_{11}\lambda & F_{12} - (G^{-1})_{12}\lambda & F_{13} - (G^{-1})_{13}\lambda & \cdots\cdots & F_{1n} - (G^{-1})_{1n}\lambda \\ F_{21} - (G^{-1})_{21}\lambda & F_{22} - (G^{-1})_{22}\lambda & F_{23} - (G^{-1})_{23}\lambda & \cdots\cdots & F_{2n} - (G^{-1})_{2n}\lambda \\ F_{31} - (G^{-1})_{31}\lambda & F_{32} - (G^{-1})_{32}\lambda & F_{33} - (G^{-1})_{33}\lambda & \cdots\cdots & F_{3n} - (G^{-1})_{3n}\lambda \\ \cdots\cdots & \cdots\cdots & \cdots\cdots & \cdots\cdots & \cdots\cdots \\ F_{n1} - (G^{-1})_{n1}\lambda & F_{n2} - (G^{-1})_{n2}\lambda & F_{n3} - (G^{-1})_{n3}\lambda & \cdots\cdots & F_{nn} - (G^{-1})_{nn}\lambda \end{vmatrix} = 0 \tag{2.37}$$

which can be expressed in a much simpler way through matrix formalism

$$|F - G^{-1}\lambda| = 0 \tag{2.38}$$

Equation (2.38) is the so-called *FG*-matrix, which upon applying matrix algebra can be re-written in various forms (Wilson et al. 1955).

The secular equations in Eq. (2.37) have t solutions for λ from which the "frequencies" (in wavenumbers) are obtained according to

$$\tilde{v} = \frac{1}{2\pi c}\sqrt{\lambda} \tag{2.39}$$

Once the eigenvalues λ_k have been evaluated, the nature of the normal mode has to be determined by evaluating the relative amplitudes A_{tk}. Using Eq. (2.35), these quantities can be normalised with respect to the potential energy such that the relative contributions of each internal coordinate t to all normal modes and the relative contributions of all internal coordinates in each normal mode sum up to one. This procedure allows for an illustrative description of the character of the normal modes in terms of the potential energy distribution (PED, given in %), e.g., x% of the stretching coordinate t_1, y% of the bending coordinate t_2, etc.

Both the *G*- and the *F*-matrix are symmetric, that is $G_{tt'} = G_{t't}$ and $F_{tt'} = F_{t't}$. This corresponds to $1/2[v(v+1)]$ different $G_{tt'}$ and $F_{tt'}$ elements for a molecule with v vibrational degrees of freedom. Whereas the $G_{tt'}$ elements can be computed readily when the structure of the molecule is known, the $F_{tt'}$ elements are not known *a priori*. Even for a simple three-atomic nonlinear molecule as depicted in Fig. 2.2, there are six different force constants: the stretching force constants F_{11} and F_{22}, referring to the bonds between the atoms 1 and 2 and the atoms 2 and 3, respectively, the bending force constant F_{33}, and the three interaction force constants F_{12}, F_{13}, and F_{23}, which are related to the interactions between the individual stretching and bending coordinates. On the other hand, there are only three normal mode frequencies that can be determined experimentally. This example illustrates the inherent problem of empirical vibrational analysis: the number of observables is always much smaller than the number of unknown force constants.

In some instances, it is possible to utilise the symmetry properties of normal modes (Box 2B) (Wilson et al. 1955; Cotton 1990). For symmetric molecules the normal modes can be classified in terms of the symmetry species of the point group to which the molecule belongs. Each point group is characterised by a set of symmetry operations, such as the reflection in a mirror plane or an n-fold rotation about an n-fold axis of symmetry. Now the individual normal modes are either symmetric or antisymmetric to these operations. For instance, a normal mode that is symmetric to all symmetry operations of the point group is denoted as a totally symmetric mode and thus belongs to the totally symmetric species of the point group. On the basis of group theory, it is possible to determine the number of normal modes for each symmetry species of the point group. This does not just facilitate computing the normal mode frequencies, because the secular determinant can be factorised. Moreover, one may predict IR and Raman activity of the individual modes taking into account the symmetry properties of the dipole moment and polarisability operator (*vide infra*) (Box 2C).

In biological systems, however, many of the molecules to be studied by vibrational spectroscopy lack any symmetry element, such that application of group theory to the analysis of vibrational spectra is restricted to only a few examples. Thus, this topic will not be covered comprehensively in this tutorial, but interested reader should consult specialised monographs (see Box 2B) (Wilson et al. 1955; Cotton 1990).

Essential support for the empirical vibrational analysis is based on isotopically labelled derivatives. A variation of the masses only alters the G-matrix and leaves the F-matrix unchanged. For the simplest case of a diatomic molecule, Eq. (2.11) shows that the frequency varies with the square root of the reciprocal reduced mass. However, for a three-atomic molecule the situation is even more complicated as the individual modes include contributions from three internal coordinates, albeit to different extents. Thus, force constants may be fitted to the experimental data set constituted by the vibrational frequencies of all isotopomers. Whereas for simple molecules with up to 10 atoms this approach has been applied with considerable success, it rapidly approaches practical limitations with an increasing number of atoms, because the synthetic efforts to produce a sufficiently large number of isotopically labelled compounds becomes enormous. Thus, the vibrational problem is inherently underdetermined.

Nevertheless, until the beginning of the 1990s, the empirical vibrational analysis was the only practicable way to extract structural information from the spectra of biological molecules such as porphyrins or retinals (Li et al. 1989, 1990a, 1990b; Curry et al. 1985). The starting point for this approach is a set of empirical force constants that have been found to be appropriate for specific internal coordinates. These force constants are derived from molecules for which a spectroscopic determination of the force field is facilitated due to the smaller size, higher symmetry, and the availability of appropriate isotopomers. Subsequently, the force constant matrix of the molecule under consideration is simplified by appropriate approximations, including the neglect of interaction force constants for internal coordinates of widely separated parts of the molecule. Finally, the normal modes are calculated for the presumed geometry (G-matrix) and adjustments of individual force constants are made to achieve the best possible agreement with the experimental data. This refinement represents the most critical step as it requires a pre-assignment of the experimentally observed bands. Inconsistencies in the assignment and substantial deviations between calculated and experimental frequencies that can only be removed by choosing unusual force constants may then be taken as an indication that the presumed geometry was incorrect. The procedure is then repeated on the basis of alternative molecular structures until a satisfactory agreement between theory and experiment is achieved. It is fairly obvious that the reliability of such a tedious procedure strongly depends on the availability of a sufficiently large set of experimental data.

2.1.4
Quantum Chemical Calculations of the *FG*-Matrix

The alternative approach is to calculate the force constant matrix by quantum chemical methods, which, due to progress in the development of the hardware and efficient and user-friendly program packages, are nowadays applicable to biological molecules, including molecules of more than 50 non-hydrogen atoms. In these methods, an initial ("guess") geometry of a molecule is set up and the Schrödinger equation is solved in self-consistent field calculations, which lead to the energy eigenvalues for this geometry. Systematic variations of internal coordinates then eventually afford the geometry of lowest energy. This energy optimisation allows determination of the force constants by calculating the second derivatives of the potential energy according to Eq. (2.17). Thus, all elements of the *F*- and *G*-matrix can be computed and the normal modes are determined as described above.

The most promising quantum chemical method is based on density functional theory (DFT), which represents an excellent compromise between accuracy and computational costs. Unlike Hartree–Fock procedures, DFT is directed to calculated electron densities rather than wavefunctions. Within this approach, the energy depends on the electron density and this dependency is included in a functional. There are various functionals that have been suggested and tested for calculating different observables. For calculations of vibrational frequencies, the B3LYP functional is widely used and it was found to reproduce experimental data in a satisfactory manner when using a standard 6-31G* basis set (Rauhut and Pulay 1995). Nevertheless, the underlying approximations cause deviations from the experimental frequencies that are approximately in the order of 4%, corresponding to a frequency uncertainty of ca. ± 60 cm^{-1} for modes between 1500 and 1700 cm^{-1}. Considering a medium-sized molecule of 25 atoms, one may expect ca. 50 normal modes in the spectral region between 200 and 1700 cm^{-1} that

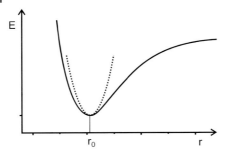

Fig. 2.4 Potential curves for a diatomic oscillator as a function of the inter-atomic distance r. The solid line is a schematic representation of a Morse potential function for an anharmonic oscillator whereas the dotted line refers to the harmonic potential function.

is usually studied by IR and Raman spectroscopy. This corresponds to an average density of modes of ca. 1 mode per 30 cm^{-1}, such that an accuracy of 4% for the calculated frequencies would not allow an unambiguous assignment for all experimentally observed bands.

The errors associated with the DFT calculations result from insufficient consideration of the electron correlation, and, more severely, from the harmonic approximation. The latter effect, illustrated in Fig. 2.4, causes an overestimation of the force constants, as the harmonic potential function is too narrow compared with an anharmonic potential function. These deficiencies of the DFT approach are systematic in nature such that they may be compensated *a posteriori*.

The simplest procedure is to correct the frequencies uniformly by multiplication with an empirical factor. This frequency scaling increases the accuracy of the calculated frequencies to ca. ± 25 cm^{-1}, which, however, is at the limit for an unambiguous vibrational assignment for molecules that include up to 50 atoms.

The most reliable procedure to correct for the intrinsic deficiencies of the quantum chemical calculations is to scale the force field directly. Using scaling factors σ_i that are specific for the various internal coordinates i, one obtains corrected force constants $(F_{ij})_\sigma$ according to

$$(F_{ij})_\sigma = \sqrt{\sigma_i}(F_{ij})\sqrt{\sigma_j} \qquad (2.40)$$

These scaling factors can be determined for small example molecules for which a sound assignment of the experimental bands is established, such that the specific correction factors can be adjusted to yield the best agreement between calculated and experimental data (Rauhut and Pulay 1995; Magdó et al. 1999). The scaling factors are characteristic of specific internal coordinates but not unique for an individual molecule. Thus, they can be transferred to the target molecule and used without any further fine tuning. This concept of global scaling factors has been shown to provide an accuracy of ca. ± 10 cm^{-1} for the calculated frequencies, even for large molecules. Attention has to be paid in the case of hydrogen bonded systems as here the 6-31G* basis set may not be sufficiently large (Mroginski et al. 2005). Applying the global scaling approach, however, requires a coordinate transformation of the force field from Cartesian to internal coordinates, which is not a routine procedure in each case (*vide supra*).

Even on the basis of scaled quantum chemical force fields, the comparison with the calculated frequencies alone does not allow for an unambiguous assignment for many biologically molecules as large as, for example, tetrapyrroles or retinals. Therefore, calculated band intensities are often required as additional assignment criteria. Calculation of IR and Raman intensities is straightforward within the software packages for quantum chemical methods used for the force field calculations. For resonance Raman intensity calculations, tailor-made solutions have to be designed (see Section 2.2.3).

2.2
Intensities of Vibrational Bands

Besides the frequencies of a normal mode, the intensity of the vibrational band is the second observable parameter in the vibrational spectrum. The intensity is simply proportional to the probability of the transition from a vibrational energy level n to the vibrational level m, typically (but not necessarily) corresponding to the vibrational ground and excited states, respectively. To understand the probabilities of transitions between different states that are induced by the interaction of the molecule with electromagnetic radiation, quantum mechanical treatments are required.

Generally, the transition probability P_{nm} is given by the square of the integral

$$P_{nm} = \langle \psi_m^* | \hat{\Omega} | \psi_n \rangle \tag{2.41}$$

where ψ_n and ψ_m are the wavefunctions for the vibrational states n and m, and $\hat{\Omega}$ is the operator that describes the perturbation of the molecule by the electromagnetic radiation. This operator is different for the physical processes in IR and Raman spectroscopy and is obtained by first-order and second-order perturbation theory, respectively.

2.2.1
Infrared Absorption

In IR spectroscopy, the transition $n \rightarrow m$ results from the absorption of a photon and thus the process is controlled by the electrical dipole moment operator $\hat{\mu}_q$, which is defined by

$$\hat{\mu}_q = \sum_\alpha e_\alpha \cdot q_\alpha \tag{2.42}$$

where e_α is the effective charge at the atom α and q_α is the distance to the centre of gravity of the molecule in Cartesian coordinates ($q = x, y, z$) (Wilson et al. 1955). The interaction with the radiation is given by the scalar product between the vector of the electric field of the radiation and $\hat{\mu}_q$. Averaging over all molecule orientations, the IR intensity for this transition is expressed by

$$I_{nm,\,\mathrm{IR}} \propto ([\mu_x]_{nm}^2 + [\mu_y]_{nm}^2 + [\mu_z]_{nm}^2) \tag{2.43}$$

where $[\mu_q]_{nm}$ is the integral

$$[\mu_q]_{nm} = \langle \psi_m^* | \hat{\mu}_q | \psi_n \rangle \tag{2.44}$$

One can easily see that a vibrational transition $n \to m$ in the IR spectrum only occurs if it is associated with a non-zero transition dipole moment $[\mu_q]_{nm}$. To decide whether or not, a normal mode is IR active, we expand $\hat{\mu}_q$ in a Taylor series with respect to the normal coordinates Q_k. Within the harmonic approximation, the series is restricted to the linear terms

$$\hat{\mu}_q = \mu_q^0 + \sum_{K=1}^{3N-6} \hat{\mu}_q^k Q_k \tag{2.45}$$

with

$$\hat{\mu}_q^k = \left(\frac{\partial \mu_q}{\partial Q_k} \right)_0 \tag{2.46}$$

With Eq. (2.41) the transition probability is then given by

$$[\hat{\mu}_q] = \langle \psi_m^* | \hat{\mu}_q | \psi_n \rangle = \mu_q^0 \langle \psi_m^* \psi_n \rangle + \sum_{k=1}^{3N-6} \hat{\mu}_q^k \langle \psi_m^* | Q_k | \psi_n \rangle \tag{2.47}$$

The first integral on the right-hand side of Eq. (2.47) is zero as the wavefunctions ψ_n and ψ_m are orthogonal. Thus, a non-zero transition probability is only obtained if two conditions are fulfilled. Firstly, the derivative of the dipole moment with respect to the normal coordinate Q_k in Eq. (2.47) must be non-zero, which requires that the normal mode is associated with a change in the dipole moment. Secondly, the integral $\langle \psi_m^* | Q_k | \psi_n \rangle$ must be non-zero, which is the situation when the vibrational quantum numbers n and m differ by one. This implies that only fundamentals are IR active within the harmonic approximation.

Equation (2.47) holds for all three Cartesian coordinates such that only one non-zero transition dipole moment $[\mu_q]_{nm}$ ($q = x, y, z$) is sufficient to account for the IR intensity of the normal mode Q_k according to Eq. (2.43) (Box 2C). Using unpolarised light and randomly oriented molecules, the experiment does not allow the conclusion to be made as to which of the components of the transition

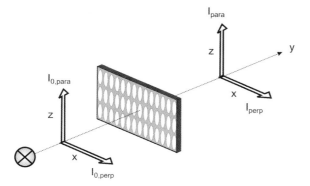

Fig. 2.5 IR dichroism of oriented molecules. The absorbance of light polarised parallel to the long molecular axis A_{para} is given by the ratio $I_{para}/I_{0,para}$ whereas the perpendicular component is defined by $I_{perp}/I_{0,perp}$.

dipole moment contributes to the IR intensity. If, however, the probe light is linearly polarised, it is possible to address the individual components of the transition dipole moments. Then IR measurements may provide additional information for the vibrational assignment, the orientation of the molecules with respect to the plane of polarisation of the incident light, or the orientation of a molecular building block within a macromolecule if the macromolecule itself is oriented.

Consider, for example, a sample of ellipsoidal molecules that are all oriented with the long axis in z-direction (Fig. 2.5). The incident light, propagating in the y-direction, can be polarised in z- or x-direction, corresponding to a parallel and perpendicular orientation of electric field vector, respectively. Parallel polarised light will thus specifically probe those vibrational modes that exhibit a transition dipole moment in the z-direction, and the absorbance A_{para} is given by

$$A_{para} \propto |\mu_{mn}|_z^2 \tag{2.48}$$

If the molecules do not exhibit a preferential orientation in the xy plane, IR absorption of perpendicular polarised light depends on both the x- and the y-component of transition dipole moment

$$A_{perp} \propto (|\mu_{mn}|_x^2 + |\mu_{mn}|_y^2) \tag{2.49}$$

The quantities A_{para} and A_{perp} are combined in the dichroic ratio d which is defined by

$$d = \frac{A_{para} - A_{perp}}{A_{para} + A_{perp}} \tag{2.50}$$

and which may vary between 1 and -1 for the limiting cases of $A_{perp} = 0$ and $A_{para} = 0$, respectively.

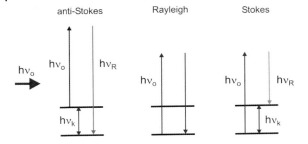

Fig. 2.6 Energy diagram representing the elastic Rayleigh scattering (centre) and the inelastic anti-Stokes (left) and Stokes (right) Raman scattering with v_0, v_R, and v_k referring to the frequencies of the incident light, the Raman scattered light, and the molecular vibration, respectively.

2.2.2
Raman Scattering

Raman spectroscopy differs principally from IR spectroscopy in that it is based on the scattering of photons by molecules rather than on the absorption of photons (Fig. 2.6). This scattering process can be illustrated readily on the basis of classical physics. Consider a molecule interacting with an electromagnetic wave with the electric field vector oscillating with the frequency v_0

$$\vec{E} = \vec{E}_0 \cos(2\pi v_0 t) \tag{2.51}$$

The oscillating electric field can induce a dipole in the molecule

$$\vec{\mu}_{ind} = \tilde{\alpha}(v) \cdot \vec{E}_0 \cos(2\pi v_0 t) \tag{2.52}$$

where $\tilde{\alpha}(v)$ is the polarisability. This quantity, which is actually a tensor, varies with time as it describes the response of the electron distribution to the movements of the nuclei that oscillate with the normal mode frequency v_k. Thus, we can express $\tilde{\alpha}(v)$ by

$$\tilde{\alpha}(v) = \tilde{\alpha}_0(v_0) + \left(\frac{\partial \tilde{\alpha}}{\partial Q_k}\right)_0 (2\pi v_k t) \tag{2.53}$$

Combining Eqs. (2.52 and 2.53) we obtain

$$\vec{\mu}_{ind} = \vec{E}_0 \left[\tilde{\alpha}_0(v_0) + \left(\frac{\partial \tilde{\alpha}}{\partial Q_k}\right)_0 \cos(2\pi v_k t)\right] \cos(2\pi v_0 t) \tag{2.54}$$

which eventually yields

$$\vec{\mu}_{ind} = \vec{E}_0 \left[\tilde{\alpha}_0 \cos(2\pi v_0 t) + \left(\frac{\partial \tilde{\alpha}}{\partial Q_k}\right)_0 Q_k \cos[2\pi(v_0 + v_k)t] \right.$$
$$\left. + \left(\frac{\partial \tilde{\alpha}}{\delta Q_k}\right)_0 Q_k \cos[2\pi(v_0 - v_k)t] \right] \tag{2.55}$$

The sum on the right side of Eq. (2.55) includes three terms corresponding to polarisabilities that depend on different frequencies, which are the frequency of the incident radiation v_0 and the frequencies $(v_0 - v_k)$ and $(v_0 + v_k)$ that differ from v_0 by the frequency of the normal mode. Scattering that leaves the frequency of the incident light unchanged is referred to as elastic or Rayleigh scattering whereas the frequency-shifted (inelastic) scattering is referred to as Raman scattering (Fig. 2.6). When the frequency of the scattered light is lower than v_0, the molecule remains in a higher vibrationally excited state ($m > n$ for the transition $n \rightarrow m$). This process is denoted as Stokes scattering whereas anti-Stokes scattering refers to $(v_0 + v_k)$ and thus to $m < n$. At ambient temperature, thermal energy is lower than the energies of most of the normal modes, such that molecules predominantly exists in the vibrational ground state and Stokes scattering represents the most important case of Raman scattering.

The energy conservation for Raman scattering is not contained in the classical treatment. It requires the quantum mechanical description of vibrational quantum states interacting with electromagnetic radiation (Placzek 1934). The operator, which according to Eq. (2.41) determines the probability of the Raman transition $n \rightarrow m$, is the polarisability $\hat{\alpha}$ with components defined by the molecule-fixed coordinates x, y, z.

$$\begin{bmatrix} \mu_{ind,x} \\ \mu_{ind,y} \\ \mu_{ind,z} \end{bmatrix} = \begin{bmatrix} \alpha_{xx} & \alpha_{xy} & \alpha_{xz} \\ \alpha_{yx} & \alpha_{yy} & \alpha_{yz} \\ \alpha_{zx} & \alpha_{zy} & \alpha_{zz} \end{bmatrix} \cdot \begin{bmatrix} E_x \\ E_y \\ E_z \end{bmatrix} \tag{2.56}$$

It is useful to define the Raman scattering cross section for the vibrational $n \rightarrow m$ transition $\sigma_{n \rightarrow m}$ by

$$I_{n \rightarrow m} = \sigma_{n \rightarrow m} I_0 \tag{2.57}$$

where I_0 is the intensity of the incident radiation and $I_{n \rightarrow m}$ is the scattered intensity integrated over all scattering angles and polarisation directions for a non-oriented sample. The Raman cross section is correlated with the Raman polarisability by

$$\sigma_{n \rightarrow m} \propto (v_0 \pm v_k)^4 \cdot \sum_{\rho, \sigma} |\alpha_{\rho, \sigma}|^2 \tag{2.58}$$

taking into account that the intensity for electric dipole radiation scales with the fourth power of the frequency. The indices ρ and σ denote the molecule-fixed coordinates.

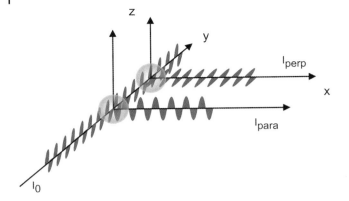

Fig. 2.7 Illustration of polarised Raman scattering in a 90°-scattering geometry. I_{perp} and I_{para} are the respective perpendicular and parallel components of the Raman scattering with respect to the polarisation of the incident monochromatic radiation.

The tensor properties of $\alpha_{\rho\sigma}$ have consequences for the polarisation of the Raman scattered light. This can easily be illustrated for molecules that are oriented with respect to the electric field vector of the incident excitation radiation, which is usually highly polarised laser light (Fig. 2.7). If we now refer the molecule-fixed coordinate system to the directions of the electric field vector (z-axis), of the propagation of the laser light (y-axis), and of the detection of the scattered light (x-axis), the intensity of the Raman scattered radiation includes two components, which are polarised parallel (I_{para}) and perpendicular (I_{perp}) with respect to the electric field vector of the incident light, and only depend on the tensor components α_{zz} and α_{yz}, respectively.

$$I_{para} \propto |\alpha_{zz}|^2 \cdot |\vec{E}_z|^2$$
$$I_{perp} \propto |\alpha_{yz}|^2 \cdot |\vec{E}_z|^2 \tag{2.59}$$

Appropriate re-alignment of the oriented sample with respect to the laser beam then allows the remaining tensor components to also be determined, taking into account that the scattering tensor is symmetric, i.e., $\alpha_{\sigma\rho} = \alpha_{\rho\sigma}$.

In the general case of non-oriented molecules, the relationship between the polarisation properties of the Raman scattered light and the scattering tensor is more complicated and the individual tensor components cannot be determined directly. For the depolarisation ratio ρ, which is defined by the ratio of perpendicular polarised and parallel polarised Raman scattered light, one obtains

$$\rho = \frac{I_{perp}}{I_{para}} = \frac{3\gamma_s^2 + 5\gamma_{as}^2}{45\bar{\alpha}^2 + 4\gamma_s^2} \tag{2.60}$$

with the mean polarisability $\bar{\alpha}$ and the symmetric (γ_s) and antisymmetric (γ_{as}) anisotropy given by

$$\bar{\alpha} = \frac{1}{3}\sum_{\rho\rho}\alpha_{\rho\rho}$$

$$\gamma_s^2 = \frac{1}{2}\sum_{\rho\sigma}(\alpha_{\rho\rho} - \alpha_{\sigma\sigma})^2 + \frac{3}{4}\sum_{\rho\sigma}(\alpha_{\rho\sigma} + \alpha_{\sigma\rho})^2 \qquad (2.61)$$

$$\gamma_{as}^2 = \frac{3}{4}\sum_{\rho\sigma}(\alpha_{\rho\sigma} - \alpha_{\sigma\rho})^2$$

In the case of symmetric molecules, determination of polarisation ratios is a useful tool for the vibrational assignment. For each symmetry species of a point group, the character table indicates which of the individual tensor components are zero and which are non-zero (Wilson et al. 1955; Czernuszewicz and Spiro 1999) (Box 2C). On this basis, it is not only possible to decide if the modes of this species are Raman-active, but also to predict the polarisation ratio, which can then be compared with the experimentally observed value.

As in IR absorption spectroscopy, a description of Raman intensities is only possible on the basis of a quantum mechanical treatment (Placzek 1934). In Raman scattering two photons are involved, hence second-order perturbation theory is required. On the basis of Kramers–Heisenberg–Dirac's dispersion theory, the scattering tensor is expressed as

$$[\alpha_{nm}]_{\rho\sigma} = \frac{1}{h}\sum_{R,r}\left(\frac{\langle nG|M_\rho|Rr\rangle\langle rR|M_\sigma|Gm\rangle}{v_{Rr} - v_k - v_0 + i\Gamma_R} + \frac{\langle rR|M_\sigma|Gm\rangle\langle nG|M_\rho|Rr\rangle}{v_{Rr} - v_k + v_0 + i\Gamma_R}\right) \qquad (2.62)$$

where $M_\sigma(M_\rho)$ is the electronic transition dipole moment in terms of a molecule-fixed coordinate system (Albrecht 1961; Warshel and Dauber 1977). The symbols v_0 and v_k denote the frequency of the excitation radiation and the normal mode Q_k, respectively (Fig. 2.6). The indices "R" and "r" refer to the respective electronic and vibrational (vibronic) states of the molecule and Γ_R is a damping constant that is related to the lifetime of the vibronic state Rr. Equation (2.62) represents a sum of integrals that describe the transitions $nG \rightarrow Rr$ and $Rr \rightarrow Gm$. The sum indicates that for the Raman transition all vibronic states have to be considered. This implies that the scattering tensor and thus the Raman intensity is controlled by the transition probabilities involving all vibronic states, even though the initial and final states refer to the vibrational ground and excited states of the electronic ground state.

In the general case, the energy of the exciting radiation hv_0 is much lower than the energy of any vibronic transition hv_{Rr}. Let us consider an example of a molecule with the first four electronic transitions E_i at 300, 240, 200, and 180 nm. Using an excitation line at 1064 nm and neglecting the damping terms, we can determine the relative weights of the individual terms in the sum of Eq. (2.62),

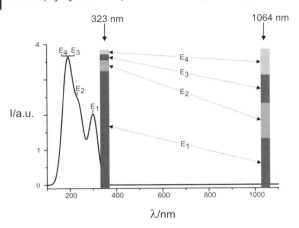

Fig. 2.8 Absorption spectrum of a molecule with four electronic transitions (E_1, E_2, E_3, and E_4). The columns represent the frequency-dependent weights of the electronic transitions to the scattering tensor [Eq. (2.62)] for excitation lines at 1064 and 323 nm. The calculations refer to a vibrational mode at 1500 cm^{-1}.

as given by the denominators, i.e., $(\nu_{Rr} - \nu_0 - \nu_k)^{-1}$ and $(\nu_{Rr} + \nu_0 - \nu_k)^{-1}$ (Fig. 2.8). One can easily see that under these conditions the contributions of the four electronic states to the scattering tensor are comparable, as shown for the normal mode ν_k at 1500 cm^{-1}. The situation is different when the excitation energy approaches the energy of an electronic transition. Then, for this specific transition the term $(\nu_{Rr} - \nu_0 - \nu_k)^{-1}$ dominates over all other terms in the sum of Eq. (2.62). This is demonstrated for $\nu_0 = 30\,960$ cm^{-1} (323 nm) for which the contribution from the first electronic transition is more than 10 times larger than that from the second transition.

2.2.3
Resonance Raman Effect

The conditions under which ν_0 is close to the frequency of an electronic transition refer to the resonance Raman (RR) effect, for which Eq. (2.62) consequently can be simplified to

$$[\alpha_{nm}]_{\rho\sigma} \cong \frac{1}{\hbar} \sum_r \left(\frac{\langle nG|M_\rho|Rr\rangle\langle rR|M_\sigma|Gm\rangle}{\nu_{Rr} - \nu_k - \nu_0 + i\Gamma_R} \right) \qquad (2.63)$$

where summation is now restricted to the vibrational states r of the resonant electronically excited state (Albrecht 1961; Warshel and Dauber, 1977). In contrast to Eq. (2.44), the wavefunctions of the integrals in Eq. (2.63) depend on the electronic and nuclear coordinates, which can be separated within the Born–Oppenheimer approximation according to

$$\langle nG|M_\rho|Rr\rangle = \langle nr\rangle\langle G|M_\rho|R\rangle = \langle nr\rangle M_{GR,\rho} \tag{2.64}$$

Here, integrals of the type $\langle nr\rangle$ represent the Franck–Condon factors that are the integrals over the products of two vibrational wavefunctions. With this approximation, Eq. (2.63) leads to

$$[\alpha_{nm}]_{\rho\sigma} \cong \frac{1}{h}\sum_r \left(\frac{\langle nr\rangle\langle rm\rangle M_{GR,\rho}M_{GR,\sigma}}{v_{Rr}-v_k-v_0+i\Gamma_R}\right) \tag{2.65}$$

The electronic transition dipole moment components $M_{GR,\rho}$ which refer to the electronic transition from the ground state G to the (resonant) electronically excited state R, can now be expanded in a Taylor series with respect to the normal coordinates Q_k.

$$M_{GR,\rho}(Q_k) = M_{GR,\rho}(Q_k^{(0)}) + \sum_k \left(\frac{\partial M_{GR,\rho}}{\partial Q_k}\right)_0 Q_k + \cdots \tag{2.66}$$

Within the harmonic approximation we neglect higher order terms and combine Eqs. (2.65 and 2.66) to obtain the scattering tensor as the sum of two terms, the so-called Albrecht's A- and B-terms.

$$[\alpha_{nm}]_{\rho\sigma} \cong A_{\rho\sigma} + B_{\rho\sigma} \tag{2.67}$$

with

$$A_{\rho\sigma} \cong \frac{1}{h}\sum_r \left(\frac{\langle nr\rangle\langle rm\rangle M^0_{GR,\rho}M^0_{GR,\sigma}}{v_{Rr}-v_k-v_0+i\Gamma_R}\right) \tag{2.68}$$

where $M^0_{GR,\rho}$ and $M^0_{GR,\sigma}$ are the components of transition dipole moment of the vertical electronic transition $G \to R$. The B-term is given by

$$B_{\rho\sigma} \cong \frac{1}{h}\sum_r \left(\frac{\langle n|Q_k|r\rangle\langle rm\rangle \left(\frac{\partial M_{GR,\rho}}{\partial Q_k}\right)_0 M^0_{GR,\sigma}}{v_{Rr}-v_k-v_0+i\Gamma_R} + \frac{\langle nr\rangle\langle r|Q_k|m\rangle \left(\frac{\partial M_{GR,\sigma}}{\partial Q_k}\right)_0 M^0_{GR,\rho}}{v_{Rr}-v_k-v_0+i\Gamma_R}\right) \tag{2.69}$$

The A- and B-terms represent different scattering mechanisms; however, common to both terms is that the dominators rapidly decrease when the frequency of the excitation line v_0 approaches the frequency of an electronic transition. Then both the A- and the B-terms and thus the RR intensity increase [Eqs. (2.57

and 2.58)], albeit to a different extent depending on the character of the electronic transitions and normal modes involved.

When the resonant electronic transition exhibits a large oscillator strength, that is a large transition dipole moment M_{GR}^0, the A-term, which scales with $|M_{GR}^0|^2$, increases more significantly than the B-term and thus becomes the leading term. Then the enhancement of a normal mode depends on the Franck–Condon factor products $\langle nr \rangle \langle rm \rangle$ (Franck–Condon enhancement). Whether or not a normal mode is resonance enhanced via the Franck–Condon mechanism, depends on the geometry of the resonant excited state.

Consider, for example, ethylene, which exhibits a strong electronic absorption band at ca. 200 nm, originating from the (first) allowed $\pi \to \pi^*$ transition of the C=C double bond. Population of the π^* orbital leads to the lowering of the bond order from 2 to 1, implying that in this excited state the C–C bond length increases. Hence, the potential energy curve for the C–C distance is displaced in this electronically excited state with respect to the ground state, as shown in Fig. 2.9, i.e., along the main internal coordinate of this normal mode. Now we must consider the Franck–Condon factors that couple the wavefunctions of the ground and excited state. According to Eq. (2.68), we have to sum up over those integrals that involve all vibrational states of the electronic excited state, with their relative weight being determined by the match with the excitation energy. For the sake of simplicity we will restrict the discussion to the RR transition $n \to m$ exclusively

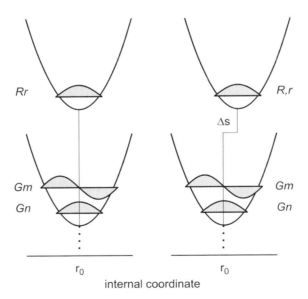

Fig. 2.9 Harmonic potential curves for the electronic ground (G) and excited state (R), illustrating the situation for an excited state displacement of $\Delta s = 0$ (left) and $\Delta s \neq 0$ (right) for RR transitions from the vibrational ground state Gn to the first vibrationally excited state Gm via coupling to the vibronic state Rr.

via the vibronic state Rr such that we only consider one term of the sum, i.e., the integrals involving the vibronic state Rr (Fig. 2.9). The wavefunctions of the first integral $\langle nr \rangle$ have the same symmetry, but due to the potential curve displacement in the direction of the internal coordinate of this mode the integral is non-zero, albeit smaller than one. For the same reason, the wavefunctions of the integral $\langle rm \rangle$ that have different symmetries are not zero either. Similar considerations hold for Franck–Condon factors involving other vibronic states. This implies that a vibrational mode including the C=C stretching coordinate exhibits non-zero Franck–Condon factor products and thus gains resonance Raman (RR) intensity via the A-term scattering mechanism (Box 2C).

The situation is different for modes including the C–H stretching coordinate (Fig. 2.9). As the C–H distance remains unaffected upon the $\pi \rightarrow \pi^*$ electronic transition, the origin of the corresponding potential curve is at the same position in the electronic ground and excited states. In this instance the integral $\langle nr \rangle$ will be non-zero but the $\langle rm \rangle$ integral will vanish, implying that a pure C–H stretching mode will not be RR active. The fact that C–H stretching modes are Raman-active instead just reflects the involvement of higher lying electronic transitions, amongst which at least one is associated with a geometry change of the C–H stretching coordinate.

The importance of excited-state displacements for the RR intensity can be generalised (Warshel and Dauber 1977). Only modes including at least one internal coordinate that changes with the electronic transition can gain resonance enhancement via the A-term mechanism. The relationship between the excited state displacement Δs and the A-term contribution to the scattering tensor can be approximated by

$$A \propto \frac{|M_{GR}^0|^2 \cdot v_k \cdot \Delta s}{(v_{Rr} - v_0 + i\Gamma_R)(v_{Rr} - v_0 + v_k + i\Gamma_R)} \quad (2.70)$$

This expression assumes that the normal modes do not change in the electronic excited state, both with respect to their composition and with respect to the frequency. Although this appears to be a severe restriction, Eq. (2.70) has been shown to provide a good basis for qualitative predictions of the RR activity of modes that are dominated by one internal coordinate, given that the nature of the resonant excited state is known. The situation is much more complicated for modes involving many internal coordinates to comparable extents, and *a priori* predictions are impossible. The contributions from different internal coordinates to Δs may have positive or negative sign, such that they can be additive or cancel each other (Mroginski et al. 2003).

Calculation of RR intensities still represents a challenge as both ground state (structure, vibrational frequencies) and excited state properties (excitation energies, structure) have to be treated. In a particularly promising concept, quantum mechanical calculations are combined with the use of experimental data. Within the framework of the so-called Transform theory, the frequency dependence of the A-term of the scattering tensor [Eq. (2.70)] may be obtained by the Kramers–

Kronig transformation of the absorption band (Peticolas and Rush 1995). The electronic transition dipole moments M_{GR}^0 and the excited state displacements of a given normal mode Δs are calculated by quantum chemical methods such as time-dependent DFT. At present, these approaches are still fairly time-consuming and, specifically, the calculation of excited state geometries, required for determining Δs, is not trivial, even for relatively small molecules.

The crucial parameter that controls resonance enhancement via the B-term scattering mechanism is the derivative of the electronic transition dipole moment with respect to the normal mode (Albrecht 1961). This derivative is large for those modes that can effectively couple to an electronic transition and thus may gain RR intensity even when the resonant electronic transition is relatively weak. This vibronic coupling enhancement may also be operative when the excitation frequency is close to the frequencies of two electronic transitions.

An alternative approach to describe the Franck–Condon-type resonance enhancement (A-term scattering) has been developed by Heller who described the scattering process in terms of wave-packet dynamics (Heller 1981). As the temporal evolution of the scattering process is much faster than the nuclear vibrations, consideration of the summation over the multitude of vibronic eigenstates is definitely not necessary. The RR intensity is induced by the force $\partial V/\partial Q_k$ exerted by the excited state potential surface on the nuclear ground state configuration. This force controls the temporal evolution of the ground state wavefunction $|n(t)\rangle$ after excitation to the excited state surface and before returning to the final state $|m\rangle$. The RR intensity for a given mode k is then given by

$$I_{RR}(k) \propto \left[\int_0^\infty |M_{GR}^2 \cdot \exp(iv_0 t - \Gamma_r t) \cdot \langle m|n(t)\rangle \, dt| \right]^2 \tag{2.71}$$

This expression is analogous to the square of Eq. (2.68). Furthermore, the treatments of the RR scattering in the time-domain by Heller (Heller 1981) and in the frequency domain via the Kramers–Heisenberg–Dirac dispersion relation are linked to each other through the Fourier transform. Also, in the time-domain treatment further simplifications can be introduced using similar assumptions as those used for Eq. (2.70). The advantage of Heller's approach is particularly evident for those instances where only a few internal coordinates exhibit large excited state displacements Δs and thus the RR spectrum is dominated only by a couple of normal modes. For these modes the RR intensities does not only depend on Δs but also on the magnitude of $\partial V/\partial Q_k$. The intensity ratio of two normal modes k and l is then given by

$$\frac{I_{RR}(k)}{I_{RR}(l)} = \frac{v_l}{v_k} \cdot \left(\frac{\partial V/\partial Q_k}{\partial V/\partial Q_l} \right)^2 \approx \frac{v_k^3 \cdot \Delta_k^2}{v_l^3 \cdot \Delta_l^2} \tag{2.72}$$

This approach has been successfully employed to analyse the RR spectra of metalloproteins upon excitation in resonance with charge-transfer transitions (Blair et al. 1985).

The expressions in Eqs. (2.67–2.72) represent approximations that can be employed when the contributions from one or two electronic transitions to the scattering tensor become dominant. Consequently, there is no sharp borderline between non-resonance Raman and RR scattering. Resonance enhancement simply means that the vibrational modes of a chromophore within a molecule selectively gain intensity when the excitation line is in resonance with an electronic transition. Such an enhancement does not necessarily require an exact frequency match of the excitation and the electronic transition. Indeed, a closer inspection of the frequency dependence of the Raman intensity reveals that a specific enhancement of these modes already occurs for excitation lines relatively far away from the maximum of the electronic transition. This is illustrated by comparing the frequency dependence of the Raman intensity under strictly non-resonance conditions [Eqs. (2.57 and 2.58)] and the RR intensity provided by the A-term scattering mechanism. In the latter, an approximate formula can be derived for the Stokes scattering under pre-resonance conditions (Albrecht and Hutley 1971), i.e., when the damping constant [Eq. (2.68)] can be neglected ($|v_R - v_0| \gg |\Gamma_R|$),

$$I_{nm, RR} \propto (v_0 - v_k)^4 \frac{(v_0 - v_k)^2 (v_0^2 + v_R^2)}{(v_R^2 - v_0^2)^2} \tag{2.73}$$

To compare the RR and Raman intensities, Eq. (2.73) has to be corrected for the "normal" v^4-dependence of the radiation intensity [Eqs. (2.57 and 2.58)]

$$\frac{I_{nm, RR}}{I_{nm, Ra}} \propto \frac{(v_0 - v_k)^2 (v_0^2 + v_R^2)}{(v_R^2 - v_0^2)^2} \tag{2.74}$$

The ratio $I_{nm, RR}/I_{nm, Ra}$ represents the resonance enhancement, which in Fig. 2.10 is plotted as a function of the frequency difference between the excitation and the the electronic transition. It can be seen that the enhancement factor solely associated with the frequency dependence of the scattering tensor strongly increases for $v_0 \rightarrow v_R$ and even for an energy gap of 5000 cm^{-1} it is nearly 500. For more rigorous resonance conditions, for which Eq. (2.74) is no longer valid, the resonance enhancement can reach 5–6 orders of magnitude, depending on the other quantities that control the scattering tensor, i.e., the Franck–Condon factor products, the square of the electronic transition dipole moment, and the damping constant [Eq. (2.68)]. As a rule of thumb, the resonance enhancement then scales with the square of the extinction coefficient of the electronic absorption band at the excitation line. Consequently, the sensitivity of Raman spectroscopy increases greatly under resonance conditions and approaches that of UV–vis absorption spectroscopy.

It should be emphasised at this point that the frequencies refer to the molecular structure in the initial electronic state, usually the electronic ground state, although the RR intensities are sensitively controlled by the properties of the electronically excited state(s). Resonance enhancement represents the gain in Raman

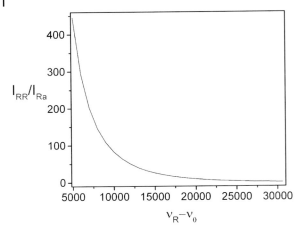

Fig. 2.10 Resonance enhancement I_{RR}/I_{Ra} via the A-term mechanism [Eq. (2.74)] as a function of the difference between the frequency of the incident light v_0 and the electronic transition v_R.

intensity for the vibrational bands of a chromophore as compared with strictly non-resonance conditions. It is restricted to that part of the molecule in which the resonant electronic transition is localised. Such a selective enhancement has enormous practical implications when we are interested in studying the vibrational spectrum of the specific constituents of a macromolecule, such as a cofactor within a protein. Choosing appropriate excitation lines will then allow probing the vibrational bands solely of the cofactor such that the non-resonant Raman bands of the apo-protein can be effectively discriminated in most instances.

2.3
Surface Enhanced Vibrational Spectroscopy

Thirty years ago it was discovered accidentally that molecules adsorbed on rough surfaces of certain metals may experience a drastic enhancement of the Raman scattering. This unexpected finding immediately prompted intensive research activities in this field. Initially, these studies were directed at elucidating the nature of this enormous enhancement, which could be of several orders of magnitude. Later, it was also recognised that this technique had a high potential for studying molecules at interfaces. Since that time, the surface enhanced Raman scattering (SERS) effect has become largely understood and this technique has found a place not only in surface and interfacial science but also in biophysics. Moreover, it was also noticed that the enhancement of vibrational bands at metal surfaces is not restricted to Raman scattering but may also take place, albeit to a smaller extent, in IR absorption, i.e., surface enhanced infrared absorption (SEIRA). SERS and SEIRA have been observed at metal interfaces with solid, liquid, or gas (vacuum) phases, but only the solid/solution interface is of relevance in biological applications. The enhancement is metal- and wavelength-specific, such that for the typical spectral range of Raman and IR spectroscopy from 400 to 10000 nm, the metals that can be employed are Ag and Au.

2.3.1
Surface Enhanced Raman Effect

The SER effect can largely be understood on the basis of classical electromagnetic theory (Kerker et al. 1980; Moskovits 1985). The starting point is the analysis of light scattering and absorption by colloidal particles, which was considered about a century ago by G. Mie (Mie 1908). In a first approximation, colloidal particles can be represented as spheres. When the size of the particles is small with respect to the wavelength of the incident light (Rayleigh limit), the electromagnetic field can effectively couple with the collective vibrations of the "free electrons" of the metal, which are denoted as surface plasmons (Kerker et al. 1980; Moskovits 1985). The incident electric field $\vec{E}_0(\nu_0)$, oscillating with the frequency ν_0, induces an electric dipole moment in the sphere and excites the surface plasmons, causing an additional electric field component $\vec{E}_{ind}(\nu_0)$, normal to the surface in the near-field of the sphere. Thus, the total electric field of the frequency ν_0 is then expressed by

$$\vec{E}_{tot}(\nu_0) = \vec{E}_0(\nu_0) + \vec{E}_{ind}(\nu_0) \tag{2.75}$$

Since $\vec{E}_{ind}(\nu_0)$ is a function of $\vec{E}_0(\nu_0)$ the enhancement of the electric field is given by

$$F_E(\nu_0) = \frac{|\vec{E}_0(\nu_0) + \vec{E}_{ind}(\nu_0)|}{|\vec{E}_0(\nu_0)|} = |1 + 2g_0| \tag{2.76}$$

The quantity g_0 is related to the dielectric properties of the metal through

$$g_0 = \frac{\tilde{\varepsilon}_r(\nu_0) - 1}{\tilde{\varepsilon}_r(\nu_0) + 2} \tag{2.77}$$

Here $\tilde{\varepsilon}_r(\nu_0)$ is the frequency-dependent dielectric constant divided by the square of the refractive index of the surrounding medium n_{solv}.

$$\tilde{\varepsilon}_r(\nu_0) = \frac{\varepsilon_{re}(\nu_0) + i\varepsilon_{im}(\nu_0)}{n_{solv}^2} \tag{2.78}$$

Equations (2.76–2.78) show that g_0, and thus the field enhancement $F_E(\nu_0)$, becomes large if the real part of the relative dielectric constant approaches -2 and the imaginary part is small. These conditions depend on the wavelength and are, within the Rayleigh limit, particularly well matched for Ag and Au colloids at ca. 400 and 560 nm, respectively.

A molecule that is located in close proximity to the sphere is excited by the electric field $\vec{E}_{tot}(\nu_0)$, which may induce all possible photophysical and photochemical processes, including Raman scattering. If the resonance conditions are fulfilled for the incident light with ν_0, this will also be approximately the case for the frequency of the Raman scattered light $\nu_0 \pm \nu_k$. Then, the electric field of the Raman scattered light, $\vec{E}_{Ra}(\nu_0 \pm \nu_k)$, of the normal mode k that is proportional to $\vec{E}_{tot}(\nu_0)$ also induces a secondary electric field component $\vec{E}_{Ra,ind}(\nu_0 \pm \nu_k)$ in the metal particle giving rise to a total electric field oscillating with $\nu_0 \pm \nu_k$ in analogy to Eq. (2.75). Thus, Eq. (2.76) holds for both the electric field of the exciting and the Raman scattered light. As the intensity of the Raman scattered light in the far field is proportional to $|\vec{E}_{Ra,tot}(\nu_0 \pm \nu_k)|^2$, the total surface enhancement factor of the Raman intensity is given by

$$F_{SER}(\nu_0 \pm \nu_k) = [(1 + 2g_0)(1 + 2g_{Ra})]^2 \tag{2.79}$$

Equation (2.79) indicates that even a field enhancement by a factor of 10 yields an enhancement of the Raman intensity of more than 10^4.

These simple considerations hold for particles much smaller than the wavelength and are thus independent of the shape and the size of the particles. The enhancement at larger particles, however, is no longer shape- and size-independent. One can show that, in general, larger spheres or ellipsoidal shapes cause a red-shift of the wavelength of maximum enhancement. This is also true for interacting particles, i.e., aggregated colloids.

A particularly large enhancement is predicted for tips of the type used in scanning probe microscopy. These predictions have indeed been confirmed and utilised in so-called tip-enhanced Raman spectroscopy (Kneipp et al. 2006). In this technique, a sharp tip, as used for scanning tunnelling or atomic force microscopy, is brought into close proximity with that part of a molecular sample that is in the focus of the incident laser beam. Only the molecules in the near-field of the tip experience an enhancement of the Raman scattering.

For biological applications, metal electrodes represent more versatile SER-active devices as they allow probing of potential-dependent processes by controlling the electrode potential (Murgida and Hildebrandt 2004, 2005). The enhancement mechanism on electrodes can be largely understood within the same theoretical framework as outlined above for the SER effect on metal colloids. A submicroscopic roughness of the electrode, typically generated by electrochemical roughening (oxidation–reduction cycles), is a prerequisite for the SER effect on metal electrodes. The scale of this roughness is comparable to the dimensions of SER-active metal colloids, such that an SER-active electrode can be approximated by an array of metal semispheres, for which a treatment of the field enhancement similar to isolated metal spheres is possible. Taking into account that the scale of roughness which is produced by electrochemical roughening of the electrode surface is approximately the same as the dimension of SER-active colloids, one may

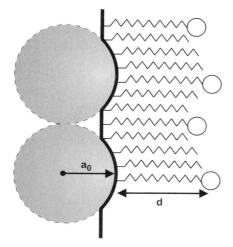

Fig. 2.11 Simplified view of a sub-microscopically rough metal surface (thick solid line) approximating the surface roughness for semi-spheres of radius a_0. Molecules contributing to the SER scattering (hollow spheres) are separated from the surface via spacers of length d.

approximate the roughness on electrodes by semi-spherical particles, or more appropriately, by connected semi-spheres (Fig. 2.11). In SER experiments, the surface roughness of electrodes is not uniform, corresponding to approximate semi-sphere structures with a wide distribution of particle radii a_0 (Fig. 2.12). Thus, the experimentally observed wavelength-dependent surface enhancement is very broad and may cover the entire spectral region from the resonance frequency at the Rayleigh limit ($a_0 \ll \lambda_0$) up to the infrared spectral region.

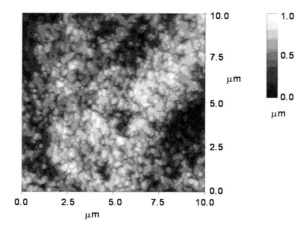

Fig. 2.12 Atomic force microscopic picture of an electrochemically roughened silver electrode.

In the relatively few instances for which the enhancement factor has been determined on the basis of the number of scattering molecules, values of 10^5–10^6 have been reported, which can be understood within the framework of the electromagnetic theory. Enhancement factors of this magnitude drastically increase the sensitivity of the Raman effect, such that SER spectroscopy represents a sensitive tool for studying molecules in the adsorbed state. The sensitivity, and, moreover, the selectivity can be further increased for molecules that exhibit an electronic absorption in the visible region. Then it is possible to tune the excitation frequency to be in resonance with both the electronic transition of the adsorbate and the surface plasmons of the metal. Under these conditions, the molecular RR and the SER effect combine (surface enhanced resonance Raman scattering – SERRS), such that it is readily possible to measure high quality spectra of molecules even if they are adsorbed at sub-monomolecular coverage. In fact, it has been shown that the effective quantum yield for the SERR process may approach unity (Hildebrandt and Stockburger 1984) and thus offers the possibility to probe even single molecules (Kneipp et al. 2006). If the chromophore associated with the resonant electronic transition is a cofactor in a protein, SERR spectroscopy displays a two-fold selectivity, as it probes selectively the vibrational spectrum of the cofactor of only the adsorbed molecules.

The electromagnetic theory of the SER effect implies that the enhancement is not restricted to molecules attached directly to the metal, although it decays according to the distance-dependence of dipole–dipole interactions. For spherical colloids of radius a_0 the decrease in the enhancement factor F_{SER} with the distance d from the surface is given by

$$F_{SER}(d) = F_{SER}(0) \cdot \left(\frac{a_0}{a_0 + d}\right)^{12} \tag{2.80}$$

where $F_{SER}(0)$ is the enhancement factor for molecules directly adsorbed onto the metal surface. For a particle radius of 20 nm, the enhancement is then estimated to decrease by a factor of ca. 10 when the molecule is separated by ca. 3.5 nm from the surface (Fig. 2.13). This prediction is in agreement with experimental findings and has also been verified for molecules immobilised on metal films and electrodes. Moreover, the consequences of this distance-dependence are important when applying SER spectroscopy to biological molecules. As direct interactions with the metal, specifically with Ag, may cause denaturation of biomolecules, it is advisable to cover the metal surfaces with biocompatible coatings that provide less harmful immobilisation conditions (see Section 4.3.6). The thickness of the coatings is typically between 1 and 5 nm, hence the surface enhancement is still sufficiently strong for the SER effect to be utilised (Murgida and Hildebrandt 2004, 2005).

In general, the electromagnetic theory provides a satisfactory explanation for the SER effect. However, there are a few documented cases that point to an additional contribution, which is commonly denoted as the "chemical effect" (Otto

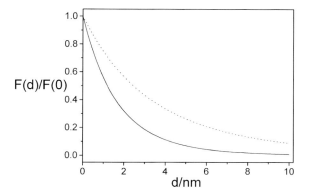

Fig. 2.13 Distance dependence of the surface enhancement for Raman scattering [solid line – Eq. (2.80)] and IR absorption [dotted line – Eq. (2.81)]. The surface roughness as defined in Fig. 2.11 is assumed to be $a_0 = 20$ nm in each instance.

2001). This effect is restricted to molecules chemisorbed on sites of atomic-scale roughness, such as ad-atoms on a surface. Such surface complexes represent essentially new compounds in which electronic transitions associated with the electronic states of the metal atom are involved. These transitions are analogous to charge transfer transitions of transition metal complexes. Excitation in resonance with such a transition leads to the enhancement of the Raman bands of the ad-atom–molecule complex similar to the molecular RR mechanism. This chemical effect has been demonstrated for small molecules under conditions where atomic-scale roughness is preserved, such as on cold-deposited metal films in an ultrahigh vacuum at low temperature. It may also play a role in so-called "hot spots" generated with metal nanoparticles and assumed to be crucial for single-molecule SER spectroscopy (Kneipp et al. 2006). The chemical effect can be neglected for those devices in which the probe molecules are not in direct contact with the metal surface, such as in SERR spectroscopy of protein-bound cofactors or for devices including biocompatible coatings. For this reason and due to the low magnitude of the enhancement (ca. 100), the importance of the chemical effect in SER (SERR) spectroscopy of biological systems is small.

2.3.2
Surface Enhanced Infrared Absorption

The electromagnetic theory also provides the conceptual framework for understanding enhanced absorption of radiation by molecules adsorbed on or in close vicinity to Ag or Au surfaces. In this instance, the optically active metal surfaces are created by electrodeless deposition of Ag or Au in the vacuum or in solution (Osawa 1997). The thin metal films that are usually deposited on an optically inert support display a morphology of interconnected islands corresponding to a roughness similar to that created on massive electrodes by electrochemical procedures. The incident broad-band radiation induces an oscillating dipole in the metal islands which, in analogy to Eq. (2.76), leads to an increase in the electric field in the near-field, and thus to an enhanced absorption by molecules that are located in close proximity to the film. As only the electric field component perpendicular to the metal surface is enhanced, SEIRA signals are not detectable for vibrations associated with dipole moment changes parallel to the surface. The enhancement is distinctly smaller than for SERS as only the field of the incident radiation is enhanced. However, for this reason the distance-dependence is not as steep because it scales only with the sixth power of the distance with respect to the centre of the island (Fig. 2.13).

$$F_{SEIRA}(d) = F_{SEIRA}(0) \cdot \left(\frac{a_0}{a_0 + d}\right)^6 \tag{2.81}$$

Even though enhancement factors are only between 10^2 and 10^3, the sensitivity gain of the SEIRA is large enough to apply this technique to biological molecules as the signals are detected in the difference mode (Ataka and Heberle 2003, 2004).

Box 2A

Quantum mechanical treatment of the harmonic oscillator

Transformation of the mass-weighted Cartesian coordinates q_i into normal coordinates Q_k is carried out such that $Q_k \cdot Q_{k'}$ cross terms are avoided in the expression for the kinetic and the potential energies [see Eq. (2.24)] (Wilson et al. 1995). Thus the kinetic energy T and the potential energy V are given by

$$T = \frac{1}{2} \sum_{k=1}^{3N-6} \dot{Q}_k^2 \tag{2.A1}$$

and

$$V = \frac{1}{2} \sum_{k=1}^{3N-6} \lambda_k Q_k^2 \tag{2.A2}$$

in analogy to Eqs. (2.15 and 2.17).
Hence, the Schrödinger equation

$$-\frac{h^2}{8\pi^2 m} \nabla^2 \psi + \hat{V}\psi = E\psi \tag{2.A3}$$

Box 2A *(continued)*

adopts the form

$$-\frac{h^2}{8\pi^2}\sum_{k=1}^{3N-6}\frac{\partial^2 \psi_\nu}{\partial Q_k^2} + \frac{1}{2}\sum_{k=1}^{3N-6}\lambda_k \cdot Q_k^2 \psi_\nu = E_\nu \psi_\nu \qquad (2.\text{A}4)$$

Solving this differential equation leads to the eigenvalues of the vibrational states given by

$$E_\nu = \left(\text{v}+\frac{1}{2}\right)h\nu_k \qquad (2.\text{A}5)$$

with the vibrational quantum number $\text{v} = 0, 1, 2, \ldots$. The eigenfunctions of Eq. (2.A4) are expressed by

$$\psi_{\text{v},k} = N_{\text{v},k} \cdot H_{\text{v},k}(Q_k) \cdot \exp\left(-\frac{1}{2}\gamma_k Q_k^2\right) \qquad (2.\text{A}6)$$

where $N_{\text{v},k}$ is the normalisation constant

$$N_{\text{v},k} = \left[\sqrt{\frac{\gamma_k}{\pi}} \cdot \frac{1}{2^{\text{v}} \cdot \text{v}!}\right]^{1/2} \qquad (2.\text{A}7)$$

The quantity γ_k is defined by

$$\gamma_k = \frac{4\cdot\pi^2 \nu_k}{h} \qquad (2.\text{A}8)$$

$H_{\text{v},k}$ denotes the Hermitian polynom, which varies with the vibrational quantum number v. Only for $\text{v} = 0$ is it equal to 1, whereas for $\text{v} \neq 0$ it becomes a function of Q_k (Atkins 2006). For $\text{v} = 1$, $H_{\text{v},k}$ is equal to $H_{\text{v},k} = \sqrt{\gamma_k} \cdot Q_k$ and the order of the polynom in terms of Q_k increases with increasing v. This corresponds to the increasing number of nodes of the wavefunction and thus affects its symmetry.

The vibrational wavefunctions actually have to be expressed as a product of the wavefunctions of the individual normal coordinates. However, the less exact treatment employed here for the sake of simplicity is appropriate to demonstrate the essential consequences for the IR, Raman, and resonance Raman selection rules (see also Box 2C). For a more thorough treatment the reader is referred to the literature (Herzberg 1945; Wilson et al. 1955).

Box 2B

Symmetry of molecules and vibrations

The symmetry of molecules is defined by a set of operations, which, when applied to the molecule, produce an identical copy (Cotton 1990). To illustrate these transformations we will consider the water molecule as a simple example (Fig. 2.B1). The three atoms of the molecule define a plane σ_{xz}, which is a symmetry plane; i.e., reflection at this plane leaves the molecule unchanged. A second plane of reflection σ_{xy} is perpendicular to σ_{xz} and bisects the bond angle. Furthermore, the intersection of both planes defines a rotational axis C_2. Upon rotation of the molecule around this axis by $2\pi/n$ ($n = 2$; two-fold axis). Other molecules may exhibit higher order rotational axes (e.g., a three-fold axis C_3 in the carbonate ion), a centre of inversion i (e.g., SF_6 with the sulfur atom being the centre of inversion), or a rotation–reflection plane S_n corresponding to the rotation around an n-fold axis and the subsequent reflection at the plane perpendicular to the rotational axis. The reflections σ, rotations C_n, rotation–reflections S_n, and the inversion I, in addition to the identity operation E which, just as a formal operation, leaves the molecule unchanged, constitute groups of symmetry operations that allows classification of all molecules into so-called point groups.

The simplest situation, which holds for many biological molecules, is given if nothing other than the identity operation applies. These molecules, which lack any symmetry, belong to the point group C_1. Nevertheless, there are also molecular groups of biological interest of higher symmetry, or whose geometry can be approximated by a point group of higher symmetry. Our example H_2O, which possesses two mirror planes and a two-fold rotational axis, belongs to the point group C_{2v}. The properties

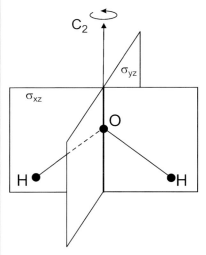

Fig. 2.B1 Symmetry operations of the C_{2v} point group, illustrated for the example of the water molecule. The two-fold rotational axis is denoted by C_2, the two mirror planes are indicated by σ_{xz} and σ_{yz}.

Box 2B *(continued)*

of these groups and the implications for (vibrational) spectroscopy become evident within the framework of group theory (Cotton 1990). Group theory by itself is a subject of pure mathematics, but it can be applied to many different fields of physics and physical chemistry. Therefore, all theorems used in the applications are based on formal derivations. Nevertheless, we consider it important to demonstrate the benefits of group theory by a more intuitive approach. Many aspects are taken from the book by Wilson, Decius and Cross, in which the application of group theory to vibrational spectroscopy is treated extensively (Wilson et al. 1955).

All elements (all symmetry operations) of a point group can be represented by matrices according to

$$\begin{pmatrix} x' \\ y' \\ z' \end{pmatrix} = \begin{pmatrix} R_{11} & R_{12} & R_{13} \\ R_{21} & R_{22} & R_{23} \\ R_{31} & R_{32} & R_{33} \end{pmatrix} \begin{pmatrix} x \\ y \\ z \end{pmatrix} \qquad (2.B1)$$

where R_{ij} constitute the elements of the transformation matrix that transform the coordinates of the molecule x, y, z into the coordinates x', y', and z'. These transformations can be illustrated by the reflection at the zy plane (Fig. 2.B2). Whereas the z- and y-coordinates remain unchanged, the x-coordinates just changes the sign such that $x' = x$. The transformation matrix is then given by

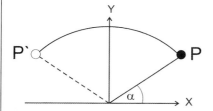

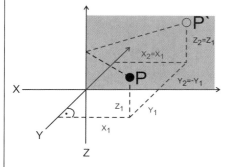

Fig. 2.B2 Transformation of P (solid circle) into P' (open circle) by rotation around the z-axis (top) and by reflection through the xz-plane (bottom).

Box 2B *(continued)*

$$R(\sigma_{v,zy}) = \begin{pmatrix} -1 & 0 & 0 \\ 0 & 1 & 0 \\ 0 & 0 & 1 \end{pmatrix} \quad (2.B2)$$

For an *n*-fold rotational axis perpendicular to the *yx*-plane (Fig. 2.B2) one obtains

$$z' = z$$
$$x' = x \cdot \cos(2\pi/n) - y \cdot \sin(2\pi/n) \quad (2.B3)$$
$$y' = x \cdot \sin(2\pi/n) + y \cdot \cos(2\pi/n)$$

such that the transformation matrix yields

$$R(C_n) = \begin{pmatrix} \cos(2\pi/n) & -\sin(2\pi/n) & 0 \\ \sin(2\pi/n) & \cos(2\pi/n) & 0 \\ 0 & 0 & 1 \end{pmatrix} \quad (2.B4)$$

These matrices describe the physical symmetry operations of the molecule.

As molecular vibrations involve distortions of the geometry, it is interesting to ask how such distortions transform under the symmetry operations of the molecule. We will take as an example the nitrate molecule, consisting of three oxygen atoms and one nitrogen atom. The symmetry is the D_{3h} point group (one 3-fold rotational axis, one horizontal mirror plane, three vertical mirror planes, three two-fold rotational axes, and the alternating 3-fold rotational axis obtained from the combination of the normal three-fold axis with the horizontal mirror plane). The deformation of the molecule is denoted by q_i with *i* running from 1 to $3N$ (for nitrate it runs from 1 to 12). The transformed displacements q_j' can be obtained for each symmetry operation by the linear equations

$$q_j' = \sum_{i=1}^{3N} R_{ji} q_i \quad (2.B5)$$

Of course, the matrices R_{ji} defined here are different from the 3×3 matrices (2.B1) describing the physical symmetry operations of the corresponding point group, although they are closely related. They are "isomorphic" to each other. The $3N \times 3N$ matrices R_{ji} (2.B5) are known as a representation of the point group, and the displacements q_i are termed the basis of this representation.

It can be shown that the matrix (2.B5) can be greatly simplified by a particular transformation of the coordinate system according to

$$\eta_i = \sum_{j}^{3N} \alpha_{ij} q_j \quad (2.B6)$$

Box 2B (continued)

Then the matrix (2.B5) transforms into

$$R'_{i'j'} = \sum_{ij}^{3N} \alpha_{i'i} R_{ij} (\alpha^{-1})_{jj'} \qquad (2.B7)$$

where α^{-1} denotes the reciprocal of the transformation (2.B6).

For each symmetry operation R of a point group a transformation α_{ij} can be found such that the transformed operation R contains only diagonal elements

$$\eta'_i = R'_{ii} \eta_i \qquad (2.B8)$$

It is generally not possible to find a single transformation by which all operations of a point group are converted into diagonal forms. However, as general derivations in group theory show, there are transformations that simplify the matrices of all the symmetry operations of a point group considerably. If the cubic groups are excluded, the corresponding matrices consist of diagonal elements and of blocks of 2×2 matrices such as:

$$\begin{pmatrix} R_{11} & & & & & & \\ & R_{22} & R_{23} & & & & \\ & R_{32} & R_{33} & & & & \\ \hline & & & R_{3N-2,3N-2} & R_{3N-2,3N-1} & & \\ \hline & & & R_{3N-1,3N-2} & R_{3n-1,3N-1} & & \\ & & & & & & R_{3N3N} \end{pmatrix} \qquad (2.B9)$$

For the cubic groups there are also blocks of 3×3, 4×4, and 5×5 matrices. It is evident that there are different matrix representations determined by the transformation (2.B6). It is straightforward to show that what all these representations have in common is that the trace of the matrices is the same, i.e.,

$$\sum_i^{3N} R'_{ii} = \sum_i^{3N} R_{ii} = \chi_R \qquad (2.B10)$$

This trace χ_R is called the character of the symmetry operation R.

Thus, in all these representations the symmetry operations are characterised by the unique value of the traces of the matrices representing the corresponding operations. If a representation is found such that all the matrices are of the form in Eq. (2.B9), the representation is referred to as being completely reduced. The corresponding new basis has the property that it can be separated into sets that do not mix with each other upon the various symmetry operations of this point group. Therefore, the transformation equations for the members of each set of the new basis can, by themselves, be

Box 2B *(continued)*

regarded as a representation of the point group. As no transformation can further simplify (reduce) the representation, it is termed an irreducible representation. Therefore, a completely reduced representation as given by the matrices such as in Eq. (2.B9) is made up of a number of irreducible representations and the character of the completely reduced representation (or of the reducible representation) is the sum of the characters of the irreducible representations, i.e.,

$$\chi_R = \sum_\gamma n^{(\gamma)} \chi_R^{(\gamma)} \tag{2.B11}$$

where χ_R^γ is the character of the irreducible representation γ with respect to the symmetry operation R, and $n^{(\gamma)}$ specifies how often the representation γ appears in the completely reduced representation. It can be shown that

$$n^{(\gamma)} = \frac{1}{g} \sum_R \chi_R^\gamma \chi_R \tag{2.B12}$$

where the sum extends over all symmetry operations of a point group and g gives the number of symmetry operations.

An additional nomenclature has to be introduced. There are symmetry operations that have the same various characters for the different irreducible representations. These operations belong to the same class. A more formal definition for a class is the following: if A, B, and X are members of a point group and if $B = X^{-1}AX$, then A and B belong to the same class. It is easy to show that the equality of the characters of A and B follows from the definition of character and from the rule of matrix multiplication. In characterising a point group, it is, therefore, not necessary to give the characters of all symmetry operations but it is sufficient to list them for each different class of symmetry operations.

Using the classes of symmetry operations, Eq. (2.B12) can be rewritten as

$$n^{(\gamma)} = \frac{1}{g} \sum_j g_j \chi_j^{(\gamma)} \chi_j \tag{2.B13}$$

where g_j is the number of elements (symmetry operations) within the class j, $\chi_j^{(\gamma)}$ is the character of the representation γ with respect to the class j, and χ_j is the character of any one of the symmetry operations within class j with respect to the completely reduced or reducible representation (as has been shown above, the character of a symmetry operation is not changed by a transformation of the coordinate system).

The representation of a point group is characterised by the character of the transformations making up this point group. Therefore, the transformation properties of the irreducible representations, i.e., the characters, can be summarised in tabular form. They are listed for the different point groups in books describing the application

Box 2B (continued)

of group theory (e.g., Wilson et al. 1955). For the D_{3h} point group, the character table is

D_{3h}	E	$2C_3$	$3C_2$	σ_h	$2S_3$	$3\sigma_v$
A_1'	1	1	1	1	1	1
A_2'	1	1	−1	1	1	−1
E'	2	−1	0	2	−1	0
A_1''	1	1	1	−1	−1	−1
A_2''	1	1	−1	−1	−1	1
E''	2	−1	0	−2	1	0

(2.B14)

The symbol in the upper left corner denotes the point group. The first column contains the various irreducible representations, in this case 6. The first row lists the various symmetry operations (unity operation E, two rotations about the 3-fold axis, one horizontal mirror plane σ_h, two alternating rotations about the 3-fold axis, and three vertical mirror planes σ_v). The second column, describing the characters of the unity operation E for the various irreducible representations, gives the dimensionality of the representation, i.e., the number of coordinates that may mix upon application of the symmetry operations. The notation for the different representations is complex (Mullican notation) and basically describes the symmetry properties upon certain basic symmetry operations of the different point groups. Here, it is sufficient to note that the letters A and B refer to one-dimensional representations, the letters E and F to two-dimensional and three-dimensional representations. Typically not all symmetry operations are listed, rather the classes together with the number of operations within the particular class.

It is easy to see that for the D_{3h} symmetry we obtain $n(A_1') = 1$; $n(A_2'') = 2$; $n(E') = 3$; $n(E'') = 1$; $n(A_2') = 1$. It is important to mention that the D_{3h} representation of the displacements as the basis does not contain the A_1'' irreducible representation.

What is the connection for these symmetry considerations with vibrational spectroscopy? It has been shown in Section 2.1.1 that the normal coordinates Q_i involve a linear transformation such as (2.B6)

$$Q_k = \sum l_{ik} q_i \qquad (2.B15)$$

It is straightforward to show that these normal coordinates form part of the irreducible representations making up the basis set of the displacements (see Wilson et al. 1955). In general, there are 6 additional representations, 3 for the translational and 3 for the rotational movements of the molecule as a whole. Thus, the normal coordinates Q_i just transform as irreducible representations, and their symmetry properties are given by the character table.

Box 2B *(continued)*

We now can further specify the Mullican notation as given in the character tables, for the C_{2v} geometry of water [Eq. (2.B16)] or for the D_{3h} symmetry of nitrate [Eq. (2.B14)]. As already mentioned the letters A and B refer to one-dimensional, E to two-dimensional, and F to three-dimensional representations. Totally symmetric representations for which all characters are 1 are denoted by A, A_1, or A_{1g}. The subscript "g" is the abbreviation for "gerade" indicating the symmetry with respect to inversion whereas "ungerade" representations (denoted by the subscript "u") are antisymmetric toward this symmetry operation.

C_{2v}	E	C_2	$\sigma_v(xz)$	$\sigma_v(yz)$
A_1	1	1	1	1
A_2	1	1	−1	−1
B_1	1	−1	1	−1
B_2	1	−1	−1	1

(2.B16)

As an example, we can now discuss the symmetry properties for the normal modes of H_2O. The three normal modes are schematically represented in Fig. 2.B3. The stretching and bending modes v_1 and v_2 are symmetric (character $+1$) with respect to all symmetry operations, i.e., the rotation and the two reflections, and thus belong to the (totally symmetric) representation A_1 [Eq. (2.B16)]. The stretching mode v_3, however, is anti-symmetric (character -1) to the rotation and the reflection with respect to the yz-plane, such that it is grouped into the B_1 representation.

Such a classification of the normal modes according to the symmetry properties can also be obtained by transforming internal coordinates into symmetry coordinates. For the water molecule, this transformation is relatively simple assuming an approximate expression for the potential energy

$$2V = F_r(r_1^2 + r_2^2) + F_\alpha \alpha^2 \qquad (2.B17)$$

where r_1 (r_2) and α are the stretching and bending coordinates, respectively. F_r and F_α denote the corresponding stretching and bending force constants, whereas interaction force constants are neglected. If we now define symmetry coordinates according to

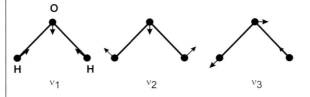

Fig. 2.B3 Illustration of the three normal modes of water. The arrows qualitatively indicate the directions of displacements of the individual atoms in the normal modes.

Box 2B (continued)

$$S_1 = \alpha$$

$$S_2 = \frac{1}{\sqrt{2}}(r_1 + r_2) \quad (2.B18)$$

$$S_2 = \frac{1}{\sqrt{2}}(r_1 - r_2)$$

the potential energy [Eq. (2.B6)] is then expressed by

$$2V = F_\alpha S_1^2 + F_r S_2^2 + F_r S_3^2 \quad (2.B19)$$

Correspondingly, the kinetic energy is given in terms of these symmetry coordinates

$$2T = M_\alpha \dot{S}_1^2 + 2\sqrt{2} M_{\alpha r} \dot{S}_1 \dot{S}_2 + (M_{rr} + M_r)\dot{S}_2^2 + (M_r - M_{rr})\dot{S}_3^2 \quad (2.B20)$$

where M_i are the reduced masses which constitute the inverse G-matrix.

With Eqs. (2.B19 and 2.B20) one obtains the secular equation

$$\begin{vmatrix} F_\alpha - M_\alpha \lambda & -\sqrt{2} M_{\alpha r} \lambda & 0 \\ -\sqrt{2} M_{\alpha r} \lambda & F_r - (M_r + M_{rr})\lambda & 0 \\ 0 & 0 & F_r - (M_r - M_{rr})\lambda \end{vmatrix} = 0 \quad (2.B21)$$

which factorises into two blocks, a 2×2 matrix referring to the normal modes of the A_1 representation, and a 1×1 matrix corresponding to the B_1 mode.

An important goal of group theory applied to molecular vibrations is to determine the number of normal modes n belonging to each irreducible representation γ of the point group. This procedure, which substantially facilitates the assignment of Raman and infrared bands (*vide infra*), can be applied on the basis of Cartesian (displacement) coordinates or internal (symmetry) coordinates. We have mentioned above that the number of given irreducible representations appears in the completely reduced representation according to Eq. (2.B13), i.e.,

$$n^{(\gamma)} = \frac{1}{g} \sum_j g_j \chi_j^{(\gamma)} \chi_j \quad (2.B13)$$

This expression can be used to determine the number of normal modes in a certain irreducible representation γ, taking into account that expressions (2.B13) also contain the 6 coordinates for translation and rotation. g_j and $\chi_j^{(\gamma)}$ are obtained from the character tables such as Eqs. (2.B14 and 2.B16). χ_j has to be determined independently, either by inspection of the effects of the symmetry operations on the displacement coordinates, or by more efficient methods described in books on the application of group theory.

Box 2B *(continued)*

Let us consider the situation when the molecular vibrations are described in terms of Cartesian displacement coordinates. It is clear that for a 3-atomic molecule χ_E equals 9 (3N). For a three-atomic molecule such as H_2O, the values of χ_j are determined to be -1, 3, and 1 for the symmetry operations C_2, $\sigma_{v,xz}$, and $\sigma_{v,yz}$, respectively. With Eq. (2.B13) we obtain

$$n(A_1) = \frac{1}{4}(1 \cdot 1 \cdot 9 - 1 \cdot 1 \cdot 1 + 1 \cdot 1 \cdot 3 + 1 \cdot 1 \cdot 1) = 3$$

$$n(A_2) = \frac{1}{4}(1 \cdot 1 \cdot 9 - 1 \cdot 1 \cdot 1 - 1 \cdot 1 \cdot 3 - 1 \cdot 1 \cdot 1) = 1 \quad (2.B22)$$

$$n(B_1) = \frac{1}{4}(1 \cdot 1 \cdot 9 + 1 \cdot 1 \cdot 1 + 1 \cdot 1 \cdot 3 - 1 \cdot 1 \cdot 1) = 3$$

$$n(B_2) = \frac{1}{4}(1 \cdot 1 \cdot 9 + 1 \cdot 1 \cdot 1 - 1 \cdot 1 \cdot 3 + 1 \cdot 1 \cdot 1) = 2$$

which can be summarised by

$$\Gamma = 3A_1 + A_2 + 3B_1 + 2B_2 \quad (2.B23)$$

Thus, altogether we have 9 modes corresponding to the $3N = 9$ degrees of freedom. To sort out the "modes" of zero frequency, we have to consider the symmetry properties of the translational and rotational degrees of freedom. Translation of the molecule in z-direction (T_z) is symmetric with respect to all symmetry operations such that T_z is assigned to the A_1 symmetry species. Translation in the x- and y-directions is symmetric with respect to $\sigma_{v,xz}$ and $\sigma_{v,yz}$, respectively, but both are antisymmetric with respect to C_2. Thus, T_x and T_y belong to the symmetry species B_1 and B_2, respectively. Correspondingly, one can show that rotation around the z-axis R_z is symmetric with respect to C_2 but antisymmetric with respect to both reflections. Likewise, we find that R_x and R_y are antisymmetric to C_2 but symmetric to $\sigma_{v,zx}$ and $\sigma_{v,zy}$, respectively, such that for these molecular motions the assignment to the symmetry species is also straightforward. Thus, we subtract the translations and rotations from the respective symmetry species to obtain the true number of vibrational normal modes. This approach is readily applicable to larger molecules and, moreover, it also holds for the internal coordinate system. In this way, it is possible to factorise the FG-matrix into blocks corresponding to the various symmetry species involved. The translational (T_x, T_y, T_z) and rotational (R_x, R_y, R_z) coordinates are usually included into the character table, so one obtains for the C_{2v} symmetry.

Box 2B *(continued)*

C_{2v}	E	C_2	$\sigma(xz)$	$\sigma(yz)$	
A_1	1	1	1	1	T_z
A_2	1	1	-1	-1	R_z
B_1	1	-1	1	-1	$T_x; R_y$
B_2	1	-1	-1	1	$T_y; R_x$

(2.B24)

For the real normal modes we obtain Γ(normal modes) $= 2A_1 + B_1$, as intuitively derived for this simple molecule.

Box 2C

Symmetry-based selection rules

Symmetry considerations allow the determination of whether a normal mode Q_k of a molecule belonging to a certain point group ($\neq C_1$) is IR- or Raman-active. This establishes if the expressions for the transition dipole moment [Eq. (2.44)] and the scattering tensor [Eq. (2.62)] are zero or non-zero. In symmetry terms this means: the product, in which $\hat{\Omega}$ is the dipole moment or polarisability operator, must lead to a symmetric function otherwise the integrals in Eqs. (2.44 and 2.62) vanish. Within the framework of group theory (Cotton 1990; Wilson et al. 1955), the symmetry properties of this product are expressed by the "direct product" of the irreducible representations

$$\Gamma = \Gamma_n \times \Gamma_\Omega \times \Gamma_m \tag{2.C1}$$

where Γ_n, Γ_m, and Γ_Ω are the irreducible representations of the wavefunctions of the initial, and the final states, and of the operator, respectively. The condition that the integral over the products in Eqs. (2.44 and 2.62) is non-zero requires that the direct product has the totally symmetric irreducible representation.

Let us first consider the wavefunction of the vibrational ground state $\psi_{n,k}$ ($\nu = 0$), which we can express by the symmetric eigenfunction of the harmonic oscillator (see Box 2A)

$$\psi_{n,k} = N_{0,k} \cdot \exp\left(-\frac{1}{2}\gamma_k Q_k^2\right) \tag{2.C2}$$

which is always totally symmetric. The actual ground state is the product of such functions belonging to the various normal modes. However, because in a fundamental vibrational transition only one normal mode is involved, it is sufficient to consider only the expression Eq. (2.C2). The situation is different if a transition to a combination state has to be treated.

For the first vibrational excited state ψ_m ($\nu = 1$), however, the eigenfunction is given by

Box 2C *(continued)*

$$\psi_{m,k} = N_{1,k} \cdot \sqrt{\gamma_k} \cdot Q_k \cdot \exp\left(-\frac{1}{2}\gamma_k Q_k^2\right) \tag{2.C3}$$

such that the symmetry of $\psi_{m,k}$ is given by the symmetry of Q_k.

Now we can investigate the consequences for the IR transitions. In this instance, the operator $\hat{\Omega}$ corresponds to the dipole moment operator [Eq. (2.44)], which has three components corresponding to the Cartesian coordinates

$$\mu_x = \sum_\alpha e_\alpha \cdot x_\alpha$$
$$\mu_y = \sum_\alpha e_\alpha \cdot y_\alpha \tag{2.C4}$$
$$\mu_z = \sum_\alpha e_\alpha \cdot z_\alpha$$

Evidently, the operator $\hat{\mu}$ transforms in the same manner as the Cartesian coordinates or as the translations in the x-, y-, and z-directions. Thus, the irreducible representations of μ_i, i.e., Γ_μ, can be derived from the character table of the individual point groups.

To determine the symmetry product according to Eq. (2.C1), we use the multiplication properties of the irreducible representations (Wilson et al. 1955). For the non-degenerate species A and B, these rules are fairly simple as $A \times A = A$ and $B \times B = A$, whereas $A \times B = B$. Likewise, for the subscripts we have $1 \times 1 = 1$, $2 \times 2 = 1$ and $1 \times 2 = 2$ (except for D_2 and D_{2h} point groups) and $g \times g = g$, $u \times u = g$ and $g \times u = u$. On this basis, we can decide which of the vibrational modes of H_2O are IR-active. For the two A_1 modes, with Eq. (2.C1) we obtain

$$\Gamma_{A_1}(z) = A_1 \times A_1 \times A_1 = A_1$$
$$\Gamma_{A_1}(y) = A_1 \times B_2 \times A_1 = B_2 \tag{2.C5}$$
$$\Gamma_{A_1}(x) = A_1 \times B_1 \times A_1 = B_1$$

implying that both A_1 modes are IR-active but only the z-component of the dipole moment operator provides a non-zero contribution to the transition dipole moment. Correspondingly, one obtains for the B_1 mode

$$\Gamma_{B_1}(z) = A_1 \times A_1 \times B_1 = B_1$$
$$\Gamma_{B_1}(y) = A_1 \times B_2 \times B_1 = A_2 \tag{2.C6}$$
$$\Gamma_{A_1}(x) = A_1 \times B_1 \times B_1 = A_1$$

Thus, in this instance only the x-component of $\hat{\mu}$ gives rise to the IR intensity.

Although the direct product formalism is general and in particular allows the treatment of more complex transitions involving overtones and combinational levels, for the fundamental transition there is an even simpler rule, which can be readily obtained.

It can be shown that the character of the direct product of two irreducible representations is the product of their characters. The condition that

Box 2C *(continued)*

$$\Gamma = \Gamma_n \times \Gamma_\Omega \times \Gamma_m \tag{2.C1}$$

contains the totally symmetric representation can be rephrased to ask how many times this representation is contained in Γ. According to Eq. (2.B12), this value, $n^{(1)}$, is given by

$$n^{(1)} = \frac{1}{g} \sum_R \chi_R^n (\chi_R^\Omega \chi_R^m) \tag{2.C7}$$

as the character of the totally symmetric representation, $\chi_R^{(1)}$, equals one for all operations R. Expression (2.C7) can be considered in such a way that $n^{(1)}$ represents the number of times Γ_n is contained in the reducible representation $\Gamma_\Omega \times \Gamma_m$.

Therefore, if Γ_n does occur, $n^{(1)}$ is greater than zero and the integral does not vanish. For the fundamental transition from the ground state (totally symmetric) to the first excited vibronic level, one therefore arrives at the very simple statement that such a transition is IR active if the corresponding normal coordinate belongs to the same representations as those of T_x, T_y, or T_z.

In the same way we determine the Raman activity on the basis of the symmetry properties of the operator for the polarisability tensor, Eq. (2.62). It can be shown (Wilson et al. 1955) that the components of this operator $\hat{\alpha}_{xx}$, $\hat{\alpha}_{xy}$, $\hat{\alpha}_{xz}$ etc. transform as do xx, xy, xz etc. Therefore, the transformation properties of the tensor can be derived in a similar way as for the x-, y-, and z-coordinates themselves. These symmetry properties are also listed in the character table, and Eq. (2.B24) is extended to

C_{2v}	E	C_2	$\sigma_v(xz)$	$\sigma_v(yz)$		
A_1	1	1	1	1	T_z	$\alpha_{xx}; \alpha_{yy}; \alpha_{zz}$
A_2	1	1	−1	−1	R_z	α_{xy}
B_1	1	−1	1	−1	$T_x; R_y$	α_{xz}
B_2	1	−1	−1	1	$T_y; R_x$	α_{yz}

(2.C8)

As shown in Chapter 2, the Raman scattering involves a change in the quantum number of a normal mode by 1, as in the infrared absorption. Therefore, the same methodology as for infrared absorption can be applied. Either the general direct product formalism can be used, or one can determine whether an element of the polarisability operator belongs to the same representation as a normal mode. We readily find that the A_1 modes are Raman-active through the tensor components α_{xx}, α_{yy}, and α_{zz}, whereas the B_1 mode gains Raman intensity through α_{xz}. By means of Eqs. (2.60 and 2.61), we can conclude that the A_1 modes exhibit a very small depolarisation ratio whereas this ratio is 0.75 for the B_1 mode. Thus, symmetry consideration do not only allow prediction for the Raman- and IR-activity of the normal modes but also provides predictions for the depolarisation ratio, which in turn is particularly helpful for the vibrational assignments.

Box 2C *(continued)*

Now we consider the RR activity. For the A-term [Eq. (2.68)], the RR intensity depends on the electronic transition dipole moments M_{GR}^0 and the Franck–Condon factor products $\langle nr \rangle \langle rm \rangle$. As the vibrational quantum number for the initial and final wavefunctions differs by one, either the left or the right integral would vanish for non-totally symmetric modes. Thus, only totally-symmetric modes gain intensity via the A-term, given that the origins of the potential curves in the ground and electronic excited state are displaced with respect to each other. The second prerequisite for A-term RR intensity is defined by the symmetry properties of M_{GR}^0

$$\Gamma = \Gamma_G \times \Gamma_\Omega \times \Gamma_R \tag{2.C9}$$

where the irreducible representation of the dipole moment operator Γ_Ω corresponds to that of the transition dipole moment as in the case of IR absorption [Eq. (2.C4)]. The electronic transition moment is non-zero if there is a component (in the x-, y-, or z-direction) for which the direct product Eq. (2.C9) is totally symmetric.

As an example, we will discuss the $\pi \rightarrow \pi^*$ electronic transitions of benzene, which belongs to the D_{6h} point group. The electronic configuration can be illustrated within the Hückel approximation of molecular orbital theory (Fig. 2.C1). In the electronic ground state, the three occupied π-orbitals are of a_{2u} and e_{1g} symmetry, leading to the $(a_{2u})^2(e_{1g})^4$ configuration. As all occupied orbitals are completely filled, the

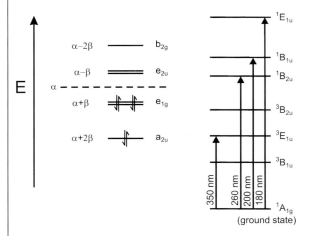

Fig. 2.C1 Electronic structure (left) and energy diagram for the electronic transitions (right) of benzene (Straughan and Walker 1976). The electronic structure is qualitatively described in terms of the Hückel molecular orbital approximation.

Box 2C *(continued)*

ground state belongs to the totally symmetric species $^1A_{1g}$ where the superscript "1" indicates the singlet state. The lowest electronic excitation promotes an electron to the e_{2u} orbital resulting in the $(a_{2u})^2(e_{1g})^3(e_{2u})^1$ configuration (Fig. 2.C1). The direct product of (e_{1g}) and (e_{2u}) then leads to symmetry properties of the three lowest excited states, i.e., $^1B_{1u}$, $^1B_{2u}$, and $^1E_{1u}$ (considering only the singlet transitions). Thus, the irreducible representation of the corresponding electronic transition dipole moments are evaluated according to Eq. (2.C9) using the irreducible representations of the dipole moment operator that are A_{2u} (z) and E_{1u} (x, y) within the D_{6h} point group. It can be verified that only the product $A_{1g} \times E_{1u} \times E_{1u}$ leads to a totally symmetric species implying that only the $^1E_{1u} \leftarrow {}^1A_{1g}$ transition, which is localised in the x,y-plane, is symmetry-allowed. This transition in fact gives rise to a strong absorption band at ca. 180 nm and excitation in resonance with this transitions provides a strong enhancement of the Raman bands originating from totally-symmetric A_{1g} modes.

Now we will consider the symmetry-based selection rules for the B-term scattering mechanism. When the excitation is in resonance with a weak electronically excited state R, the Herzberg–Teller approximation (Albrecht 1961) applies and Eq. (2.69) is transformed into

$$B \cong \frac{1}{h} \frac{M^0_{GS} M^0_{GR} H_{RS}}{\nu_S - \nu_R} \sum_r \left(\frac{\langle n|Q_k|r\rangle\langle rm\rangle + \langle n|r\rangle\langle r|Q_k|m\rangle}{\nu_{Rr} - \nu_k - \nu_0 + i\Gamma_R} \right) \qquad (2.C10)$$

where H_{RS} is the vibronic coupling integral describing the mixing between the resonant excited state R and a second (allowed) excited state S, and ν_S and ν_R are the frequencies of the corresponding $0 \to 0$ excitations.

Again, we restrict the discussion to the transition from the vibrational ground state n (with $v = 0$) to the first vibrational excited state m (with $v = 1$), and inspect the individual terms in Eq. (2.C10) separately. The vibrational integrals that include the normal coordinate Q_k, i.e., $\langle n|Q_k|r\rangle$ and $\langle r|Q_k|m\rangle$, are only non-zero if the symmetry product of Q_k with $\psi_{m,k}$ or $\psi_{r,k}$ is totally-symmetric. In addition, the integrals $\langle n|r\rangle$ and $\langle r|m\rangle$ must be non-zero, which is only fulfilled for $\psi_{r,k}$ having the same vibrational quantum number as $\psi_{n,k}$ or $\psi_{m,k}$. Finally, the integral H_{RS} has to be large, which is true for modes that can effectively couple the electronically excited states R and S. Combining these three conditions, maximum enhancement via the B-term mechanism is expected:

(i) for modes of the symmetry species given by the direct product of the two electronic transitions that are involved
(ii) when the excitation line is in resonance with the weak $0 \to 0$ or $0 \to 1$ vibronic transition Rr
(iii) when the energy difference with respect to a second electronically excited state S is small.

References

Albrecht, A. C., **1961**, "On the theory of Raman intensities", *J. Chem. Phys.* **34**, 1476–1484.

Albrecht, A. C., Hutley, M. C., **1971**, "On the dependence of vibrational Raman intensity on the wavelength of incident light", *J. Chem. Phys.* **55**, 4438–4443.

Ataka, K., Heberle, J., **2003**, "Electrochemically induced surface-enhanced infrared difference absorption (SEIDA) spectroscopy of a protein monolayer", *J. Am. Chem. Soc.* **125**, 4986–4987.

Ataka, K., Heberle, J., **2004**, "Functional vibrational spectroscopy of a cytochrome *c* monolayer: SEIDAS probes the interaction with different surface-modified electrodes", *J. Am. Chem. Soc.* **126**, 9445–9457.

Atkins, P., **2006**, "*Physical Chemistry*", Oxford University Press, Oxford.

Blair, D. F., Campbell, G. W., Schoonover, J. R., Chan, S. I., Gray, H. B., Malmström, B. G., Pecht, I., Swanson, B. I., Woodruff, W. H., Ch., W. K., English, A. M., Fry, H. A., Lunn, V., Norton, K. A., **1985**, "Resonance Raman studies of blue copper proteins: Effect of temperature and isotope substitutions. Structural and thermodynamic implications", *J. Am. Chem. Soc.* **107**, 5755–5766.

Cotton, F. A., **1990**, "*Chemical Applications of Group Theory*", Wiley, New York.

Curry, B., Palings, I., Broek, A. D., Pardoen, J. A., Lugtenburg, J., Mathies, R., **1985**, "Vibrational analysis of the retinal isomers", in *Adv. Infrared and Raman Spectrosc.* **12**, Clark, R. J. H., Hester, R. E. (Eds.), chap. 3, Wiley, Heyden, New York.

Czernuszewicz, R. S., Spiro, T. G., **1999**, "IR, Raman, and resonance Raman spectroscopy", in *Inorganic Electronic Structure and Spectroscopy*, Solomon, E. I., Lever, A. B. P. (Eds.), Vol. 1, Wiley, New York, pp. 353–441.

Fogarasi, G., Zhou, X., Taylor, P. W., Pulay, P., **1992**, "The calculation of *ab-initio* molecular geometries: Efficient optimization by natural internal coordinates and empirical corrections of offset forces", *J. Am. Chem. Soc.* **114**, 8191–8201.

Heller, E. J., **1981**, "The semi-classical way to molecular spectroscopy", *Acc. Chem. Res.* **14**, 368–375.

Herzberg, G., **1945**, "*Molecular Spectra and Molecular Structure: II, Infrared and Raman Spectra of Polyatomic Molecules*", Van Nostrand Reinhold, New York.

Hildebrandt, P., Stockburger, M., **1984**, "Surface enhanced resonance Raman spectroscopy of Rhodamine 6G adsorbed on colloidal silver", *J. Phys. Chem.* **88**, 5935–5944.

Kerker, M., Wang, D. S., Chew, S., **1980**, "Surface enhanced Raman scattering (SERS) by molecules adsorbed on spherical particles", *Appl. Opt.* **19**, 4159–4273.

Kneipp, K., Moskovits, M., Kneipp, H. (Eds.), **2006**, "Surface-enhanced Raman scattering: Physics and applications", *Topics Appl. Phys.* **103**, Springer, Berlin.

Li, X.-Y., Czernuszewicz, R. S., Kincaid, J. R., Spiro, T. G., **1989**, "Consistent porphyrin force field. 3. Out-of-plane modes in the resonance Raman spectra of planar and ruffled nickel octaethylporphyrin", *J. Am. Chem. Soc.* **111**, 7012–7023.

Li, X.-Y., Czernuszewicz, R. S., Kincaid, J. R., Su, Y. O., Spiro, T. G., **1990a**, "Consistent porphyrin force field. 1. Normal mode analysis for nickel porphine and nickel tetraphenylporphine from resonance Raman and infrared spectra and isotope shifts", *J. Phys. Chem.* **94**, 31–47.

Li, X.-Y., Czernuszewicz, R. S., Kincaid, J. R., Stein, P., Spiro, T. G., **1990b**, "Consistent porphyrin force field. 2. Nickel octaethylporphyrin skeletal and substituent mode assignments from ^{15}N, meso-d$_4$, and methylene-d$_{16}$ Raman and infrared isotope shifts", *J. Phys. Chem.* **94**, 47–61.

Magdó, I., Nemeth, K., Mark, F., Hildebrandt, P., Schaffner, K., **1999**, "Calculation of vibrational spectra of linear tetrapyrroles. Global sets of scaling factors for force fields derived by *ab-initio* and density functional theory methods", *J. Phys. Chem. A* **103**, 289–303.

Mie, G., **1908**, "Beiträge zur Optik trüber Medien, speziell kolloidaler Metallösungen", *Ann. Physik* **25**, 377–445.

Moskovits, M., **1985**, "Surface-enhanced spectroscopy", *Rev. Mod. Phys.* **57**, 783–826.

Mroginski, M. A., Hildebrandt, P., Kneip, C., Mark, F., **2003**, "Excited state geometry calculations and the resonance Raman spectrum of hexamethylpyrromethene", *J. Mol. Struct.* **661–662**, 611–624.

Mroginski, M. A., Németh, K., Bauschlicher, T., Klotzbücher, W., Goddard, R., Heinemann, O., Hildebrandt, P., Mark, F., **2005**, "Calculation of vibrational spectra of linear tetrapyrroles. 3. Hydrogen bonded hexamethylpyrromethene dimers", *J. Phys. Chem. A* **109**, 2139–2150.

Murgida, D. H., Hildebrandt, P., **2004**, "Electron transfer processes of cytochrome c at interfaces. New insights by surface-enhanced resonance Raman spectroscopy", *Acc. Chem. Res.* **37**, 854–861.

Murgida, D. H., Hildebrandt, P., **2005**, "Redox and redox-coupled processes of heme proteins and enzymes at electrochemical interfaces", *Phys. Chem. Chem. Phys.* **7**, 3773–3784.

Osawa, M., **1997**, "Dynamic processes in electrochemical reactions studied by surface-enhanced infrared absorption spectroscopy (SEIRAS)", *Bull. Chem. Soc. Jpn.* **70**, 2861–2880.

Otto, A., **2001**, "Theory of first layer and single molecule surface enhanced Raman scattering (SERS)", *Phys. Stat. Sol. A* **188**, 1455–1470.

Peticolas, W. L., Rush, T., **1995**, "*Ab-initio* calculation of the ultraviolet resonance Raman spectra of Uracile", *J. Comp. Chem.* **16**, 1261–1270.

Placzek, G., **1934**, "Rayleigh-Streuung und Raman-Effekt", in *Handbuch der Radiologie*, Vol. VI, Marx, E. (Ed.), chap. 3, Akademische Verlagsanstalt, Leipzig.

Rauhut, G., Pulay, P., **1995**, "Transferable scaling factors for density functional derived vibrational force fields" *J. Phys. Chem.* **99**, 3093–3100.

Straughan, B. P., Walker, S. (Eds.), **1976**, "*Spectroscopy*", Chapman and Hall, London.

Warshel, A., Dauber, P., **1977**, "Calculations of resonance Raman spectra of conjugated molecules", *J. Chem. Phys.* **66**, 5477–5488.

Wilson, E. B., Decius, J. C., Cross, P. C., **1955**, "*Molecular Vibrations: The Theory of Infrared and Raman Vibrational Spectra*", McGraw-Hill, New York.

3
Instrumentation

This chapter is devoted to the experimental set-ups of IR and Raman experiments suitable for studying biological systems. We focus on the "hardware", the various types of spectrometers, light sources, detectors, and other components. The goal of this chapter is to present guidelines for setting up IR and Raman workplaces according to specific types of experiments. Even though the hardware and its specifications are very different in IR and Raman spectroscopy and, therefore, are treated separately, there are also aspects that are common to both techniques. As an example, we mention interferometers and the Fourier transform (FT) method, which are essential for modern IR spectroscopy and thus are covered in the "IR spectroscopy" section. However, also in Raman spectroscopy the interferometric detection is an option, albeit not the most versatile one. Instead of duplicating paragraphs, cross-links in the individual paragraphs will guide the reader specifically interested in either IR or Raman spectroscopy. The chapter also describes the principles of time-resolved IR techniques as some of these (step-scan, rapid-scan) are based on hardware components that are integrated into the interferometer. Conversely, time-resolved Raman spectroscopic methods are employed with the same spectrometers as used for stationary Raman spectroscopy and, therefore, these approaches will be treated in Chapter 4.

3.1
Infrared Spectroscopy

At first sight it might be surprising for the reader not familiar with IR spectroscopy that a lengthy chapter is devoted to infrared instrumentation. One might think that is not much different from that used for UV–vis spectroscopy. However, the first difficulty one encounters is that of infrared transparent material, which might be used as windows for cuvettes. The spectral range in mid-infrared spectroscopy used for vibrational spectroscopy covers the range from 4000 to 200 cm^{-1}, corresponding to the wavelength range from 2.5 to 50 µm. The choice of window material will be discussed later in this chapter, however, the difficulty of finding infrared-transparent material also has consequences for the infrared radiation source.

Vibrational Spectroscopy in Life Science. Friedrich Siebert and Peter Hildebrandt
Copyright © 2008 WILEY-VCH Verlag GmbH & Co. KGaA, Weinheim
ISBN: 978-3-527-40506-0

From Planck's radiation law one might think that the hotter the source the higher the infrared intensity and a halogen lamp would certainly be a good choice, if one could find a material equivalent to quartz from which the bulb could be manufactured. So far, this has not been possible. Thus, in all instruments, the source is just in air at ambient temperature. Therefore, the temperature has to be limited, and usually it is around 1300 K. This constraint restricts the infrared power of the source substantially, although the intensity emitted on the long-wavelength side of the emission maximum increases only linearly with the temperature. The emission maximum of a source kept at room temperature is ca. 10 μm. This has the severe consequence that detectors sensitive towards infrared radiation at room temperature will exhibit large amounts of noise. Here noise is the stochastic signal produced by thermal events in the detector. It is desirable that the signal produced by the infrared radiation gives a signal in the detector larger than the noise fluctuations. Thus, in IR spectroscopy one is dealing with low-intensity sources and noisy detectors, hence it is not easy to record spectra with high precision.

Up to the middle of the 1970s, IR spectrophotometers consisted of an infrared source, and, as in UV–vis spectroscopy, of a monochromator and a detector. In order to increase the total power of the radiation emitted by the infrared source, its area is usually fairly large (up to 1 cm^2). Therefore, mirrors and gratings used for the monochromator and other parts of the instrument had to be relatively large in order to collect and transfer as much as possible of this radiation. This makes the optics fairly expensive. However, despite the special optics used, a large fraction of the radiation is lost due to the entrance and exit slits of the monochromator, which have to be narrow (approximately 2–5 mm) to provide the required spectral resolution of typically 2–4 cm^{-1} (see Section 3.2.2.1). All these factors contributed to IR spectroscopy based on monochromators being very tedious and scanning a spectral range from 1000 to 2000 cm^{-1} took several hours.

The introduction of the FT principle to IR spectroscopy revolutionised this field. Nowadays, practically every IR spectrophotometer, it being a routine or a research instrument, is based on this principle. As the measuring time for, e.g., a 1000-cm^{-1} wide spectral range is drastically reduced without introducing any new disadvantages, this approach has completely replaced the old dispersive technique over the last 25 years. Therefore, it is more than appropriate to explain this principle in greater detail. In addition, as we will see later, biomolecular applications of FT IR spectroscopy are very demanding and not routine tasks. A deeper understanding of the principle will allow improving the measuring conditions by choosing the correct parameters and sampling forms.

3.1.1
Fourier Transform Spectroscopy

The heart of an FTIR instrument is a Michelson interferometer, as depicted in Fig. 3.1. In its classical form, it consists of two plane mirrors oriented perpendic-

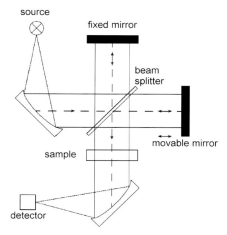

Fig. 3.1 Michelson interferometer. The radiation from the IR source is collimated and passed through the interferometer, consisting of the beam splitter and two plane mirrors. The beam splitter divides the IR beam into two parts and subsequently recombines them and directs the combined beam to the output. With the movable mirror the optical path difference between the two beams is varied. Before the output beam is focussed onto the detector, it passes through the sample.

ularly to each other and a beam splitter, which ideally transmits 50% of the incoming radiation to one of the mirrors and also reflects 50% to the other mirror. One of the mirrors is movable, changing the optical path length in one arm of the interferometer. The beams reflected by the two mirrors are recombined via the beam splitter, which also divides the combined beam into two parts, one leaving the interferometer into the direction of the incoming beam, the other into the exit port of the interferometer. One might think that each of the two beams carries half the intensity of the beam entering the interferometer (neglecting any losses). However, this applies only if interference within each of the two beams is not taken into account.

Let us first assume that the distances of the two mirrors from the beam splitters are exactly equal within a small fraction of the shortest wavelength of the radiation. In order to consider interference, one has to sum the amplitudes of the electromagnetic field of the radiation. For an ideal beam splitter, each reflection reduces the amplitudes by a factor of $\sqrt{2}$. The same reduction applies for the transmitted beam. Thus, the amplitudes of the two beams interfering with each other is $1/2 E_0$ each, with E_0 being the amplitude of the radiation entering the interferometer. As the optical pathlengths of the two interfering beams are equal, there is no phase difference, such that the amplitude of the resulting beam is just E_0, and the intensity is proportional to the square of the amplitude, i.e., E_0^2. Thus, the intensity leaving the interferometer through one arm is equal to the intensity entering the interferometer. However, under these conditions, the same arguments would also apply to the other arm of the interferometer, implying that the intensity of the radiation leaving the interferometer would be twice as high as the incident intensity.

In order to resolve this contradiction, one has to take into account the phase shifts at the beam splitter, which influence the interference and differ for the two beams leaving the interferometer. Usually, for the mid-infrared spectral range, the beam splitter is made from a thin layer of a material with high refractive index, such as Ge or Si, deposited on a substrate with a low refractive index, such as KBr or CaF_2. The optical properties can be calculated using Fresnel equations. In most commercial instruments the beam splitter is designed in such a way that the situation described above essentially applies to the beam leaving the interferometer through the exit port, i.e., perpendicularly to the entering beam. In this instance, the beam splitter must cause a phase shift of π between the two beams interfering to form the other beam that leaves through the entrance port. This phase shift causes destructive interference and no doubling of the intensity occurs. For the following discussion of FT spectroscopy we will assume a beam splitter with such optical properties and focus on the beam leaving through the exit port, i.e., where no phase shifts occur.

In order to understand the principle of FT spectroscopy in an intuitive way, we now assume that the radiation entering the interferometer is monochromatic with the wavelength λ_0. If the movable mirror is displaced by $\lambda_0/4$, the optical path difference between the two interfering beams is $\lambda_0/2$, corresponding to a phase difference π, which leads to destructive interference. If the mirror moves with constant velocity v, the phase difference between the two beams increase linearly in time by $4\pi vt/\lambda_0$, i.e., the intensity varies sinusoidally with the frequency $2v/\lambda_0$, or, using the wavenumber notation v_0, $2vv_0$. Correspondingly, the signal from the detector shows the same time dependence.

If instead of monochromatic radiation a beam with a spectral distribution $S(v)$ ($\sim E_0^2(v)$) enters the interferometer, each spectral component will contribute its own oscillation frequency, and the signal from the detector is the superposition of all the components weighted with the corresponding spectral sensitivity of the detector. This signal is a function of mirror position, or more precisely, of the optical path difference between the two interferometer arms. If the optical path difference is zero, all spectral components superimpose constructively in the beam leaving the interferometer through the exit port and thus, maximum intensity is observed. Therefore, this position of zero path difference is also known as the centre burst. If the mirror moves out of this position the intensity decreases, reaching values even smaller than one would observe in the absence of any interference. Finally, for positions further away from the centre burst, the oscillation amplitudes get smaller and the signal settles around this "no-interference" intensity. Thus, the detector "sees" two components, one of which corresponds to the integrated intensity without interference, whereas the other is induced by the movement of the mirror. Therefore, this second component is sometimes called the ac-part of the signal (ac = alternating current, only changes are monitored), and, correspondingly, the first component is called dc-part of the

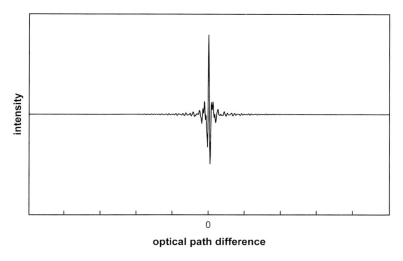

Fig. 3.2 IR interferogram. Only the central part is shown in order to resolve the oscillations near the zero optical path difference better.

signal (dc = direct current, no variation in time). The ac-part represents the so-called interferogram, which decays to zero for mirror positions far from the centre burst. A typical IR interferogram (or at least the central part) is shown in Fig. 3.2. It is interesting to note that the phase difference π between the two output beams mentioned above for the centre burst condition (maximum intensity versus zero intensity) generally holds, independent of the mirror position. This guarantees the conservation of total intensity.

In order to obtain the spectral information, one has to decompose the interferogram into its frequency components. The decomposition corresponds to the mathematical operation "Fourier transform". The spectral information $S(v)$ is the Fourier transform of the interferogram (here, and in the following discussions, we will neglect the spectral dependence of the detector). More thorough mathematical derivations are given in the Boxes 3A and 3B.

Thus, we have shown that, if all the steps described above can be established in practice, the desired spectral information can be obtained with an interferometer instead of a monochromator. The name "Fourier transform spectroscopy" derives from the mathematical operation needed to obtain the spectrum.

3.1.1.1 Interferometer

The most critical part of the FTIR set-up is, of course, the interferometer and particularly the movement of the moveable mirror. To ensure interference, the mirrors and the beam splitter must be precisely adjusted such that the deviations from the ideal arrangement are considerably smaller than the shortest wavelength of the spectrum, i.e., <2 µm. The same requirement also holds for the movement of the mirror. Here it is not only required that the adjustment remains unchanged, but also the absolute mirror position must be known precisely up to a fraction of the wavelength. This requirement can be intuitively appreciated from the fact that variations in the mirror position smaller than 1/4 of the shortest wavelength to be covered still cause intensity variations at the detector, and these variations need to be resolved in order to determine this spectral component.

The oscillatory modulation of a monochromatic beam described above is used to monitor the movement of the mirror. For this purpose, a monochromatic beam from an HeNe laser (632.8 nm) is directed into the interferometer, parallel and very often coaxial to the infrared beam. The centre of the beam splitter is specifically designed for the wavelength of the laser, and the detector for the laser, a small photodiode, is placed into the centre of the exiting infrared beam. By counting the zero crossings of the laser modulation, i.e., the laser interferogram, a very precise record-keeping of the mirror positions can be established.

In addition to controlling the movement of the mirror, the HeNe signal detected by the photodiode serves to control the digitisation of the infrared interferogram, which is needed to apply the mathematical Fourier transform operation in a computer. Details of this are described in the Box 3B, but a few comments might be helpful in this context. The common algorithm used for fast Fourier transform requires that the digital values of the interferogram are sampled equidistant with respect to the optical path difference, and this equidistant spacing has to be determined with high precision. In the set-up described above, the sinusoidal signal from the HeNe laser provides this information, and the zero crossings are used to precisely sample the interferogram and trigger the digitisation. From information theory (Nyquist criteria, see Box 3B) we know that Fourier transformation resolves a frequency component v only if the signal is sampled at a time spacing determined by $\Delta t = 1/(2 \cdot v)$. Thus, if each zero crossing is used to sample the interferogram, the highest frequency that can be resolved is that of the HeNe laser, i.e., 15802.8 cm^{-1}. If only each second zero crossing is used, the limit is 7901.4 cm^{-1}, or for each fourth the limit is 3950.7 cm^{-1}. Thus, the implementation of an HeNe laser into the interferometer was essential for the practical development of FT spectroscopy.

The required digitisation of the interferogram naturally asks questions of the required resolution. To understand the underlying principles in a qualitative manner, we have to take into account that, unlike for monochromators, the detector only "sees" the spectrally integrated intensity that is modulated by the interference (Fig. 3.2). For the determination of small changes of a single absorption band, not only its contribution to the total intensity but also its influence on the interference modulation must be detected. Thus, the requirements for the digitisation accuracy are demanding.

Another aspect is related to the digitisation resolution of the noise level. To obtain the full information content of a signal, the accuracy must be high enough to resolve the noise. However, it is often necessary to co-add interferograms for improving the signal-to-noise ratio in the spectrum. This signal averaging reduces the effective noise level and it is, therefore, limited by the boundary condition that the noise must still be visible in the averaged interferogram for a full extraction of the spectral information from the interferogram. For long a time, the dynamic range of digitisation has been 16 bit, limited by the requirement that a single interferogram is obtained within a few milliseconds. This restriction determines the electronic bandwidth to approx. 200–500 kHz. The noise reduction by signal averaging has specific consequences for the most sensitive liquid N_2-cooled semiconductor detectors (see Section 3.1.1.2). To exploit this sensitivity and the potential of signal averaging, the digitisation accuracy has to be increased, and new instruments often offer a 24-bit dynamic range digitisation.

However, the design of the experiments can also be optimised to improve the accuracy. Very often one is interested in only a limited spectral range, e.g., between 1900 and 600 cm^{-1}. In such a situation, it might be useful to block the unwanted spectral part from the detector. This results in an increase in the relative contribution of a spectral feature to the interferogram.

3.1.1.2 Infrared Detectors

The properties of infrared detectors are important for the performance of FTIR spectroscopy. As we shall see below, the advantages of FTIR spectroscopy only become evident if the spectral range to be covered is not too small, thus broad-band detectors have to be used. Another requirement concerns the speed (or electronic bandwidth) of the detector. As we have seen in the discussion of the principles of FT spectroscopy, the detector must be able to resolve the modulation of the radiation caused by the movement of the movable mirror, and this frequency can be as high as 200 kHz, depending on the speed of the mirror and the highest wavenumber of the radiation. Thus, detectors must be sufficiently fast. Finally, as the total integrated intensity almost always reaches the detector, it must tolerate this high intensity and nevertheless exhibit, as far as possible, a linear response. Despite this, detectors must have a high sensitivity to be able to resolve small spectral features. It is obvious that not all these requirements can be fulfilled by one detector and compromises have to be made taking into account the demands of the experiments.

There are two classes of detectors, thermal and photon detectors. Pyroelectric detectors are the commonly used thermal detectors in FTIR spectroscopy. They are operated at room temperature and exhibit an excellent linear performance, at a somewhat limited speed, but are considerably limited in sensitivity. The principle is based on the pyroelectric effect, which induces changes in the surface charge density of special materials if the temperature is changed. The radiation being absorbed by the detector causes an increase in temperature. As only changes in the surface charge can be detected by the subsequent preamplifier, the detector monitors changes in intensity, and thus, the ac-part of the intensity, i.e., the interferogram directly. Typical materials for the pyroelectric detectors encompass triglycine sulfate (TGS), deuterated TGS (DTGS), and $LiTaO_3$. The relatively low speed of these detectors is caused by the thermal properties of the materials.

The photon detectors are semiconductor detectors operating at liquid N_2 temperature. Through the absorption of infrared photons, charge carriers are generated in the material, and thus the conductivity is increased. The increased conductivity is monitored by the subsequent preamplifier as an increase in current flowing through the detector element if a constant voltage is applied. These detectors are characterised by excellent sensitivity, and high speed, but they become nonlinear even at intensities common for mid-infrared FTIR spectroscopy. The detector sees both the dc- and the ac-part of the intensity, and the interferogram is obtained by ac-coupling of the signal either within the preamplifier or after the preamplifier. This class of photon detectors includes photoconductive and photovoltaic detectors. The first group refers to homogeneous semiconductors, whereas the second group are photodiodes such as Si photodiodes for the detection of visible and near-IR radiation. The sensitivity of the photovoltaic detectors are comparable to that of the photoconductive ones, however, they are considerably faster, i.e., 30 versus 700 ns response time and they provide a better linear performance. Photoconductive and photovoltaic detectors are mainly fabricated from HgCdTe. Through variation of the material composition, the wavenumber of peak sensitivity of photoconductive detectors can be altered, typically between 1400 and 600 cm^{-1}. Such a sensitivity variation does not seem to be possible for photovoltaic detectors, which exhibit maximum sensitivity between 1000 and 800 cm^{-1}. All semiconductor detectors are characterised by a cut-off at lower wavenumbers, that is between 1200 and 400 cm^{-1}. The sensitivity usually decreases at higher wavenumbers according to $1/\nu$, as the sensitivity is related to the intensity and not to the photon flux. In addition, the sensitivity might be further influenced by the IR absorption properties of the specially treated detector material and of the window of the dewar for the liquid N_2-cooled detector element.

3.1.2
Advantages of Fourier Transform Infrared Spectroscopy

In Section 3.1.1, it was emphasised that in FTIR experiments the measuring time is drastically reduced, making measurements which could not be performed previously on the basis of monochromatic detection now possible. For optimal measurements it is useful to understand the reasons for this improvement.

The main reduction in intensity using a monochromator takes place at the entrance and exit slits, which must be kept narrow to guarantee sufficient spectral resolution. In Fourier transform spectroscopy an aperture is needed for collimating with a parabolic mirror the beam entering the interferometer. It can be shown that relatively large apertures can be tolerated, which reduce the optical losses at the entrance of the interferometer substantially. This effect is called Jacquinot's advantage, the extent of which strongly depends on the actual optical design, the upper frequency limit and the spectral resolution: a larger upper frequency and a better resolution require a better collimated beam (Box 3C).

However, the main advantage is caused by a different feature of FT spectroscopy that is based on the fact that the signal of the detector contains both the optical signal and the noise. The noise has its own frequency spectrum, and thus, the noise power is distributed over the complete spectral range by the Fourier transformation. In IR spectroscopy, the noise associated with a signal is determined by the detector noise. In spectrophotometers using a monochromator, there is no Fourier transformation of the signal, and thus, the full noise power is present at each spectral data point. This can be expressed in a quantitative way by taking into account the spectral resolution and the time-constant of the measurement. The signal-to-noise (S/N) ratio of spectra measured on an FT spectrometer will be greater by a factor of $\sqrt{M}$ (M is the number of resolution elements) than the S/N ratio of the spectrum measured in the same time and at the same resolution on a grating instrument with the same source, detector, optical throughput and time-constant. As a typical example we will consider a spectrum measured over a range of 2000 cm^{-1} with a spectral resolution of 4 cm^{-1}. The number of resolution elements is, therefore 500, and thus, the improvement in the S/N ratio is 22. Conversely, the measuring time with the grating instrument must be increased by a factor of 500 to afford the same S/N ratio as with the interferometric set-up. This advantage is called multiplex-advantage or Felgett's advantage.

In FT spectroscopy, the noise is not independent of the resolution. With increasing length of the interferogram, more noise is sampled, and thus, the noise in the spectrum is also increased. However, it increases only by $\sqrt{1/\Delta v}$ whereas for a grating instrument it is proportional to $1/(\Delta v)$.

Another advantage of the FT technique is the absolute wavenumber accuracy originating from the precise control of the interferometer and of the data acquisition by the HeNe laser interferogram. This is very important for high-resolution FTIR spectroscopy in the gas phase, and is also helpful for studying samples in the condensed phase, i.e., also biological samples, although the bandwidth of the absorption bands is relatively large and the required spectral resolution is lower.

3.1.3
Optical Devices: Mirrors or Lenses?

For the construction of an IR spectrophotometer it is necessary to collimate the radiation from the IR source into the interferometer and subsequently focus it onto the sample. Such collimation and focussing can be achieved with lenses or with special mirrors (spherical mirrors, parabolic, or elliptical mirrors). The main advantage of mirrors is that their performances are independent of the infrared spectral range, given that absorption losses are avoided by appropriate coatings. In contrast, lenses have only a limited spectral range, where they are transparent, in the infrared. In addition, there is a considerable variation in the refractive index for the materials usable for lenses. This results in a spectral dependence of their optical performance (e.g., variation of the focal length). Both limitations are particularly severe, as IR spectroscopy must cover a very broad spectral range, typically from 500 to 4000 cm^{-1}. Finally, the materials that have the broadest transparent spectral range are either very expensive, such as Ge, ZnSe, or they are fairly hygroscopic, such as NaCl, KBr or CsBr. Therefore, in almost in every respect mirrors are superior to lenses. Today they are fabricated by diamond milling machines with very high precision, hence the prices have dropped considerably.

3.1.4
Instrumentation for Time-resolved Infrared Studies

The elucidation of a reaction mechanism often requires studying the molecular properties of intermediates of the reaction. If such intermediates cannot be stabilised, e.g., by cryogenic trapping, time-resolved techniques have to be employed. There are various approaches to measure time-resolved IR spectra, but there is not a single one that covers the entire time range of interest in biology from the femtosecond to second time scale.

We distinguish between pulsed and continuous-wave (cw) techniques. For time-resolutions longer than 10 ns, the monitoring IR beam is cw, requiring fast IR techniques and detector electronics for acquisition of the signals reflecting the time-resolved absorbance changes. Time-resolutions shorter than 10 ns are based on pulsed IR irradiation and, currently, 100 fs can be reached by dedicated techniques.

3.1.4.1 Time-resolved Rapid-scan Fourier Transform Infrared Spectroscopy
As explained in the description of the FTIR method, it takes only 5–20 ms for the movable mirror of the interferometer to travel over the required distance (typically 0.5 cm or less). Thus, a complete interferogram is measured within that time. Most manufacturers of FTIR instruments provide the means to synchronise the data acquisition of the first interferogram to an external trigger linked to an event in the sample, that is, the initiation of a chemical reaction or a physical process. Thus, spectral changes induced by the event can be monitored in a time-resolved manner on the basis of successively measured interferograms. The spectral changes are calculated with respect to the spectrum before the event or with respect to the spectrum a long time after the event when the final state of the sample has been reached. Both reference spectra can be measured with a high S/N ratio as they refer to stable states of the sample. The time-resolution is not only determined by the travelling time of the mirror but also by the time the mirror needs to change its direction. Usually, data can be collected during both directions of movement. Because of the rapid movement of the mirror, this method is called time-resolved rapid-scan FTIR spectroscopy. The principle is shown in Fig. 3.3. However, the S/N ratio of such a single measurement is not sufficient to detect the spectral changes. Therefore, this procedure has to be repeated several times (50–100). If the reaction triggered by the event is reversible, the repetitive accumulation of interferograms is straightforward. For irreversible processes, however, a fresh sample has to be used after each measurement, requiring a correspondingly larger amount of material, which is not always available.

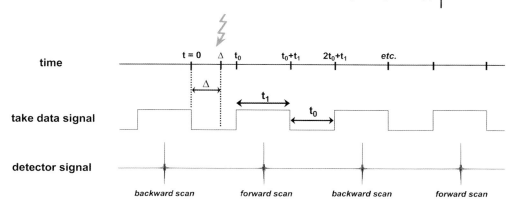

Fig. 3.3 Principle of the rapid scan technique. The electronic signal (take data) from the interferometer electronics is synchronous with the acquisition of the interferogram. The falling edge of the signal is used as time zero. After a delay Δ, which is shorter than the time elapsed before the acquisition of the next interferogram, the reaction is triggered. Thus, the interferogram directly after the trigger records the spectral changes within the time-resolution. Time-resolution is essentially determined by the time t_1 needed by the movable mirror to move over the path difference required for the acquisition of the interferogram. Subsequent interferograms are stored separately and contain the spectral information at later times. t_0 is the time between two successive interferograms and is mainly determined by the time the movable mirror needs to reverse its direction. The timing is shown for interferogram acquisition during both directions of the mirror movement.

For longer reaction times that allow reduction of the time resolution, several successive spectra can be co-added, thereby increasing the S/N ratio. A practicable way to achieve such a progressive slowing down of the effective time resolution is the implementation of a quasi-logarithmic time base. For example, the first 8 spectra are linearly spaced, for the next 16 scans, 2 consecutive interferograms are averaged, resulting again in 8 linearly spaced spectra, however with a doubling of the time-spacing; and for the next 32 scans 4 consecutive interferograms are averaged. Such a quasi-logarithmic time base is often used in time-resolved studies. In addition to the improvement of the S/N ratio for longer times it allows a much larger time range to be covered with a still manageable number of data points. For time-resolved events determined by linear reaction kinetics this reduction of data points is essentially accomplished without loss of information.

There are two basic methods for triggering a reaction, these are the initiation (i) of a photochemical or photophysical process by a light pulse, and (ii) of a chemical reaction via rapid mixing of the reactants. For rapid mixing, time-resolution is limited by the mixing process, which takes a few milliseconds using commercial devices, and by the time required for the mixed solution to be transferred into the IR cuvette, which also takes a few milliseconds. Thus, only processes that are faster than 10 ms can be probed and for this time-scale rapid-scan FTIR is the appropriate approach. A detailed description of rapid mixing techniques is given in Section 4.4.2. Shorter time scales are accessible when light pulses, typically laser pulses, can be employed to trigger reactions. Then other time-resolved IR spectroscopic techniques are required.

3.1.4.2 Time-resolved Studies Using Tunable Monochromatic Infrared Sources

In the UV–vis spectral range, the most common experimental set-up for time-resolved spectroscopy is based on a broad-band light source, a monochromator for selecting the wavelength of the probe beam, and a fast detector and electronics to record the time-resolved intensity changes. This is usually achieved by a transient recorder or a digital oscilloscope. The complete time-resolved spectrum is obtained by changing the monochromator setting point by point to the desired wavelengths. In principle, it should be possible to extend this technique also to the infrared spectral region. However, owing to the experimental difficulties associated with weak light sources and noisy detectors (*vide supra*) this approach does not represent a practical solution in IR spectroscopy, although a couple of applications have been published (Barth et al. 1995; Siebert et al. 1981). In a modified version, tunable lasers can be employed as probe beam sources. With the availability of cheaper plane array detectors, the disadvantage of weak thermal sources may be compensated by the multichannel advantage. This aspect, also in connection with time-resolved IR spectroscopy, has been thoroughly discussed (Pellerin et al. 2004). The higher intensity of the laser light results in a considerably shorter acquisition time as signal averaging can be reduced. At present, the only sufficiently tunable laser sources are the lead-salt laser diodes, which are commercially available (Mäntele and Hienerwadel 1990). These lasers are operated at low temperature (10–120 K), and the tuning range varies between 50 and 200 cm^{-1}. Therefore, the method is particularly suitable if one is interested in spectral data of only a limited range.

3.1.4.3 Time-resolved Fourier Transform Infrared Spectroscopy Using the Step-scan Method

It is evident from the discussion of the advantages of FT spectroscopy that the method becomes increasingly powerful with an increasing spectral range to be covered. Therefore, it is desirable to develop a method for time-resolved spectroscopy that preserves the advantages of FTIR spectroscopy but overcomes the limited time-resolution of the rapid-scan technique. The basic idea of such an improvement of time-resolved FTIR spectroscopy is simple, but it requires substantial modifications of the interferometer. It somehow uses the principle of the method with tunable IR sources. However, here it is not the source that is tuned in discrete steps, but the position of the movable mirror. At each fixed position the reaction is triggered, and the time-resolved change of the interferogram at this mirror position is recorded. The principle is shown in Fig. 3.4, where the time-dependent changes of the interferogram evoked by the laser flash are shown for a number mirror positions. It is necessary to control the mirror in such a way that a step-wise movement from one sampling position to the next one can be achieved until the complete length of the interferogram is covered. As outlined

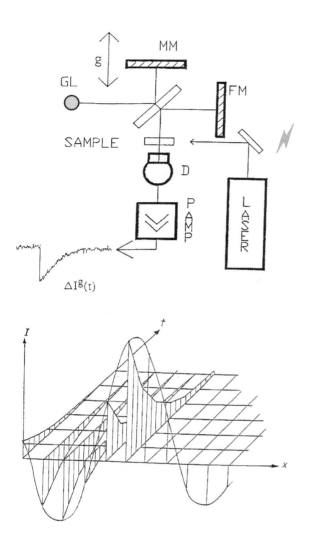

Fig. 3.4 Principle of the step-scan technique. Upper part: the principle of an FTIR instrument is shown. The movable mirror (MM) is held in a fixed position and the reaction is triggered by a laser flash. The detector records the time-dependent change of the interferogram after the flash. Lower part: 3D representation of the time-dependent changes of the interferogram at a number of mirror positions, evoked by the laser flash. It assumes that the reaction returns to the initial state. Abbreviations: GL globar, FM fixed mirror, D detector, PAMP preamplifier.

above, both the spacing between two successive positions and the number of stepping points, i.e., the length of the interferogram, are determined by the FT principle. The construction of an interferometer capable of this stepwise mirror movement is not straightforward, as positioning and the mechanical stability has to be very precise. There are several companies that have now implemented the step-scan feature in the normal rapid-scan interferometers. As with tuneable monochromatic sources, the time-resolution is only limited by the speed of the detector and of the electronics (amplifier, transient recorder, digital oscilloscope).

After recording the time-resolved changes of the interferogram at each position, the data are rearranged so that interferograms at each time slice of the time traces are obtained. After Fourier transformation the corresponding spectral changes are obtained. From the measuring principle, it is clear that the reaction has to be triggered at least as often as there are sampling points of the interferograms, typically 540. Taking into account the need for signal averaging, the number of trigger events amounts to several thousand. This is an important prerequisite that has to be taken into account when planning such measurements.

Manufacturers of interferometers that are capable of step-scan movements provide software to perform the measurement and to calculate the time-resolved spectra. For more details, in particular for difficulties one often encounters, the reader is referred to the original literature (Uhmann et al. 1991, Rödig and Siebert 1999).

3.1.5
Time-resolved Pump-probe Studies with Sub-nanosecond Time-resolution

Pushing the time-resolution of the step-scan technique or of the tunable monochromatic set-up toward the short nanosecond time scale is associated with increasing technical difficulties. The electronic components are not sufficiently fast. Furthermore, the intensity of the IR probe beam provided by conventional light sources also becomes too weak. Because the photon flux, i.e., the number of incident photons per second, is constant, the number of photons per sampling interval Δt decreases with decreasing Δt, such that eventually the signal becomes too weak to exceed the noise level. Thus, alternatives are required for approaching the sub-nanosecond time scale. Such techniques, which have been developed along with the availability of (sub-)picosecond laser pulses, are used to study photoinduced processes. Based on Ti:sapphire lasers, light pulses with a duration between 50–100 fs have become almost routine nowadays. These short-time pulses are used both for initiating the reaction and probing the intermediates. This so-called pump-probe technique is well established in transient absorption spectroscopy in the visible range. The exciting pulse (pump laser) and the probe pulse are derived from the same laser and a variable time difference between the two pulses can be adjusted by an optical delay line, as described in detail in Chapter 4. Measuring the absorbance at the probe wavelength prior and subsequent to the pump event, the absorption changes can be monitored as a function of the delay time with respect to pump event. As time resolution is controlled by the optical delay line, fast electronics are not required. Furthermore, the energy of the probing pulse is measured such that the detector does not need to temporally resolve the signal, and therefore, the response time is not important. However, it must be guaranteed that the response of the detector with respect to the energy of the

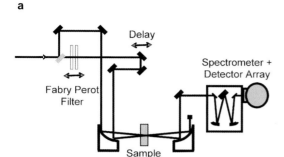

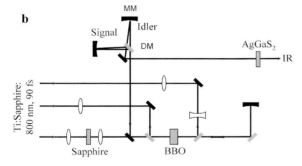

Fig. 3.5 (a) IR/IR pump/probe set-up for femtosecond time-resolved measurements. The femtosecond broad-band IR beam is split into two beams. One beam is converted by a tunable Fabry–Pérot filter into a narrow-band beam and is used as the pump beam. The other is passed through a variable delay line. After passing through the sample, the probe beam is directed to the monochromator and detected by an array detector. 2D-IR spectra are obtained by tuning the Fabry–Pérot over the range of the broad-band beam. (b) Principle of femtosecond mid-infrared pulse generation. Details are given in the text (after Hamm et al. 2000).

ultra-short light pulses is linear. This principle of pump-probe experiments also holds when this approach is extended to the IR region, albeit modifications of the experimental design are necessary. In particular, frequency converters have to be employed to generate the infrared probing pulse from the master laser that usually emits in the UV–vis or near-infrared spectral range. A typical set-up for IR-pump/IR-probe spectroscopy is shown in Fig. 3.5(a), which has been used for 2D-IR experiments on peptides.

In this scheme, a narrow-band infrared pump pulse and a broad-band probe pulse are used. The pump pulse is derived from the probe pulse by a beam splitter and a Fabry–Pérot filter. The timing between pump and probe pulses is set by the delay line. The broad-band pulse is dispersed by the monochromator and detected by a 2×32 element array detector covering an infrared spectral range of ca. 200 cm^{-1}. This set-up has been used to probe the peptide bonds in proteins. In such experiments, one C=O oscillator of a single peptide group is excited by the strong pump pulse and the probe pulse is used to monitor its influence on the neighbouring C=O groups via dipolar coupling (Wouterson and Hamm 2002) (see Section 5.4).

In the more general situation, this approach is employed to study light-induced reactions, using a pump pulse in the UV–vis spectral range for initiating the process. Pulses at these wavelengths can be obtained by second-harmonic generation of a fraction of the 800-nm pulse provided by the Ti:sapphire laser. This 400-nm pulse may be used either directly or converted into pulses at longer wavelengths between 450 and 750 nm, using an optical parametric generator (OPG) or amplifier (OPA).

The generation of infrared pulses is shown in Fig. 3.5(b) (Hamm et al. 2000). It is somewhat more complicated, but for this tutorial it is sufficient to explain the basics of this design. Three pulses are split from the 800 nm Ti:sapphire laser pulse. The 2.5 µJ pulse is focussed onto a sapphire plate, from which a continuum pulse is generated both in the visible and in the near-infrared spectral range. The near-infrared continuum is used to seed the OPA. The OPA is constructed on the basis of a BBO crystal, which is double passed (back and forth) with 3.5- and 200-µJ pulses from the Ti:sapphire laser, respectively. The amplified signal is separated from the idler and recombined by the dielectric mirror DM. By adjusting the delay of one of the two pulses with the help of MM, they are brought into time coincidence and focussed onto an $AgGaS_2$ crystal. The phase-matching properties of this crystal favour difference frequency generation of these two pulses, and a broad-band (200 cm^{-1}) mid-infrared pulse in the spectral range from 3 to 10 µm is generated.

Because with one excitation pulse only one time slice is obtained, at least as many exciting pulses have to be applied as time slices are required, which are typically between 100 to 1000. In addition, in this technique the signal has to be averaged, hence the number of excitation pulses increases correspondingly. Therefore, the applicability of this approach and of all other time-resolved methods is most easily performed with systems undergoing a reversible processes. For other system, the sample has to be replaced after each shot (in time-resolved infrared spectroscopy of biological systems, due to the small absorbance changes, a large fraction of the sample is required to be excited). The availability of broad-band IR probe pulses and array detectors thus reduces the requirements on the measuring time and sample amount substantially as compared with monochromatic detection. Furthermore, both the pump and probe pulses originate from a laser. Hence the beams are well collimated and, therefore, can be focussed onto a small spot, such that the irradiated volume is very small and the amount of material needed for these experiments is further reduced. These technical improvements are, without any doubt, a prerequisite for applying these techniques to biological systems.

It is interesting to ask whether the combination of broad-band source, focal plane monochromator, and array detector could also be used for slower time-resolved studies. In fact, it is possible to use a thermal source for the infrared radiation, as in a conventional spectrophotometer. The array detectors can be gated, implying that they are active only during a certain period of time, which is controlled by the length of the gate pulse. The gate pulse is synchronised with the trigger event, and in this way, broad-band spectral information is obtained within a single time-slice. As compared with the single-wavelength method described above, this approach represents a trade-off between broad-band spectral information and temporal information. Furthermore, the lower energy input into the monochromator as compared with the interferometer also prevails here. Such a device has been described recently, and one has to gain more experience on the practical performance (Pellerin et al. 2004).

3.2
Raman Spectroscopy

A Raman spectroscopic set-up is based on five essential components, a monochromatic light source (i.e., a laser), an appropriate arrangement of the sample (see Section 4.2), an analyser for the scattered light which may either be a spectrometer or a spectrograph, a device for converting photons into electrical signals, and a control unit for data acquisition (Fig. 3.6).

Nowadays, compact Raman spectrometer systems are available in which all these components are integrated. Such systems offer the advantages and disadvantages of typical "black box" systems: they are easy to handle but cannot readily be modified, such that they are difficult to adapt to non-standard applications. Dedicated and specifically tailored systems may overcome these restrictions, but setting up and maintainance require significant skills and knowledge of optics.

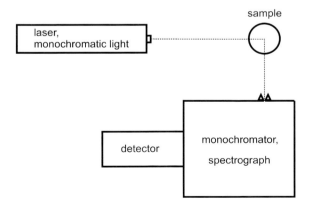

Fig. 3.6 Block diagram of a Raman set-up including the essential components.

3.2.1
Laser

Excitation lines used for Raman spectroscopy cover the spectral range from the near-infrared (1064 nm) to the deep-UV (ca. 180 nm). These excitation lines are obtained by various types of lasers (Silvfast 1996). The most commonly used lasers are noble gas ion lasers that emit a large number of discrete lines, mainly in the visible spectral range. These lasers are based on well-established technologies and are easy to handle. However, these ion lasers generally require demanding water-cooling systems, which, in addition to relatively high maintenance costs, represent serious drawbacks. Emerging alternatives are solid state lasers, among which the most widespread system is the already well-established Nd:YAG laser system. Although this laser possesses only one fundamental line at 1064 nm, further lines are available upon frequency doubling (532 nm), tripling (355 nm), or quadrupling (266 nm). Furthermore, commercially available diode lasers operate in the red to near-infrared region, thereby filling a gap between Kr ion and Nd:YAG lasers. Up to now such diode lasers cannot fully replace Kr or Ar ion lasers when wavelengths below 600 nm and high laser powers are required, but they may become alternatives in the future in view of the rapid technological developments in this field.

Whereas all these lasers operate at discrete and well-defined emission lines, the output wavelength of dye and Ti:sapphire lasers is tunable in a relatively wide range. Dye lasers cover the entire range from the near-UV to the red spectral region but set-up and operation are relatively demanding. Firstly, the dye laser must be optically pumped by a high power laser (Ar ion or frequency-doubled or -tripled Nd:YAG laser) at a wavelength below the desired emission line of the dye. Secondly, each laser dye is suitable only for a wavelength range of 20–50 nm, such that scanning over a larger spectral range requires time-consuming replacement of the dyes and optics. As an alternative, the Ti:sapphire laser has been introduced, which is in fact easier to handle. Also, this laser has to be pumped by an Ar or a frequency-doubled Nd:YAG laser. The emission of the Ti:sapphire laser is restricted to the range between 700 and 800 nm, but the output is sufficiently high for frequency doubling such that lines between 350 and 400 nm can also be obtained.

Frequency multiplying is generally needed to obtain excitation lines in the deep-UV region below 300 nm. In most instances, deep-UV lines are only accessible in the pulsed mode as high intensities are required for generating higher order harmonics, such as the quadrupled output of the Nd:YAG laser fundamental at 266 nm. To obtain further lines in the deep-UV, the stimulated Raman effect of gases such as hydrogen, deuterium, or methane has been utilised (Fodor et al. 1986). Excitation of these molecules with high laser powers can lead to a high population of vibrationally excited states. Once inversion of energy levels is achieved, these molecules can act as molecular lasers that emit with a frequency ν_{stim}

$$\nu_{stim} = \nu_{exc} + n \cdot \nu_{mol} \tag{3.1}$$

where ν_{exc} and ν_{mol} are the frequencies of the primary laser excitation and the fundamental molecular vibrational mode, respectively, and n is the vibrational quantum number. The fundamental, the second, third, or the fourth harmonics of the Nd:YAG laser can serve as excitation frequencies, such that a large number of lasing emissions can be obtained upon progressive excitation of the vibrational energy level. The only examples for cw-laser lines in the deep-UV are based on intracavity frequency doubling, which can be achieved for the 514-, 488-, and 413-nm lines of Ar and Kr ion lasers.

3.2.1.1 Laser Beam Properties

Besides the wavelength, there are various parameters that characterise the quality of laser radiation which make it is suitable for Raman spectroscopic experiments. Firstly, the spectral bandwidth must be distinctly smaller than the widths of the vibrational bands to be probed, which are typically between 8 and 15 cm^{-1} for biological systems. Ion lasers operating in the cw-mode easily fulfil this requirement. This is not *a priori* the situation for pulsed lasers and dye lasers, for which, however, optional devices (etalons) are available to reduce the bandwidth below 1 cm^{-1}. Furthermore, the spectral bandwidths of pulsed lasers are associated with a principle restriction, the transform limit, which is ca. 3.5 nm for 1-ps pulse. For an excitation line at 500 nm, this would result in a spectral bandwidth of 14 cm^{-1}, given that this limit is reached. Then, 1-ps pulses lead to a substantial broadening of the Raman bands and a decrease in the peak intensity corresponding to a lowering of the S/N ratio. Thus, short time Raman spectroscopic experiments in the picosecond regime are usually associated with a trade-off between temporal resolution and spectral information. An exception is the broad-band stimulated Raman scattering discussed in Section 4.4.1.2.

Specifically for biological applications, the photon flux at the sample is a critical parameter. For a cw-laser emitting at a wavelength of $\lambda = 532$ nm with a power of $P = 0.1$ W, the number of photons per second is given by

$$N_p = \frac{P_0 \cdot \lambda}{h \cdot c} = 2.7 \cdot 10^{17} \text{ s}^{-1} \tag{3.2}$$

The same number is obtained for pulsed laser emission with 10-ns pulses with a peak energy of 10 mJ and a repetition rate of 10 Hz. However, in this instance one tenth of the photons are emitted within 10 ns, corresponding to a temporary photon flux that is higher by seven orders of magnitude than with cw-excitation. Such high peak energies, therefore, bear the risk of inducing unwanted photo-induced reactions and photodamage of the biomolecule. Lowering the pulse energy, however, decreases the average photon flux per second and, thus, also the number of Raman scattered photons. For this reason pulse laser excitation leads to Raman spectra with poorer S/N ratios and, hence, should be avoided if possible.

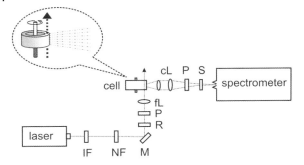

Fig. 3.7 Optical set-up for a Raman experiment using a rotating cell in a right-angle scattering geometry. The optical components include an interference filter (IF), a neutral-density filter (NF), a mirror (M), a focussing lens (fL), and two collecting lenses (cL). Additional components required for polarisation-dependent Raman measurements are a Faraday rotator (R), a polariser (P), and a scrambler (S).

In general, the laser beam is focussed onto the sample, hence the relevant parameter is the photon flux (power) per unit area (energy flux) rather than the photon flux itself. Diameters of laser beams that exhibit a Gaussian profile are defined by the full-width of the beam at $1/e$ of the peak intensity. Typical values of these diameters are between 50 and 150 µm when 5- to 20-cm focussing lenses are used in conventional set-ups. For the example given in Eq. (3.2), a diameter of 50 µm then corresponds to a photon flux density of $3.4 \cdot 10^{25}$ s^{-1} m^{-2}. In confocal Raman spectrometers, in which the laser beam is focussed onto the sample by a microscope objective, beam diameters down to 1 µm can be obtained, corresponding to an increase in the energy flux by more than three orders of magnitude. To avoid photodamage of the sample, the laser power has to be attenuated.

In the optical pathway from the laser to the sample compartment, a variety of optical elements are needed for "cleaning", proper alignment, and focussing of the beam (Fig. 3.7). In Raman spectroscopic experiments lasers usually operate in the single-line mode, however, specifically in the UV region, multi-line operation may be unavoidable. In these instances, prisms or simple prism spectrographs and pinholes have to be inserted into the optical pathway to filter out unwanted laser lines. For closely adjacent laser lines, such as the 333.6-, 334.4-, and 335.8-nm lines of the Ar ion laser, passage through a quartz prism yields sufficient spatial separation only at relatively long distances.

In both single- and multi-line operation, the laser output includes non-lasing emission lines that are much weaker than the lasing line(s) but that are still sufficiently intense to interfere with the Raman spectrum when they reach the detector. These lines can easily be identified as their half-widths are much smaller than those of Raman bands of molecules in the condensed phase and thus they appear as sharp spikes in the spectrum. These lines can be removed by appropriate interference filters, which are available for a large variety of laser lines.

In addition to the removal of unwanted spectral components, "cleaning" of the laser beam also aims at achieving a high polarisation ratio. Laser light is polarised perpendicularly to the propagation direction and the laser is usually mounted such that the electric field vector (polarisation) is oriented perpendicular to the plane of the optical table. For cw-ion lasers, the polarisation ratio is typically higher than 100:1, which may be sufficient for many applications. However, optical elements such as mirrors and lenses may lower this ratio such that determination of polarisation ratios of Raman bands is aggravated. In this instance, insertion of a polariser and a Faraday rotator in the optical pathway, preferentially close to the sample, are required to obtain a beam polarisation of variable but well-defined orientation. Because the gratings used in dispersive spectrometers exhibit a polarisation-dependent sensitivity, a scrambler should be placed just in front of the spectrometer entrance.

3.2.1.2 Optical Set-up

Each mirror, lens, or filter in the set-up may degrade the beam quality, affecting the mode pattern or the polarisation and, furthermore, causing optical losses that could add up considerably, depending on the complexity of the optical layout. In the visible range, even high quality lenses and mirrors may be associated with losses of up to 5%, such that ten of these components may reduce the laser intensity to nearly 50% at the sample, while in the near-infrared and in the UV spectral regions losses may be even higher. Appropriate coatings can improve reflectivity and transmission of the mirrors and lenses, respectively.

In conventional Raman set-ups, precise and stable alignment of the focussed beam onto the sample is crucial for an efficient collection of the Raman scattered photons by the spectrometer. This is specifically true for a $90°$-scattering geometry when the image of the scattered light on the entrance slit of the spectrometer is usually magnified for an optimum illumination of the slit height. Then even subtle displacements or a tilting of the exciting laser beam with respect to the slit, which may be easily introduced by small misalignments of lenses and mirrors, can drastically reduce the intensity of the detected signal.

Pump-probe time-resolved Raman experiments impose even higher demands on the precision and stability of the optical set-up as the probe beam has to be aligned both with respect to the entrance slit of the spectrometer and with respect to the pump beam. Clearly, the performance of dedicated and laboratory-designed Raman spectroscopic set-ups depends sensitively on the mechanical stability, which in turn requires high quality optical mounts and positioning systems and appropriate optical tables. Actively damped tables represent the best choice. Such tables provide a platform for the set-up that is isolated against vibrations from the surroundings and offer a high degree of flexibility for mounting the individual components.

Pump-probe experiments not only require precise and stable alignments but also high quality beam profiles (i.e., a TEM_{00} mode pattern). Low divergence of the laser beams as a prerequisite for well-defined illumination conditions of the sample is particularly important in time-resolved pump-probe Raman measurements (see Section 4.4.1). These conditions are usually fulfilled by the fundamentals of cw-lasers. As a general rule, the number of optical components should be kept as low as possible to reduce intensity losses, lowering of the polarisation ratio, and the deterioration of the mode profile.

3.2.2
Spectrometer and Detection Systems

Analysis and detection of the scattered light is possible in two different modes, the monochromatic and polychromatic detection (Schrader 1995). Systems that are based on dispersive elements are principally similar to those used in fluorescence spectroscopy but exhibit a distinctly higher spectral resolution.

3.2.2.1 Monochromators

Monochromatic detection utilises spectrometers equipped with holographic gratings serving as dispersive elements (Fig. 3.8). The light scattered by the sample is focussed onto the entrance slit of the spectrometer and falls onto the grating, which disperses the light. Thus, only light of a small wavenumber increment leaves the monochromator through the exit slit that lies in the focal plane of the dispersed light. The size of this increment is given by the geometrical width of the

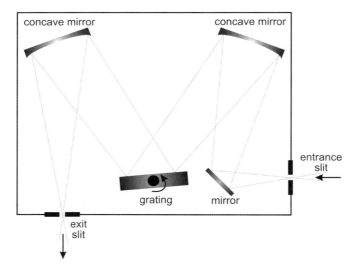

Fig. 3.8 Optical lay-out of a Czerny–Turner-type monochromator with a plane reflection grating. The position of the grating controls the spectral segment that is focussed onto the exit slit.

entrance and exit slits and the lateral dispersion $d\lambda/dx$ of the spectrometer, which in turn depends on the focal length f of the spectrometer and the groove to groove distance (g) of the grating.

$$\frac{d\lambda}{dx} = \frac{g}{f} \cos\beta \tag{3.3}$$

For a 1800 lines mm^{-1} grating and a focal length of 1 m, a typical working angle of $\beta = 30°$ yields a lateral dispersion of ca. 0.48 nm mm^{-1} such that at 500 nm a spectral range of 20 cm^{-1} is spread over 1 mm. A high focal length and a large groove density provide a high dispersive power as a prerequisite for a good spectral resolution, which in addition, depends on the band pass determined by the finite spectrometer slit widths ($\Delta\lambda_{\text{slit}}$) and unavoidable aberrations and diffraction effects ($\Delta\lambda_{\text{sys}}$). For a Gaussian band profile with a natural bandwidth of $\Delta\lambda_{\text{band}}$, the measured bandwidth $\Delta\lambda_{\text{exp}}$ is approximately given by

$$\Delta\lambda_{\text{exp}} \approx \sqrt{(\Delta\lambda_{\text{slit}})^2 + (\Delta\lambda_{\text{sys}})^2 + (\Delta\lambda_{\text{band}})^2} \tag{3.4}$$

Typical bandwidths of molecules in the condensed phase are not less than 10 cm^{-1}, which corresponds to $\Delta\lambda_{\text{band}} = 0.265$ nm at 500 nm. Then even a spectral band pass of the slit of 0.1 nm (ca. 4 cm^{-1}), which is distinctly larger than $\Delta\lambda_{\text{sys}}$, would only cause a "broadening" of the measured band profile of less than 10%.

For a spectrometer with a focal length of 1 m and a 1800 lines mm^{-1} grating, a spectral slit width of ca. 4 cm^{-1} at 500 nm corresponds to a geometrical slit width of ca. 300 µm, which is sufficiently large for collecting the scattered light, with the appropriate entrance optics. Continuous rotation of the grating leads to the displacement of the spectrum of the scattered light along the exit slit such that each position of the grating can be correlated with a specific spectral increment of the scattered radiation that leaves the monochromator.

The photons that pass the exit slit are converted into electric signals by means of photomultipliers, which multiply the number of primary photoelectrons by several orders of magnitude. Adjusting the high operating voltage of the photomultiplier, one may find a compromise between signal amplification and noise enhancement. Photomultipliers offer the advantage of low dark currents and a wide and linear dynamic range. For this reason, very weak and very strong signals can be detected upon scanning the grating through a spectral range of interest. Usually, the photomultiplier is set-up for single photon detection, which further reduces the dark noise, as it allows discrimination of the (larger) electrical pulses evoked by single photons against the (smaller) pulses thermally evoked from dynodes. Subsequently, the single photon pulses are converted into normalised standard pulses, which can be summed digitally (photon counting devices). This last step is essential for a computer-based data acquisition and its synchronisation with the controlled movement of the grating(s) in the monochromator.

In most Raman spectrometers, two monochromator stages (double-monochromator) are combined and both gratings are moved synchronously to reduce the stray light at the exit slit, i.e., the light that does not follow the theoretical optical pathway. This stray light mainly originates from reflected and elastically scattered light, which is more intense than the Raman light by more than 10 orders of magnitude. Thus, an effective suppression of the stray light is essential for reducing the overall background and for detecting low frequency Raman bands that would otherwise be obscured by the wings of the Rayleigh line.

3.2.2.2 Spectrographs

Modern Raman spectrometers are based on polychromatic detection. In this case, no exit slit is required as the "entire" spectrum is focussed onto the detector, which consists of a large number of active elements (Fig. 3.9). CCD (charge coupled devices) cameras are nowadays the most widely used detector arrays for Raman spectroscopy as they exhibit a very high quantum yield and a low dark current. In these respects, each photoactive element exhibits a performance that is comparable to a sensitive photomultiplier. Also in this instance, the entrance slit width, the focal length of the spectrometer, and the type of gratings control the spectral resolution; however, equally important is the dispersive power and the number of photoactive detector elements that finally gives the wavenumber increment per pixel. For a spectrometer with a focal length of 1 m and a 1800 lines mm^{-1} grating, the lateral dispersion is ca. 0.48 nm mm^{-1} (*vide supra*). Taking into account the typical size of a CCD array with a length of ca. 25 mm including ca. 1000 individual photoactive elements, the wavenumber increment per pixel at 500 nm is 0.5 cm^{-1} and the total range to be covered is 500 cm^{-1}. Then, besides the constraints due to the spectral slit width, in principle, bands

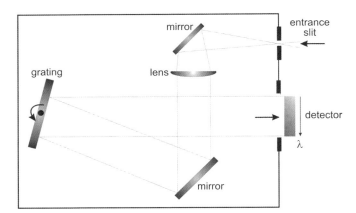

Fig. 3.9 Optical lay-out of a single-stage spectrograph. The light is dispersed onto a photoactive array that is positioned into the focal plane of the spectrograph.

can be resolved that are separated by at least three pixels which, in the present example, corresponds to 1.5 cm^{-1}. The size of the spectral range that is covered by the CCD array increases with decreasing number of grooves of the grating, such that the choice of the grating is a trade-off between spectral coverage and spectral resolution.

The sensitivity of gratings and detectors used in both the polychromatic and monochromatic detection mode exhibits a distinct wavelength dependence, which has to be taken into account in designing the Raman spectroscopic set-up for the desired applications. Clearly, if only for the time gain of a factor of ca. 100, polychromatic detection is superior to monochromatic detection. In addition, frequency accuracy in the scanning mode is principally lower due to the mechanical movement of the gratings. The only exception refers to Raman difference spectroscopy, which allows a very precise determination of the frequency differences in two samples (see Section 4.2.2.1).

With the availability of high quality CCD cameras, spectrometers that were originally designed as (double-)monochromators have been modified to operate as spectrographs. These modifications cannot correct the original spectrometer optics that are optimised for focussing the dispersed light onto the narrow exit slit of less than 1 mm rather than onto an array with a length of more than 20 mm. This deficiency causes aberrations at the edges of the array, which reduces the total usable number of CCD elements to ca. 75% or even less.

A critical aspect of polychromatic detection is stray light rejection. Single-stage monochromators operating as spectrographs limit the accessible spectral range to measurements above ca. 700 cm^{-1} when studying condensed matter. An additional monochromator stage can lower this limit to ca. 300 cm^{-1}. In "true" spectrographs stray light rejection is achieved via a filter stage that may be either a small-frame double-monochromator ("triple system") or a notch filter. A double-monochromator represents the most versatile solution inasmuch as it allows measurements very close (<5 cm^{-1}) to the Rayleigh line upon operation in the subtractive mode. Furthermore, a grating-based filter stage can be employed for a wide range of excitation lines. This mode of stray light rejection, however, reduces the optical throughput due to the additional gratings and mirrors, which is disadvantageous when very weak signals have to be measured.

As an alternative, holographic notch filters are employed that are selective for a specific excitation wavelength and afford a much better optical throughput. These filters efficiently cut off the reflected and elastically scattered light with a band pass that may allow approaching the excitation line down to 200 ("notch filter") and 40 cm^{-1} ("super notch filter"). If, however, many different excitation lines have to be employed, this option may become fairly costly.

3.2.2.3 Confocal Spectrometers

In view of the inherently low sensitivity of Raman spectroscopy, great care has to be taken to transfer the scattered light most efficiently from the sample to the spectrometer. Conventional spectrometers possess an entrance slit, which, for the

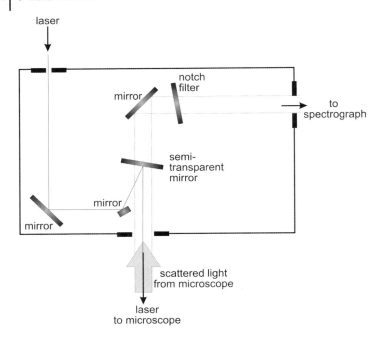

Fig. 3.10 Optical lay-out of the entrance stage of a confocal spectrometer. The laser is focussed onto the sample via a microscope objective, which in turn also collects the scattered light. The reflected laser light and the Rayleigh light is removed from the scattered light by a notch filter prior to entering the spectrograph stage.

sake of an acceptable spectral resolution, has to be kept small and thus optical losses are unavoidable. This problem is elegantly overcome in confocal Raman microscopy in which the exciting laser beam is focussed onto the sample via a microscope objective, which simultaneously serves to collect the scattered light in the backscattering configuration (Fig. 3.10). In this arrangement, no entrance slit is required but the scattered light enters the spectrometer through a pinhole, which ensures an optimal optical throughput. Prior to focussing onto the grating, the scattered light passes a notch filter to remove the reflected and elastically scattered light. In terms of optical throughput, confocal Raman spectrometers are clearly superior to conventional spectrographs and monochromators and, in addition, they offer the advantage of combining microscopic and Raman spectroscopic measurements (see Section 4.4.4). However, these advantages can only be fully exploited for those samples that are appropriate for back scattering detection, i.e., solid states, surfaces, or concentrated solutions. For diluted solutions that require a large irradiated volume, and for special techniques such as pump-probe experiments, conventional Raman spectrometers and 90° scattering geometries may afford at least comparably good results despite the poorer light conductance from the sample to the detector.

3.2.2.4 Fourier Transform Raman Interferometers

As an alternative to polychromatic detection based on dispersive elements, the interferometric detection of Raman scattered light has been developed following the well-established methodology in FTIR spectroscopy. The optical layout of the Michelson-type interferometer is the same as described in Section 3.1.1 (Fig. 3.1), but it includes an additional filter stage to remove the reflected laser radiation and Rayleigh scattered light. As FT Raman and FTIR detection are based on the same interferometer, FTIR instruments are offered with an optional Raman module. Stand-alone FT Raman spectrometers, on the other hand, are compact and rugged and thus may be used for routine measurements. However, the commercially available FT Raman instruments are restricted to excitation in the near-infrared using the fundamental output of a Nd:YAG laser at 1064 nm.

The FT Raman technique offers all of the advantages of other types of FT spectroscopy, (i.e., Jaquinot and Felgett advantage, high intrinsic frequency accuracy; see Section 3.1.2). Furthermore, the 1064-nm excitation line is sufficiently far shifted to the near-infrared such that in most instances the energy gap with respect to electronic transitions is large enough to avoid interference with fluorescence or photochemical processes. However, the price is a weaker (pre-)resonance enhancement for biological chromophores, which in addition to the v^4-law leads to distinctly lower Raman intensities as compared with excitation in the visible or UV region. Furthermore, the performance of detectors in interferometers is distinctly worse than that of CCD detectors, which today can be operated to wavelengths up to 1000 nm. Thus, CCD-based spectrometers may represent an attractive alternative to FT Raman spectrometers when excitation wavelengths up to 850 nm are used. On the other hand, one has to consider that the poorer detector performance in the near-infrared is at least partially compensated by the Jaquinot and Felget advantages of FT spectroscopy (see Section 3.1.2).

Box 3A

Theory of Fourier transform infrared spectroscopy

Let E be the amplitude of the radiation field entering the interferometer. Then the time (t) and space (r) dependence of the electromagnetic radiation is described by

$$E = E_0 \exp[i(\omega t - kr)] \tag{3.A1}$$

where the circular frequency is given by

$$\omega = 2\pi \nu \tag{3.A2}$$

and k is defined by

$$k = \frac{2\pi}{\lambda} \tag{3.A3}$$

At the output of the interferometer, two fields $E_1 \{= (1/2) \cdot E_0 \exp[i(\omega t - kr_1)]\}$ and $E_2 \{= (1/2) \cdot E_0 \exp[i(\omega t - kr_2)]\}$ superimpose. The path difference γ, brought about by the movable mirror,

$$\gamma = r_2 - r_1 \tag{3.A4}$$

then results in a phase difference expressed by

$$k(r_2 - r_1) = \frac{2\pi\gamma}{\lambda} = 2\pi\gamma\nu \tag{3.A5}$$

Thus, the total radiation field is given by the sum of the two amplitudes, i.e., $E_1 + E_2$, and the intensity, which is registered by the detector, is proportional to $|E_1 + E_2|^2$ according to

$$|E_1 + E_2|^2 = |E_1|^2 + |E_2|^2 + |E_1 E_2^*| + |E_2 E_1^*| = \frac{1}{2} E_0^2 [1 + \cos(2\pi\gamma\nu)] \tag{3.A6}$$

when using the expression (3.A1) and its conjugate complex form.

The total intensity is obtained by integration of Eq. (3.A6) over the spectrum, i.e.,

$$\frac{1}{2} \int_0^{\nu_{max}} S(\nu) \, d\nu + \frac{1}{2} \int_0^{\nu_{max}} S(\nu) \cos(2\pi\nu\gamma) \, d\nu = \frac{1}{2} I_0 + I(\gamma) \tag{3.A7}$$

with I_0 being the total integrated intensity entering the interferometer. $S(\nu)$ is the spectral intensity or the spectrum with ν_{max} as the upper limit of the spectrum. For $\gamma = 0$, $I(0) = I_0$, i.e., the total incoming intensity is directed to the output arm of the interferometer. Non-ideal performance of the interferometer (mirror, beam splitter,

Box 3A *(continued)*

alignment) and of the detector including signal electronics will result in a smaller value of $I(0)$ (see below).

$I(\gamma)$ is the ac-part of the detector signal, and thus, represents the interferogram. $I(\gamma)$ can be rewritten by formally extending the wavenumber scale to negative values and setting $S(-\nu)^* = S(\nu)$:

$$I(\gamma) = \frac{1}{4}\int_{-\infty}^{+\infty} S(\nu) e^{i2\pi\nu\gamma}\, d\nu \tag{3.A8}$$

Thus, $I(\gamma)$ and $S(\nu)$ are formally related by the mathematical operation of the Fourier transformation. This means that $S(\nu)$ can be obtained from $I(\gamma)$ by the inverse Fourier transformation, i.e.:

$$S(\nu) = 4\int_{-\infty}^{+\infty} I(\gamma) e^{-i2\pi\nu\gamma}\, d\gamma \tag{3.A9}$$

As $I(\gamma)$ is symmetric, $S(\nu)$ is automatically a real quantity, as required.

In Eq. (3.A9), the integration extends from $-\infty$ to $+\infty$, which is, of course not possible, as the movable mirror can only span a limited path length, resulting in a restricted range of γ. We will now demonstrate the consequences of this limitation.

We assume that the movement of the mirror extends from $-\gamma_{max}$ to $+\gamma_{max}$; thus the integration in Eq. (3.A9) only refers to this range. The unlimited integration can be brought back into Eq. (3.A9) by multiplying $I(\gamma)$ with a rectangle function, which is 1 for $-\gamma_{max} < \gamma < +\gamma_{max}$ and zero elsewhere. This function is known as $R^{\gamma_{max}}(\gamma)$.

The observed spectrum obtained with a real interferometer is thus given by

$$S_{obs}(\nu) = 4\int_{-\infty}^{+\infty} R^{\gamma_{max}}(\gamma) I(\gamma) e^{-i2\pi\nu\gamma}\, d\gamma \tag{3.A10}$$

We now use a theorem well-known in Fourier transformation. The Fourier transform of the product of two functions is equal to the folding of the Fourier transform of the two functions, such that one obtains

$$S_{obs}(\nu) = FT[R^{\gamma_{max}}(\gamma)] * FT[I(\gamma)] \tag{3.A11}$$

Here, the asterisk denotes the folding operation and $FT[I(\gamma)]$ is equal to the "true" spectrum $S(\nu)$.

The Fourier transform of the rectangle function $R^{\gamma_{max}}(\gamma)$ is given by the so-called dif-function, i.e:

$$FT[R^{\gamma_{max}}(\gamma)] = 2\gamma_{max}\, dif(2\pi\nu\gamma_{max}) \tag{3.A12}$$

Box 3A *(continued)*

with

$$\text{dif}(x) = \frac{\sin(x)}{x} \tag{3.A13}$$

This function is shown in Fig. 3.A1. It represents a main peak with oscillating side lobes. The half-width of the main peak is given by

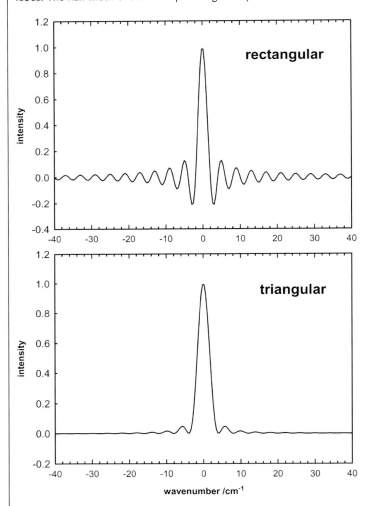

Fig. 3.A1 Rectangular apodisation $\text{dif}(2\pi v \gamma_{max})$ (top); and triangular apodisation $\text{dif}^2(\pi v \gamma_{max})$ (bottom). A realistic value has been used for γ_{max} in order to provide a half-width of approx. 4 cm^{-1} for the *dif*-function, a common value in the spectroscopy of proteins.

Box 3A (continued)

$$\Delta\nu = \frac{1.25}{2\gamma_{max}} \tag{3.A14}$$

such that we obtain

$$S_{obs}(\nu) = 8\gamma_{max} \int_{-\infty}^{+\infty} S(\nu) \, dif[2\pi\gamma_{max}(\nu - \nu')] \, d\nu' \tag{3.A15}$$

The folding of the true (or theoretical) spectrum $S(\nu)$ with the *dif*-function as described in Eq. (3.A15) results in line-broadening with the half-width given by Eq. (3.A14). In addition, the observed spectrum is somewhat distorted by the oscillating side lobes. This will be of particular importance for spectra with well-defined narrow bands. The oscillation is mainly caused by the sharp edges of $R^{\gamma_{max}}(\gamma)$. This can be reduced by multiplying a so-called apodisation function onto the measured interferogram, which reduces the sharp edges.

The simplest function is the triangular function $T^{\gamma_{max}}(\gamma)$, varying linearly from zero at $\pm\gamma_{max}$ to 1 at $\gamma = 0$. The Fourier transform of this function is also shown in Fig. 3.A1. As the triangular function is obtained by folding two rectangular functions, again using the folding theorem in Fourier transformation, the Fourier transform of $T^{\gamma_{max}}(\gamma)$ is given by a squared *dif*-function, i.e.:

$$FT[T^{\gamma_{max}}(\gamma)] = \gamma_{max} \, dif^2(\pi\nu\gamma_{max}) \tag{3.A16}$$

It is obvious that the oscillations are reduced, however at the expense of a broadening of the main peak. The half-width $\Delta\nu$ now amounts to

$$\Delta\nu = \frac{0.91}{\gamma_{max}} \tag{3.A17}$$

The most important consequence of the limited interferogram is the reduction in spectral resolution, as the larger γ_{max} is, the better the spectral resolution is [Eq. (3.A17)]. With a maximum optical path difference of 0.25 cm and triangular apodisation, $\Delta\nu$ amounts to 3.6 cm^{-1}. This broadening is acceptable in FTIR spectroscopy of biological systems for which "true" band widths are usually larger than 10 cm^{-1}.

There are numerous other apodisation functions, which may make a better compromise with suppression of oscillations and reduction of resolution. These functions are implemented in all commercial FTIR instruments.

Under ideal conditions assumed in the derivation of the equations of Fourier spectroscopy, one concludes from Eq. (3.A7) that $I(\gamma)$ is a symmetric function. However, in general this does not hold exactly. Typical origins for deviations of the symmetry are the thickness of the beam splitter with a refraction index depending on the wavenumber, and a frequency-dependent phase-shift of the detector/amplifier system. Such factors will introduce a wavenumber-dependent phase shift in Eq. (3.A7). Thus, the interferogram is expressed as

Box 3A *(continued)*

$$I(\gamma) = \frac{1}{2}\int_0^{\nu_{max}} S(\nu)\cos[2\pi\nu\gamma + \phi(\nu)]\,d\nu$$

$$= \frac{1}{2}\int_0^{\nu\,max} S(\nu)[\cos(\phi(\nu))\cos(2\pi\nu\gamma) + \sin(\phi(\nu))\sin(2\pi\nu\gamma)]\,d\nu \qquad (3.A18)$$

The introduction of a non-symmetric interferogram into Eq. (3.A18) leads to a complex spectrum, with the real part $S^r(\nu)$ given by $\cos\phi(\nu)S(\nu)$ and the imaginary part $S^i(\nu)$ by $\sin\phi(\nu)S(\nu)$. These quantities are obtained by the complex Fourier transform of Eq. (3.A9).

There are two ways to obtain the physical spectrum $S(\nu)$, either using

$$S(\nu) = \sqrt{S^{r2}(\nu) + S^{i2}(\nu)} \qquad (3.A19)$$

or

$$S(\nu) = \frac{S^r(\nu)}{\cos[\phi(\nu)]} \qquad (3.A20)$$

where $\phi(\nu)$ is given by

$$\phi(\nu) = \arctan\left[\frac{S^i(\nu)}{S^r(\nu)}\right] \qquad (3.A21)$$

Although the two methods appear to be equivalent, there are small differences. In Eq. (3.A20), there is only one sign for $S(\nu)$, either positive, or negative (usually positive, as the spectral intensity is positive), whereas in Eq. (3.A21) changes to the sign can be recovered. Of course, this is of no consequence when only the spectral intensity is measured. However, there are particular measurements, e.g., difference spectroscopy, where indeed $S(\nu)$ changes its sign. Furthermore, in judging the noise in a spectrum it is often useful to take a spectral range where there is no intensity. Using the method of Eq. (A3.21), the noise oscillates around zero with a mean value of zero, whereas in Eq. (3.A20) the negative noise values become positive, and the mean value of the noise is no longer zero. Therefore, generally, the second method expressed by Eq. (3.A21) is preferred. Thus, first $S^r(\nu)$ and $S^i(\nu)$ are obtained, then $\phi(\nu)$ and subsequently $S(\nu)$ are calculated. In commercial instruments, in addition to $S(\nu)$, the other quantities can also be recovered.

Box 3B

Practical implementation of the Fourier transform

As the spectra are calculated by means of computer-based programs, the interferogram must be obtained in a digital form. For practical reasons it is no longer a continuous but a discrete function with values only at predetermined sampling points. Thus, one has to ask what the consequences are for this loss of information. A theorem in information theory (Nyquist criteria) deals with this issue. It states that the frequency of a sinusoidal signal can be determined unequivocally if this signal is sampled at least twice per cycle. Thus, specifically, in order to determine the spectrum contained in a signal, in this instance the interferogram, loss of information is avoided if the signal is sampled at time intervals

$$\Delta t = \frac{1}{2 v_{max}} \quad (3.B1)$$

where v_{max} is the highest frequency component in the spectrum.

Conversely, if the interferogram is only determined for $\gamma < \gamma_{max}$, the corresponding spectrum is fully described by discrete points at the spectral spacing:

$$\Delta v = \frac{1}{2 \gamma_{max}} \quad (3.B2)$$

These relationships reflect the reciprocity of Fourier transformation.

In order to perform the Fourier transformation [Eq. (3.A9)] using a computer, the integral has to be replaced by a sum. The following derivations are adapted from Gronholz and Herres (Gronholz and Herres 1984, Herres and Gronholz 1984).

$$S(k\Delta v) = \sum_{n=0}^{N-1} e^{i 2\pi k n/N} I(n\Delta \gamma) \quad (3.B3)$$

Here, the continuous interferogram $I(\gamma)$ has been replaced by the discrete function $I(n\Delta\gamma)$. For simplicity, the factor 4 in front of the integral has been omitted, and the sum begins from 0 to $N-1$, instead of from $-N/2$ to $+(N/2-1)$. This shift in the γ-scale is possible as we perform the complex Fourier transform.

As outlined above, the discrete interferogram can be used without loss of information provided v_{max} is limited according to the Nyquist criteria. Equally, because the maximum optical path difference is given by $1/2N\Delta\gamma$, the spectrum can be depicted as a discrete function $S(k\Delta v)$, again fulfilling the Nyquist criteria. This is equivalent to the discussion on the effect of a limited interferogram on the spectrum obtained (Box 3A). Using the so-called Cooley–Tukey algorithm, the summation in Eq. (3.B3) is effectively performed. The requirements for its application are that N is a power of 2, and that the number of sampling points in the interferogram is equal to the number of sampling points in the spectrum. This is equivalent to the Nyquist criteria, i.e., $\Delta v = 1/N\Delta\gamma$, and $\Delta\gamma = 1/N\Delta v$.

Box 3B *(continued)*

It can be seen that the spectrum obtained by Eq. (3.B3) is symmetric

$$S[(N-k)\Delta v] = S(-k\Delta v) \tag{3.B4}$$

and S has the periodicity of N, i.e.:

$$S[(k+mN)\Delta v] = S(k\Delta v) \tag{E5}$$

and thus $S(k\Delta v) = S(-k\Delta v)$.

This, however, further means that the spectrum calculated according to Eq. (3.B3) has a mirror symmetry at the so-called Nyquist wavenumber $(N/2)\Delta v = 1/(2\Delta \gamma)$. This symmetry property provides another explanation of the Nyquist criteria. If the spectrum extended over the critical Nyquist wavenumber, there would be contributions from the mirrored spectrum. This is equivalent to folding back the spectral part extending over the Nyquist wavenumber, and therefore, the spectra are distorted.

The most important consequence of the discrete interferogram is the limitation of the spectral range, and if necessary, the spectrum must be limited by a low-pass filter in order to fulfil the Nyquist criteria. The spectrum obtained by the sum in Eq. (3.B3) using the Cooley–Tukey algorithm is characterised by N data points, but the actual spectrum has only $N/2$ points. The spacing of the spectral data points corresponds to the maximum optical path difference $\gamma_{max} = (N/2)\Delta \gamma$.

Box 3C

Beam collimation

It can be shown that the maximum tolerable solid angle Ω of radiation entering the interferometer is given by:

$$\Omega = \frac{2\pi \Delta v}{v_{max}} \tag{3.C1}$$

This equation is obtained directly by calculating the actual optical path lengths in the interferometer for a beam entering it parallel to the central beam and for a beam entering it at an angle α. For the latter, displacement of the mirror by Δx causes a larger optical path difference $\Delta \gamma$. In order to obtain a spectral resolution of Δv, the phase difference at any mirror displacement for two beams with frequencies v and $v + \Delta v$ must be larger than the phase difference between the two beams entering the interferometer at zero angle and at an angle α, taken at the same mirror displacement.

References

Barth, A., Hauser, K., Mäntele, W., Corrie, J. E. T., Trentham, D. R., **1995**, "Photochemical release of ATP from 'caged ATP' studied by time-resolved infrared spectroscopy", *J. Am. Chem. Soc.* **117**, 10311–10316.

Fodor, S. P. A., Rava, R. P., Copeland, R. A., Spiro, T. G., **1986**, "H_2-Raman shifted YAG laser ultraviolet Raman spectrometer operating at wavelengths down to 184 nm", *J. Raman Spectrosc.* **17**, 471–475.

Gronholz, J., Herres, W., **1984**, "Datenverarbeitung in der FT-IR Spektroskopie Teil 2: Einzelheiten der Spektrenberechnung", *Comp. Anw. Lab.* **6**, 418–425.

Hamm, P., Kaindl, R. A., Stenger, J., **2000**, "Noise suppression in femtosecond mid-infrared light sources", *Opt. Lett.* **25**, 1798–1800.

Herres, W., Gronholz, J., **1984**, "Datenverarbeitung in der FT-IR-Spektroskopie, Teil 1: Datenaufnahme und Fourier-Transformation", *Comp. Anw. Lab.* **6**, 352–356.

Mäntele, W., Hienerwadel, R., **1990**, "Application of tunable infrared lasers for the study of biochemical reactions: time-resolved vibrational spectroscopy of intermediates in the primary process of photosynthesis", *Spectrosc. Int.* **2**, 29–35.

Pellerin, C., Snively, C. M., Chase, B., Rabolt, J. F., **2004**, "Performance and application of a new planar array infrared spectrograph operating in the mid-infrared (2000–975 cm^{-1}) fingerprint region", *Appl. Spectrosc.* **58**, 639–646.

Rödig, C., Siebert, F., **1999**, "Error and artifacts in time-resolved step-scan FT-IR spectroscopy", *Appl. Spectrosc.* **53**, 893–901.

Schrader, B. (Ed.), **1995**, "*Infrared and Raman Spectroscopy*", VCh-Verlag, Weinheim.

Siebert, F., Mäntele, W., Kreutz, W., **1981**, "Biochemical application of kinetic infrared spectroscopy", *Can. J. Spectrosc.* **26**, 119–125.

Silfvast, W. T., **1996**, "*Laser Fundamentals*", Cambridge University Press, New York.

Uhmann, W., Becker, A., Taran, Ch., Siebert, F., **1991**, "Time-resolved FT-IR absorption spectroscopy using a step-scan interferometer", *Appl. Spectrosc.* **45**, 390–395.

Wouterson, S., Hamm, P., **2002**, "Nonlinear two-dimensional vibrational spectroscopy of peptides", *J. Phys. Condens. Matter* **14**, R1053–R1062.

4
Experimental Techniques

In this chapter we discuss the practical problems of IR and Raman spectroscopy that are relevant to applications in life sciences. In particular, we outline various "non-standard" methods that have become valuable tools in studying biological systems. These methods include time-resolved techniques in addition to surface-enhanced Raman and IR approaches. The objective of this chapter is to provide a sound practical and theoretical manual for adapting these techniques to specific biochemical and biophysical problems.

4.1
Inherent Problems of Infrared and Raman Spectroscopy in Life Sciences

Applying vibrational spectroscopy to biological systems is associated with some difficulties that the reader should be aware of. These difficulties are associated with the systems to be studied and the techniques themselves. Firstly, biological systems are rather sensitive objects and experimental conditions have to be chosen to minimise the risk of damaging the sample during the experiments. Secondly, IR absorption and Raman scattering are based on relatively weak effects such that the desired signals of the biomolecules may be obscured by much stronger signals from the medium or by competing photophysical effects. To overcome these drawbacks, various strategies have been established, which are described in the following paragraphs.

4.1.1
The "Water" Problem in Infrared Spectroscopy

Spectroscopy in life sciences means spectroscopy of molecular systems in an aqueous environment. This is a particularly stringent prerequisite as many biological samples only remain functionally intact in the presence of water. Very often, proteins denature and are irreversibly damaged when water is exchanged by an organic solvent. Furthermore, reduction of the water content by drying may also cause degradation of the system. For UV–vis spectroscopy the aqueous solution is not a problem as water is sufficiently transparent over the spectral

Vibrational Spectroscopy in Life Science. Friedrich Siebert and Peter Hildebrandt
Copyright © 2008 WILEY-VCH Verlag GmbH & Co. KGaA, Weinheim
ISBN: 978-3-527-40506-0

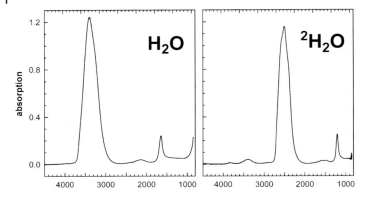

Fig. 4.1 Absorption spectra of H_2O (left) and ^2H_2O (right). An aqueous film is produced by squeezing an aqueous drop between two infrared windows. From the absorbance at 1650 cm^{-1} one can estimate a film thickness of approx. 1 µm.

range from 180 to 1400 nm. For infrared spectroscopy, however, water represents a major obstacle. Water has strong, broad absorption bands around 3400 cm^{-1} (stretching vibrations), 2200 cm^{-1} (combination band of OH bending mode and libration mode of associated water molecules), 1650 cm^{-1} (bending vibration), and below 900 cm^{-1} (libration modes). In addition, water also exhibits a so-called continuum absorption below 2000 cm^{-1}. Figure 4.1, left, shows a spectrum of water at a path length of approximately 1 µm. For this spectrum, the two IR windows have just been squeezed together and the aqueous film is formed in between, and the sample thickness is estimated from measurements using a cuvette of 4-µm spacing, taking the band at 1650 cm^{-1} for calibration. The band at 3300 cm^{-1} can, however, not be faithfully displayed at this cuvette spacing. It is clear that even at a 4-µm cuvette thickness the absorption of water is considerable, and a larger path length cannot be tolerated. In Fig. 4.1, right, the corresponding spectrum of heavy water (^2H_2O) is shown, again using a film squeezed between two windows. As expected from the larger mass, absorption bands are shifted down in frequency. In addition, quantitative measurements with the 4-µm cuvette show that the absorption strength is somewhat reduced. Thus, by changing the solvent from H_2O to ^2H_2O, the measuring conditions can be somewhat improved. However, one has to take into account that in biological molecules many hydrogens are exchangeable upon dissolving in ^2H_2O, and the replacement by deuterium will, therefore, affect some of the vibrational modes one is interested in. As discussed in Section 2.1.3, this is not necessarily a disadvantage as isotopic shifts may also support the vibrational assignment and the identification of the contribution of specific molecular groups to the spectrum.

In any case, the spectra in Fig. 4.1 demonstrate that one has to employ very thin cuvettes. This requirement implies the use of very concentrated solutions in the 1–10 mM range. For proteins this is sometimes difficult to establish due to the irreversible formation of aggregates, which may influence the functional properties of the biomolecule. Here, the lower extinction coefficient of ^{2}H$_2$O may circumvent this difficulty as a larger path length of up to 25 μm can be tolerated and the concentration may be reduced correspondingly according to the Lambert–Beer law.

4.1.2
Unwanted Photophysical and Photochemical Processes in Raman Spectroscopy

Interaction of electromagnetic radiation with molecules may induce a variety of photophysical and photochemical processes in addition to Raman scattering. This is particularly true when the energy of the incident light matches that of an electronic transition (Fig. 4.2) (Turro 1991). The primary event is absorption of a photon such that the molecule is promoted to the electronically and vibrationally excited state. In the excited *vibronic* state, a variety of radiative and non-radiative processes may occur. Some of these processes are fairly annoying for Raman spectroscopists inasmuch as they obscure the Raman bands (fluorescence) or alter the molecular system to be studied (photochemical reaction).

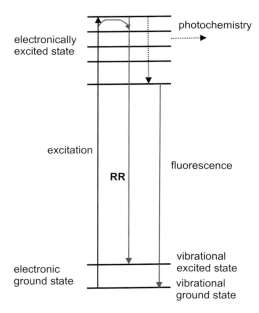

Fig. 4.2 Jablonski diagram for illustrating the photoinduced processes competing with RR scattering. The solid lines represent radiative processes whereas the dotted lines indicate the non-radiative reaction channels.

4.1.2.1 Fluorescence and Raman Scattering

After light absorption has promoted the molecule to a vibronic state, it may fall directly back to the ground state under emission of a photon, which has exactly the same energy as the incident photon. This resonance fluorescence should therefore not be confused with Rayleigh scattering, although the energy of the detected photons is the same. In contrast to Rayleigh scattering, resonance fluorescence represents a sequential absorption and emission of a photon, implying that the resonant excited vibronic state exhibits a finite lifetime.

However, more probable than resonance fluorescence is the emission of a photon from the vibrational ground state of the electronically excited state, which is reached via rapid radiationless transitions (internal conversion) from the originally excited level. This process gives rise to a fluorescence band at energies lower than the excitation energy and thus it falls into the region of Stokes Raman scattering. Fluorescence quantum yields can vary substantially from almost one (e.g., laser dyes, chlorophylls) down to values $< 10^{-6}$ such that, owing to the broad band shape, the fluorescence hardly exceeds the background noise. These variations depend on the efficiency of competing processes, such as intermolecular energy transfer, internal (radiationless) conversion into the electronic ground state, or intersystem crossing to a (lower lying) triplet state, which may convert into the ground state either via intersystem crossing (radiationless) or upon emission of a photon (phosphorescence) of much lower energy than in fluorescence. The yields of these processes depend on the molecular environment (e.g., energy transfer) or on the involvement of heavy atoms that open radiationless decay channels (e.g., intersystem crossing). Thus, metalloporphyrins, which include open shell transition metal ions, exhibit a fluorescence quantum yield lower than 10^{-6}. Finally, a molecule in the electronically excited state can undergo photochemical reactions such as isomerisations, proton or electron transfer. Upon resonant excitation, all these processes may possibly compete with RR scattering for which the quantum yield may be at best in the order of 10^{-5}.

Even for a compound that exhibits a fluorescence quantum yield as low as 10^{-3} and thus is considered to be weakly fluorescent, the RR signals can be largely obscured. This is also true for RR or Raman spectra of non-fluorescing compounds in samples that contain fluorescent impurities. Annoying fluorescent contaminations in biological samples are flavins or chlorophylls. These chromophores have fluorescence quantum yields larger than 0.1 and thus contaminations of even 0.1% or less may obscure the Raman or RR signals of the compound of interest. However, there are several possibilities to circumvent the interference with intrinsic or impurity fluorescence.

Firstly, the excitation line may be shifted either to the low or to the high-energy side of the fluorescing transition, which usually originates from the lowest excited state. In the latter, the excitation line may be chosen to be in the proximity of the second electronically excited state, which thus provides resonance enhancement of the Raman bands without interference by fluorescence. Alternatively, one may select an excitation line sufficiently separated from the first electronic transition

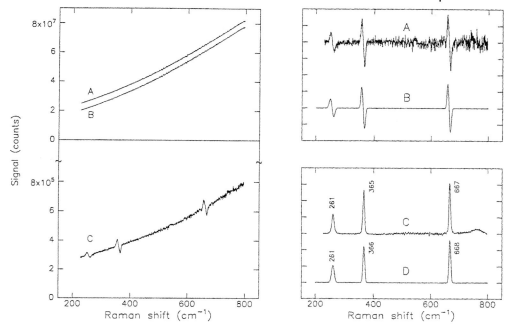

Fig. 4.3 Raman spectroscopy in the presence of a strong fluorescence background using the shifted excitation difference technique. The left panel shows the spectra of chloroform solution including ca. 40 μM LDS-925, a highly fluorescing dye. Spectrum A was obtained with 725-nm excitation whereas for spectrum B the excitation line was shifted by 10 cm^{-1} to longer wavelengths. Trace C displays the difference "A" minus "B", indicating the Raman signals of the solvent. The right panel shows a background correction applied to the difference spectrum (trace A). Trace B, which represents a fit to trace A, is used to generate the absolute Raman spectrum of the solvent (trace D), which is compared with the measured spectrum of neat chloroform (trace C). The figure is reproduced from Shreve et al. (1992) by permission.

on the low energy side. Under these conditions, pre-resonance enhancement is distinctly lower (see Section 2.2.3) but fluorescence can largely be avoided.

An alternative approach that has been demonstrated to be applicable also for strongly fluorescent molecules utilises the fact that fluorescence bands are very broad and poorly structured (Fig. 4.3) (Shreve et al. 1992). Thus, intensity and band shape of the fluorescence do not differ for spectra obtained with slightly different excitation lines v_{exc} and $v_{exc} + \Delta v_{exc}$ with Δv_{exc} being smaller than the band width of a Raman band (e.g., $\Delta v_{exc} = 10$ cm^{-1}). Conversely, the absolute positions of the Raman bands are shifted by the increment Δv_{exc}. In the difference spectrum constructed from both spectra measured under otherwise identical conditions the fluorescence background cancels whereas the Raman signals that can not be identified in the original spectra become visible as derivative-type features on top of a noisy but relatively straight line. The final step is to reconstruct the

Raman bands by simulation of the difference signals on the basis of Lorentzian band shapes. The reliability of the procedure is limited by the total noise of the difference spectrum, which increases with the height of the fluorescence background in the absolute spectra.

Yet another strategy employs optoelectronic switches (Kerr cells) that open the time window for detection solely of the instantaneously scattered Raman photons, whereas the fluorescence photons following with a delay time are rejected (Matousek et al. 2001). This technique is limited to those instances when the fluorescence lifetime is in the long picosecond time range.

Quenching fluorescence represents another approach. It requires energy acceptors brought into contact with the electronically excited molecules. Early RR studies of highly fluorescent chromophores in flavoproteins exploited the quenching effect of iodide that promotes the non-radiative decay of excitation energy (Benecky et al. 1979). To achieve the desired effect, however, high iodide concentrations are required, which may cause unwanted perturbations of the protein structure. Also, metal surfaces can be appropriate for fluorescence quenching as the high density of states in the metal provide efficient decay channels for the excitation energy (Hildebrandt and Stockburger 1984). Here, the only requirement is a sufficiently close proximity of the molecule to the surface. Thus, the suppression of fluorescence is a highly welcome side effect of SER spectroscopy, which, like the surface enhancement of the Raman scattering, is not restricted to the molecules directly attached to the metal but is also operative for molecules in the near-field of the metal surface.

4.1.2.2 Photoinduced Processes

Electronically excited molecules can undergo a variety of chemical processes, such as electron or proton transfer, or double bond isomerisation. These processes may be unwanted side reactions or natural reaction steps of biomolecules, i.e., photoreceptors. In any case, RR spectroscopic experiments should be designed such that the exciting laser beam does not induce chemical reactions, in order to avoid accumulation of reaction intermediates or products in the irradiated sample element. Photosensitive molecules, therefore, have to be probed (i) by reducing the photochemical rate constant l_0 as much as possible, and (ii) by flowing the sample through the exciting laser beam such that the dwell time Δt of a volume element in the laser beam is small (Schneider et al. 1989). The product of both quantities determines the degree of photoconversion according to

$$[A(t)] = [A_0] \exp(-\Delta t \cdot l_0) \tag{4.1}$$

where $[A_0]$ and $[A(t)]$ are the respective concentrations of the molecule in the parent state prior to irradiation and in the volume element irradiated for the time interval Δt. The photochemical rate constant l_0 not only depends on the power of the laser beam P_0 but also on the photophysical properties of the compound, which are the extinction coefficient at the excitation wavelength λ_0 (in nm), ε (in L mol^{-1} cm^{-1}), and the quantum yield for the photochemical reaction γ. Assuming a Gaussian laser profile with a beam radius of r_0 (in m) one obtains

$$I_0 = 4.81 \cdot 10^{-10} \left(\frac{\gamma \cdot \varepsilon \cdot P_0 \cdot \lambda_0}{r_0^2} \right) \tag{4.2}$$

where the P_0 is expressed in J s^{-1}.

To keep the portion of photoinduced reaction products that are formed during the irradiation interval Δt below 10%, the photoconversion parameter $\Delta t \cdot I_0$ [Eq. (4.1)] should be smaller than 0.11. For 514.5-nm excitation (0.003 J s^{-1}) with a beam radius of 40 µm, an extinction coefficient of 30 000 L mol^{-1} cm^{-1}, and a quantum yield for the photochemical process of 0.3, this condition is fulfilled with a dwell time Δt of ca. 27 µs. Thus, the sample has to pass the laser beam with a velocity of ca. 3 m s^{-1}, which can be achieved by pumping the solution through a capillary flow system, or by using a rotating cuvette.

In photo-inactive and non-fluorescent molecules, the major portion of the incident radiation energy that is absorbed is converted into thermal energy via radiationless transitions. For photon fluxes typically used in cw-experiments, the energy is rapidly dissipated to the environment, which may cause a local heating of the irradiated volume of the solution. As a consequence, degradation of the biomolecules may occur. For strongly absorbing samples, i.e., at resonance conditions, the light-to-heat conversion may lead to a substantial temperature increase in a stationary sample. The effects of laser-induced heating can be even more severe in pulse-laser experiments. If the pulse energies are very high and energy transfer to the solvent is not sufficiently efficient, local heating may occur even within the light-absorbing molecule. Also for photo-inactive molecules, a moving sample device is therefore recommended to avoid laser-induced heating. Additionally, cooling the sample is advisable although for aqueous solutions the accessible temperature range is fairly small due to the relatively high freezing temperature. In any case, however, the energy flux P_0/r_0^2 should be kept as small as possible, which is achieved by reducing the laser power and by increasing the beam focus. Both measures, however, also reduce the number of Raman photons and the transfer of the scattered light to the spectrometer. Thus, a successful Raman spectroscopic experiment of photo- and temperature-sensitive materials is always based on a compromise between preservation of the sample integrity and good spectral quality. Consequently, the optimised collection of the scattered light by the spectrometer is essential in Raman spectroscopy. In this respect, confocal spectrometers are superior but one has to take into account that for the same laser power, the energy flux is typically higher by three orders of magnitude as compared with conventional Raman set-ups.

4.2
Sample Arrangements

One of the main advantages of vibrational spectroscopic methods is their flexibility by which they can be adapted to various sample arrangements, which in turn are related either to the properties of the system under consideration or the specific questions to be addressed. In most instances one is interested in studying biological molecules in aqueous solution at ambient temperature. Low-temperature studies may be needed for biomolecules that are not sufficiently stable during the Raman or IR experiment at ambient temperature. Furthermore, for redox proteins it may be desirable to combine spectroscopic and electrochemical investigations, such that specific devices have to be developed that simultaneously allow for both types of measurements. In this section, we will discuss the most common devices in addition to specific arrangements that are particularly useful for novel techniques.

4.2.1
IR Spectroscopy

4.2.1.1 Sandwich Cuvettes for Solution Studies

To overcome the main difficulty in IR spectroscopy of biological systems, i.e., the water absorption bands, thin cuvettes have to be employed that are capable of reducing the optical pathlength below 25 µm (see Section 4.1.1). The simplest method is to use two flat windows and separate them by a spacer of the corresponding thickness. Such spacers can be easily prepared from appropriate foils, e.g., of aluminium or polyethylene. After smoothly attaching the spacer to one window, a drop of the concentrated solution is put onto this window, and the second window is overlaid, squeezing the solution over the opening of the spacer. Excess solution spills over the spacer. By pressing the two windows tightly together, an adequate seal is usually achieved. The extent of the sealing critically depends on the flatness of the windows and how smoothly the spacer is placed onto the window. Tiny amounts of the solution left from the spilling over may cause problems, both with respect to the exact cuvette thickness and with respect to the cuvette sealing. Such a sample device is termed a sandwich cuvette. There are also commercially available cuvettes that can be refilled through holes near the rim of the window. However, in practice, two major problems are encountered. Firstly, the cuvette is difficult to clean, and secondly, a considerable amount of excess solution is required as compared with the small volume necessary to match the focus of the IR beam.

Because the sealing is often not perfect, the solution in the cuvette may dry out, especially during longer measuring periods. In order to circumvent this problem, a different type of sandwich cuvettes has been designed (Vogel and Siebert 2003). The pathlength is determined by grinding a very shallow trough (2–6 µm) into one window. This trough is surrounded by a groove with a width and depth of approximately 0.5 mm. This groove, being in contact with the sample layer, serves as a reservoir. The sample volume amounts to approximately 0.150 µL, whereas the volume of the groove is approximately 8 µL. This reservoir also guarantees that the composition of the sample solution is determined by this reservoir, and, therefore, can be precisely controlled and kept reasonably constant over a prolonged measuring time. The design of this cuvette is shown in Fig. 4.4(A). The sample preparation is the same as for the sandwich cuvette but considerably more biological material is needed because of the reservoir. If, however, the bio-

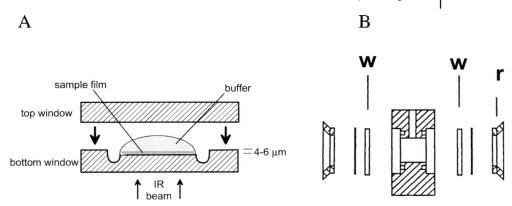

Fig. 4.4 IR cuvettes for biological samples. A, Specially designed sandwich cuvette. B, Cuvette for hydrated films: w, infrared windows; r, metal ring to fix the windows; washers between the rings and the windows protect the windows; via the hole of the cuvette body a drop of H_2O or ^2H_2O can be placed onto the bottom, allowing the hydration of the film deposited onto one of the windows via the vapour phase.

logical system tolerates moderate temporary drying, the sample can be deposited onto the actual window surface. A layer from the appropriate buffer having the volume of the reservoir is overlayed, and subsequently squeezed into the groove by the second flat window. The high viscosity of the concentrated sample considerably slows down diffusion of the biological material into the groove. If the material consists of natural or artificial membrane vesicles or membrane sheets that are dried onto the window surface from a suspension, it often sticks firmly onto the window and rehydration by the buffer layer does not spill it into the groove. Given that the surfaces of the windows are sufficiently plane, satisfactory sealing of the cuvette is possible. This sampling device is also called a sandwich cuvette.

In some instances, the S/N ratio is critical, such that the loss of IR intensity is too large even for the thin aqueous layer in the types of cuvettes described so far. For membrane systems, and sometimes also for soluble proteins, hydrated films may offer a solution. Here, the samples are dried onto the surface of one window, which is placed into another sample holder, as shown in Fig. 4.4(B). The second window is separated by a thick spacer (1–2 mm). By placing a drop of water into the hole of the spacer, the dried film is hydrated via the water vapour phase. The amount of water adsorbed onto the film is considerably less than the amount of water present in the sandwich cuvettes, improving the IR transmission significantly. There are two main disadvantages: (i) in some instances it has been shown that the hydration is not sufficient to guarantee full functionality, and the functionality has to be controlled by complementary methods; and (ii) as the amount of water adsorbed by the film cannot be controlled precisely, the composition of the solvent (pH, salt concentration, etc.) is only poorly defined. On the other hand, this approach, given that it does not perturb the biological function, allows the same sample to be used several times. The film can be dried again and rehydrated. In this way, ^{1}H/^{2}H exchange can also be performed on the same film.

All sample devices described so far have the disadvantage that after the second window is in place and the cell is sealed, the composition of the sample cannot be changed further. Thus, no reaction can be initiated just by adding a substrate, a ligand or an activator, which is common practice if biochemical reactions are followed by UV–vis spectroscopy. This is a particularly severe restriction as the strongest potential for IR spectroscopy of biological systems is associated with reaction-induced difference spectroscopy. For the cuvettes described above, this technique is only applicable if the reaction is triggered by light absorption. For all other types of "reactions" alternative sample arrangements have to be chosen.

4.2.1.2 The Attenuated Total Reflection (ATR) Method

To study molecular mechanisms of receptor functions it is often sufficient to monitor the molecular changes between the ligand-free and ligand-bound states. There are several other examples for which it is not the kinetics of the reaction that are important but the molecular differences between the initial and final states that are the main interest. The difference spectrum between two different samples could perhaps be produced, one spectrum representing the initial, i.e., ligand-free state, the other the final, i.e., the ligand-bound state. Of course, small variations in concentration have to be compensated for. However, experience has shown that only large bands can be reliably identified in this way. Therefore, it is desirable to add the component (substrate, cofactor, ligand) to the same sample from which a spectrum of the initial state has been taken. For this purpose, the attenuated total reflection (ATR) method offers a solution. The principle of this technique is shown in Fig. 4.5.

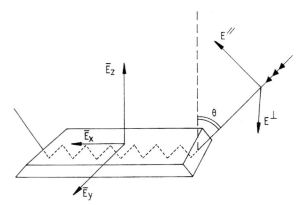

Fig. 4.5 Principle of ATR spectroscopy. The ATR plate is shown together with the radiation entering the plate, and the definition of the coordinate system together with the polarisation vectors; 13 total reflections are shown, 7 on the bottom surface, and 6 on the top surface. In most instances, the sample is placed onto only one surface (Goormaghtigh et al. 1999).

The IR beam enters an IR transparent slab of high refractive index, such that total reflection occurs at its surface. In the example shown in Fig. 4.5, approximately six reflections take place at each surface. If the surfaces are clean, the IR beam leaves the slab without appreciable attenuation. However, if one of the surfaces is in contact with material absorbing in the IR region, the IR beam is attenuated at the frequencies of the absorption bands. How can this happen when total reflection takes place within the slab? Here we only want to discuss the basic aspects of the ATR method. Knowledge of these principles is important to understand the advantages of this sampling device. For a more detailed treatise and applications to biological systems, the reader is referred to the literature (Harrick 1987; Goormaghtigh et al. 1999).

We will first concentrate on the situation for a single-reflection element. Using the Maxwell equations it can be shown that upon total reflection an evanescent wave penetrates into the optically less dense or rarer medium (i.e., possessing a lower refractive index) with an amplitude exponentially decaying in the direction normal to the surface (i.e., in z-direction), and carrying no energy. If, however, an absorbing medium is placed onto the surface, it couples with the electric field and energy is drawn out of the radiation, i.e., the total reflection is attenuated. The exponentially decaying field amplitude

$$E = E_0 \exp\left(-\frac{z}{d_p}\right) \tag{4.3}$$

is characterised by the penetration depth d_p

$$d_p = \frac{\lambda_1}{2\pi\sqrt{\sin^2\theta - n_{21}^2}} \tag{4.4}$$

Here λ_1 is the wavelength of the radiation in the optically denser medium with refractive index n_1, θ is the angle of incidence in the denser medium, and n_{21} is the ratio of the refractive indices n_2/n_1 where the index "2" refers to the rarer medium. As expected, the penetration depth becomes infinite as θ approaches the critical angle $\arcsin(n_{21})$ when light begins to be transmitted into the medium of lower refractivity. The minimum penetration depth is reached at grazing incidence, i.e., at $\theta = 90°$.

The underlying idea for exploiting the ATR principle in biological applications is based on the fact that the penetration depth is rather small. Thus, a bulk layer of the biological material can be deposited onto the surface. Even if the layer thickness is much larger than the penetration depth, the attenuation of the totally reflected IR beam only depends on the penetration depth and on the extinction coefficient of the material. Of course, the optical element must be transparent to the IR beam, as the effective path length within the optical element can be fairly large, i.e., 1 cm or more. As an example, we will consider ZnSe, which has a refractive index of 2.4, whereas the average index of refraction of biological material in an aqueous environment is 1.4 (Goormaghtigh et al. 1999). Thus, the critical angle is around 35°. If we assume that θ is 45, Eq. (4.4) yields a ratio of the penetration depth with respect to the wavelength in vacuum of

$$\frac{d_p}{\lambda} = 0.166 \tag{4.5}$$

For the range between 5 and 10 µm (2000–1000 cm^{-1}) the penetration depth then varies between 0.83 and 1.66 µm. Thus, it is much smaller than the thickness of the sandwich cuvettes. However, it is not straightforward to correlate quantitatively the attenuation of the reflected beam to the absorption properties of the material measured in transmission. For an exact treatment, one has to calculate the penetrating electric field strength, which decreases exponentially in the optically less dense medium according to Eq. (4.3). It should be mentioned that the penetration depth d_p is the same for both directions of polarisation of the electric field.

In transmission spectroscopy the absorption coefficient α is defined by

$$I = I_0 \exp(-\alpha d) \tag{4.6}$$

where d is the sample thickness, i.e., the cuvette path length.

Note that the commonly used definition for the absorption coefficient refers to

$$I = I_0 \cdot 10^{-\alpha_{10} d} \tag{4.7}$$

but both quantities can be easily converted into each other by

$$\alpha = \alpha_{10} \ln 10 \tag{4.8}$$

For small values of $\alpha \cdot d$, Eq. (4.8) may be approximated by

$$\frac{I}{I_0} = 1 - \alpha d \tag{4.9}$$

In ATR spectroscopy, the quantity that is measured is the attenuated reflection R as defined by

$$R = \frac{I}{I_0} \tag{4.10}$$

Conversely, the energy that is taken out of the radiation is related to

$$I_0(1 - R) = a I_0 \tag{4.11}$$

where a is termed the reflection loss. The correlation between α and a is complicated. Formally, one can define an effective thickness d_e, which relates the reflection loss to the absorption coefficient

$$a = d_e \alpha \tag{4.12}$$

The meaning of the effective thickness becomes clearer if Eq. (4.9) (low absorption approximation) is compared with Eq. (4.11). The effective thickness represents the actual thickness of a film of the absorbing material that would be required to obtain the same absorption in a transmission experiment compared with the ATR experiment using infinitely extended bulk material deposited onto the surface of the slab. Only for low absorption in bulk media or for very thin films does a direct proportionality between a and α hold [Eq. (4.12)]. In the more general situation, d_e itself is a function of α.

Using the Fresnel equation for the totally reflected beam and expanding it with respect to α, the factor of the linear term is the effective thickness d_e, which is expressed by

$$d_e = \frac{n_{21}}{\cos \theta} \int_0^\infty E^2 \, dz = \frac{n_{21} E_0^2 d_p}{2 \cos \theta} \tag{4.13}$$

For this derivation it has been used that the evanescent wave in the rarer medium shows the exponential z-dependence of Eq. (4.3). The factor $1/\cos \theta$ is due to the angle dependence of the sampled area; n_{21} describes the index of refraction matching, and E_0 is the field amplitude in the rarer medium at the surface, i.e., at $z = 0$. E_0 is given by a complicated expression and depends on n_{21} and θ (Harrick 1987).

Of particular practical importance is the fact that E_0 is different for the two polarisation directions, meaning that d_e is always smaller for parallel polarisation. For a semi-quantitative interpretation of ATR spectra using ATR elements of ZnSe ($n = 2.4$) and Ge ($n = 4$) and typical angles of θ between 40° and 60°, d_e does not differ strongly from d_p as defined in Eq. (4.3). Additionally, the low absorption approximation used here holds for values of α up to 10^4 cm^{-1}. In general, the values are considerably smaller in the IR and even the strong IR absorber water has a value of only ca. 2800 cm^{-1}. Thus, with the penetration depth of 0.83 to 1.66 μm (see above), only very low attenuation is obtained with a single reflection. At least qualitatively, the penetration depth can, therefore, be taken as the effective thickness of the sample. Because of the low attenuation, a multi-reflection element is often used, as depicted in Fig. (4.5). For multiple (N) reflections R amounts to

$$R = (1 - \alpha d_e)^N \tag{4.14}$$

If $\alpha \cdot d_e$ is small, Eq. (4.14) can be approximated by

$$R = 1 - N \alpha d_e \tag{4.15}$$

The effective thickness is now defined by $N \cdot d_e$. Thus, for a reflection element allowing five or six reflections, an absorbance of an aqueous biological sample is reached that is comparable to that of the sandwich cuvette. For the Ge ATR element, and the same 45° geometry, the number of reflections has to be higher as the penetration depth is smaller according to Eq. (4.4). However, the advantage as compared with the sandwich cuvette is clear. Because the thickness of the biological sample in the z-direction can be arbitrarily larger than the penetration depth without affecting the attenuation, it is possible to overlay the biological sample with an extended aqueous layer of water, i.e., the bulk phase of the aqueous solution. This guarantees a full hydration of the sample.

Furthermore, ATR represents an ideal device for probing the effect of ligand binding to a receptor. Now the same sample can be measured in the absence of the ligand and directly compared with the sample, after addition of the ligand to the aqueous solution. The only prerequisite is that the sample sticks firmly to the surface of the ATR element. For natural or artificial biological membranes including membrane bound receptors, this can be accomplished by drying them onto the surface (see Section 6.3).

FTIR instruments usually possess a circular aperture. Therefore, an ATR element with a square surface is more suitable as compared with flat plates with rectangular surfaces matching the entrance slits of old monochromator instruments. A ZnSe element with a square cross section has, typically, the dimensions $5 \times 5 \times 60$ mm^3. The entrance and exit faces are inclined by 45°, allowing a perpendicular entrance of the IR beam for 45° total reflection. The optical set-up is shown in Fig. 4.6. It is remarkable that, if properly placed, only flat mirrors are needed to reflect the IR beam into the element and to collect the radiation from the exit face. It should be mentioned that complete optical set-ups for this and for other designs are commercially available.

The system described in Fig. 4.6 has the disadvantage that the area onto which the sample has to be deposited is large. Although in principle only a thin layer of the order of the penetration depth needs to be deposited, it usually turns out that the corresponding films are fairly inhomogeneous. Therefore, in order to guarantee a homogeneous absorption over the complete area, a considerably

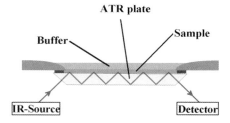

Fig. 4.6 ATR sample cell using a plate with a square cross section. With two plane mirrors, the IR radiation is directed to the entrance of the ATR element and collected from the exit, respectively. The element must be correspondingly displaced from the axis of the normal radiation path in the sample chamber, such that entrance and exit of the plate are in the focus of the two focussing mirrors. The cross section of the element approximately matches the size of the focus.

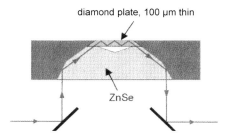

Fig. 4.7 Diamond ATR cell (Resultec Analytical Equipment, Garbsen, Germany). The diamond plate has a diameter of approximately 5 mm and is 100 µm thick. The ZnSe optical element is used to guide the IR radiation into the thin diamond plate. The required amount of material is considerably less as compared with that needed for the cell shown in Fig. 4.6.

larger amount of sample is needed as compared with the sandwich cuvette. This difficulty has been overcome by a new design that is now commercially available. When the slab gets thinner, the same number of reflections can be achieved over a shorter length of the slab. Using a 0.1 mm thick diamond plate, the active area is reduced to a circle of 4 mm with ca. seven reflections. The IR beam is guided into this plate by a type of waveguide made from ZnSe, which matches the index of refraction of diamond. This device is depicted in Fig. 4.7. Except for the spectral region where diamond absorbs, the device shows a very good performance. Alternatively, an Si plate is offered for such an arrangement, which it is also applicable in the region of absorption of diamond.

Many ATR devices are also available as flow cells, allowing either the measurement of liquid samples or the change of the buffer solution in experiments while the sample sticks to the surface. The diamond set-up can be changed to a microdevice where the volume of the liquid is less than 1 µL. Thus, the ATR method has many advantages and could, in principle, replace the "classical" cuvettes. However, in all the ATR devices a considerable fraction of IR intensity is lost and for high-sensitivity measurements requiring a very good S/N ratio, the sandwich or the hydrated film cuvettes are superior. Nevertheless, the ATR technique is the method of choice for many biological applications as it avoids unwanted structural distortions due to extensive dehydration, and furthermore allows flexible control of the environmental parameters (buffer, pH, ionic strength) and of probing molecular processes induced by binding of molecular reactants.

4.2.1.3 Electrochemical Cell for Infrared Spectroscopy

There are many important biological reactions that involve electron transfer processes of redox proteins. These processes can be triggered electrochemically. For membrane-bound redox systems, this can be achieved by changing the concentrations of the redox partners in the surrounding buffer (Rich and Breton 2002), and the reaction is then probed by the ATR-IR technique (see Section 4.2.1.2).

A more general approach is based on electrochemical cells. In such a cell, electrons are transferred between the redox-active protein and the electrode, which is controlled and triggered by the electrode potential. The oxidation/reduction process is usually monitored by measuring the current flow (e.g., in cyclic voltammetry) but it can also be followed spectroscopically. This technique, denoted as spectroelectrochemistry, is well established for monitoring the UV–vis spectral range. Implementation for IR spectroscopy is considerably more complicated due to the water absorption in aqueous solutions.

In general, the electrochemical cell consists of a three-electrode device, which is controlled by a potentiostat [Fig. 4.8(a)]. The electron transfer reaction of interest takes place at the working electrode (WE). The potential of the working electrode is measured with respect to a reference electrode (RE), which is itself a redox couple and in contact with the reaction volume via a diffusion barrier (e.g., a mem-

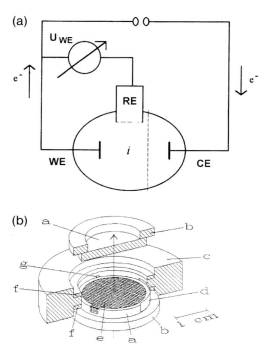

Fig. 4.8 (a) Principle of an electrochemical cell. The potential at the working electrode (WE) is measured with respect to the reference electrode (RE). The voltage is applied with respect to the counter electrode (CE) such that a current flows until the potential at the working electrode equals a pre-set value at the potentiostat. This potential corresponds to the redox-equilibrium in the cell. (b) Electrochemical cell for IR spectroscopy: a, IR windows; d, counter electrode made of platinum; e, gold grid, serving as working electrode and as spacer; g, capillary providing connection to the Ag/AgCl reference electrode (Baymann et al. 1991).

brane or a frit). Under high input-resistance conditions, the reference electrode allows a precise measurement of the potential between the working electrode and the reaction volume. For practical purposes, Ag/AgCl electrodes are used, which exhibit a redox potential that is 200 mV negative to the normal hydrogen electrode (NHE), the reference potential ($= 0.0$ V) to which all tabulated redox potentials are referred. Thus, the electrode potentials measured with an Ag/AgCl reference electrode have to be increased by 200 mV to obtain potentials on the NHE scale. Unless indicated otherwise, all potentials listed in this book refer to the NHE. The potential at the working electrode is applied with respect to the counter electrode (CE), and this causes a current to flow. The counter electrode is separated from the reaction volume by a diffusion barrier, which allows the flow of electric current via electrolytes but inhibits the diffusion of the active components to the electrode. Normally, there is only an Ohmic current according to the applied potential. If the potential of the working electrode is such that electrons can be transferred to or from the redox-active system, an additional current is observed. Concomitantly, the concentration changes of the redox active compounds in solution can be probed spectroscopically. When the redox equilibrium is established corresponding to the potential at the working electrode (U_{WE}), again only the Ohmic current is observed and the composition of reaction partners in solution remains unchanged. Regardless of the mode of how the redox processes are monitored, there are three boundary conditions imposed on the design of the cell (Heinze 1984). Firstly, the counter electrode should be positioned equidistant from all spots on the surface of the working electrode to ensure a mainly homogeneous current flow through the solution. Secondly, the reference electrode should be at the shortest possible distance to the working electrode to reduce the Ohmic potential drop through the solution. Thirdly, the diffusion distance should be kept as small as possible to establish equilibrium conditions in a short time.

Using IR spectroscopy for monitoring the redox processes, additional constraints are imposed on the design of the cell. The layout most frequently used is shown in Fig. 4.8(b) (Baymann et al. 1991). The working electrode consists of a gold grid, which, after filling the cuvette with the electrolyte solution including the redox protein, is squeezed between two infrared windows (e.g., CaF_2). The gold grid also serves as a spacer for this type of sandwich cuvette. The thickness of the grid is between 5 and 10 µm, and the IR transmission is ca. 55%. A small part of the grid is bent over the rim of one window ensuring electrical contact. A strip of platinum, serving as the counter electrode is attached to another part of the rim, and it is in electrical contact with the grid only via the electrolyte. The reference electrode is in contact with the reaction volume via a capillary hole in the body of the cell. Although the assembling and handling of this cuvette requires some experience, it has been successfully applied to the investigation of various redox proteins (see Chapters 7 and 8). Moreover, spectroelectrochemical cells of this type can also be employed for probing the reactions by RR spectroscopy.

4.2.2
Raman and Resonance Raman Spectroscopy

The sample arrangements for Raman and RR spectroscopy that can be chosen for the study of biological systems depend on the properties of the biomolecule and on the available spectrometer (see Section 3.2.2).

4.2.2.1 Measurements in Solutions

The standard experiments refer to measurements of biomolecules in aqueous solutions. For diluted solutions and non-absorbing species, the 90° scattering geometry represents the best arrangement that can be easily adapted to flowing sample devices. In an RR experiment, there is an optimum value for the concentration of the absorbing compound, which corresponds to an optical density of ca. 1.4 for a 1-cm optical pathlength. Increasing the concentration further will attenuate the intensity of the exciting laser beam, such that the irradiated volume element is reduced and, despite the higher concentration of scattering molecules, the detected RR intensity does not increase further. For a chromophore with an extinction coefficient of 50 000 L mol^{-1} cm^{-1}, this optimum concentration is ca. 30 µM.

Among the flowing sample devices, rotating cylindrical cells represent the most versatile and simple sample containers (Fig. 4.9). These cells can be designed such that the required sample volume is at low as 50 µL, which for the above example would correspond to a total amount of the biomolecules of less than 2 nmol. Such rotating quartz cells, which are tailor-made by optical companies and machined with a high precision and quality, are relatively expensive. However, they have a much better performance than the NMR, EPR, or capillary tubes that are frequently used as cheaper variants. The latter tubes cannot be operated

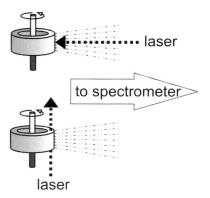

Fig. 4.9 Rotating cells for 180° scattering (backscattering, top) and 90° scattering (bottom). The thick arrows indicate the incident laser light whereas the dashed–dotted lines represent the scattered light to be collected by the spectrometer.

in the 90° scattering geometry but instead require a backscattering arrangement. This geometry is associated with a higher efficiency for collecting scattered photons and is ultimately optimised in the confocal arrangement; but it requires concentrations that are usually higher by one order of magnitude as compared with the 90° scattering geometry. Specifically, in confocal spectroscopy the higher concentration requirements can be compensated by the volume reduction.

The 90° scattering geometry facilitates the sample exchange in the exciting laser beam. For an aqueous solution of an absorbing biomolecule (30 µM) in a rotating cylindrical cell with a diameter of 1 cm, a laser beam focussed into the close to the cell wall with a beam radius of 40 µm, produces an irradiated volume of ca. 25 µL per rotation, corresponding to $7.5 \cdot 10^{-10} \cdot N_A$ irradiated molecules. In the backscattering experiment with a rotating 2-mm NMR tube, including a solution of a ten-fold higher concentration, the number of irradiated molecules is ca. 100-times smaller, assuming the same radius of the focussed laser beam. Thus, in this case the average number of photons per molecule is larger by two orders of magnitude and the risk of photodamage is correspondingly higher than using a rotating cylindrical cell.

Specially designed rotating cylindrical cells that include two equally sized chambers have been introduced for comparative measurements of two samples contained in the separated compartments (Kiefer 1973). This technique is applicable for scanning spectrometers and requires a photon counting detection system with two separate channels. The cell is placed on a mechanical rotator equipped with a chopper and a photodiode emitter and receiver pair that produces an electrical trigger signal each time the laser passes from one cell compartment to the other. This trigger signal is used to gate the photomultiplier output to either one or the other channel of the photon counting device corresponding to the origin of the photons from the individual compartments. Subtracting the spectra of two different samples, which are measured quasi-simultaneously, from each other allows for a highly precise determination of frequency differences far beyond the spectral resolution of the absolute spectra and the accuracy of the spectrometer movement.

4.2.2.2 Solid State and Low-temperature Measurements

Raman spectroscopy of solid samples has to be carried out in the backscattering mode (Fig. 4.9). Solid samples may be crystals or powders of pure compounds, solid mixtures involving mainly Raman-inactive diluents such as KBr, frozen solutions, or glasses. Low-temperature Raman experiments require specially designed cryostats that allow adaptation of the focussing and collecting optics. Such cryostats are also commercially available for the confocal arrangement and may operate at liquid nitrogen or liquid helium temperatures. With such set-ups, however, it is very difficult to study photosensitive materials as the laser is permanently focussed onto the same spot unless the entire cryostat is moved. To overcome this drawback, laboratory-built cryostats have to be employed in which, for instance, the sample is attached to a metal spring (Eng et al. 1985). Using a magnetic switch to induce a tuning-fork type motion of the spring, the sample then oscillates between two positions, thereby increasing the total illuminated area of the sample considerably. In another set-up, the sample is deposited onto a rotating cold finger (Braimann and Mathies 1982).

4.3
Surface Enhanced Vibrational Spectroscopy

The enhancement of signals associated with vibrational transitions on nanoscopically rough metal surfaces was first discovered and utilised for Raman scattering (SER, SERR) whereas the potential of surface enhanced infrared absorption spectroscopy (SEIRA) was recognised much later (see Section 2.3). SER and SERR spectroscopy was initially applied to biomolecules as an easy approach to measure Raman or RR spectra molecules that were hardly accessible by conventional Raman/RR spectroscopy due to the high fluorescence or the low quantities of the compounds being available. Because of the high sensitivity, SER/SERR spectroscopy was thought to have a high potential as an analytical technique. In addition, it has been suggested that the metal/electrolyte interface may be considered as a model for biological interfaces such that SER/SERR spectroscopy could provide novel insight into interfacial processes of biomolecules (Cotton et al. 1980). However, along with an increasing number of systematic experimental studies and an improved understanding of the nature of the surface enhancement of radiation fields, the problems associated with surface enhanced vibrational spectroscopy have become evident and led to a more realistic view about the potential of this technique.

Firstly, direct adsorption of biomolecules onto bare metal surfaces may cause structural changes that range from local perturbations of the binding site to global structural rearrangements. These changes can be reversible or irreversible (denaturation) but in any case convert the biomolecule into a conformational state that is different from the reference state, the biomolecule in solution. Moreover, there are no complementary techniques that are sufficiently sensitive to provide structural data about the biomolecules which are usually adsorbed at submonolayer coverages. Thus, possible adsorption-induced structural perturbations have to be identified on the basis of the SER/SERR and SEIRA measurements themselves and only in a few instances can additional indirect information be extracted by other techniques, such as electrochemical methods. In this respect, it was a substantial methodological improvement when surface enhanced vibrational spectroscopies were extended to metal surfaces covered with biocompatible coatings. Thus, the direct contact of the biomolecules with the metal is avoided and the risk of irreversible denaturation processes is largely reduced. Moreover, choosing appropriate coatings leads to much better models for biological interfaces than a bare metal surface.

Secondly, the coupling of the surface plasmons with the electromagnetic radiation does not only produce an enhancement of the Raman scattering but may also enhance photoinduced processes, such as photoinduced electron transfer or photochemical reactions. Here, low photon fluxes, short sampling times, and the movement of the sample through the exciting laser beam are the only means of preventing these unwanted side reactions.

Thirdly, roughened metal surfaces that are required for SER/SERR and SEIRA spectroscopy represent non-equilibrium states. Thus, the morphology of the surface, which is a crucial determinant of the absorption and scattering enhancement, may be subject to time-dependent changes, and moreover, may differ from preparation to preparation. Hence, the *magnitude* of the enhancement is difficult to reproduce in subsequent surface preparations and it may change with time during the individual experiments. For these reasons, SER/SERR spectroscopy may be well suitable for a qualitative but not for a quantitative analysis of specific compounds.

However, despite these limitations and taking into account some essential experimental precautions, SER/SERR and SEIRA spectroscopies are not only powerful analytical tools but, what appears to be even more attractive, represent approaches that allow studying structure and dynamics of biomolecules at interfaces that can be designed to mimic fairly closely biological membranes (Murgida and Hildebrandt 2005).

For biological applications, basically, only Ag and Au are used as metal supports as they exhibit particularly strong enhancements at wavelengths above 400 and 600 nm, respectively. Rough Ag and Au surfaces are either provided by colloidal suspensions, massive electrodes, or chemically deposited or evaporated metal films. All of these systems are suitable for SER and SERR spectroscopy. SEIRA studies on biological molecules, however, are essentially restricted to thin metal films on ATR elements, due to the strong absorption of water in colloidal suspensions and electrochemical cells. Because the wavelength dependence of the enhancement sensitively depends on the nanostructured surface morphology, the various materials can be optimised empirically for operation in the visible and in the infrared spectral region. Unlike colloidal suspensions, electrochemical cells for SER/SERR and SEIRA measurements represent fairly demanding devices that are not easy to fabricate, but the underlying technical efforts are particularly justified if potential-sensitive biological systems are to be studied, such as electron-transferring proteins or redox enzymes. Then, the electrode does not only function as an amplifier of the spectroscopic signals but also serves as a reaction partner that delivers or accepts electrons to or from the attached redox-active molecules. Moreover, in conjunction with the potential-jump technique, SERR and SEIRA on electrodes can also be operated in the time-resolved mode (see Section 4.4.3).

4.3.1
Colloidal Suspensions

Soon after the discovery of the SER effect for molecules adsorbed on roughened electrode surfaces, it was shown that colloidal metal particles in aqueous suspensions also possess comparable signal amplifying properties. In fact, the first classical electromagnetic treatment of the origin of the SER effect was developed for colloidal particles assuming spherical shapes with diameters distinctly smaller than the wavelength of the incident laser light (Kerker et al. 1980). Colloidal Ag and Au hydrosols are easy to prepare by chemical reduction of $AgNO_3$ and $KAuCl_4$ with borohydride or citrate. Specifically, the citrate-reduced preparations are very stable and can be stored as stock solutions for many months. In each instance, the colloidal suspensions are polydisperse and include particle aggregates. This heterogeneity is reflected by a broad and asymmetric absorption band with an extended wing towards longer wavelengths. Addition of electrolyte causes aggregation of the particles, which decreases the stability of the hydrosol but leads to stronger SER signals particularly upon excitation at longer wavelengths. It has been proposed that aggregation is associated with the formation of so-called "hot spots" where enhancement of Raman scattering is significantly larger than at "normal" adsorption sites (Hildebrandt and Stockburger 1984; Kneipp et al. 2006). Thus, it may be that the experimentally observed SER signals predominantly result from molecules adsorbed at such "hot spots", which may be located in between two adjacent colloidal particles. Electrolyte-induced aggregation of colloids, specifically Ag, brings the sensitivity of the SER effect to a level that, in combination with the molecular RR effect, allows single-molecule detection of chromoproteins by SERR spectroscopy (Kneipp et al. 2006). Colloidal suspensions can be measured using the same devices that are employed for experiments with aqueous solutions. Also, in this instance, rotating cells represent preferable arrangements in order to avoid the consequences of the enhanced photosensitivity of the adsorbates in addition to laser-induced desorption of the molecules.

4.3.2
Massive Electrodes in Electrochemical Cells

Roughening of electrodes to provide the required nanostructured surface morphology is achieved by sequential oxidation and reduction of the metal at sufficiently high currents such that re-deposition of the metal ion does not lead to perfect surface lattices. Details of the protocol depend on the geometry of the electrochemical cell used for roughening and may be empirically optimised on the basis of the measured SER intensities. Although correlations between the charge transferred during a reduction step, the surface morphology, and the SER intensity have not yet been explored in detail, one may consider a charge of ca. 25 $C\,cm^{-2}$ as a guideline for good surface enhancements for Ag. Under these conditions, the average surface roughness is in the order of 50–100 nm, depending on the rate of reduction and the type of supporting electrolyte. The most durable SER activity has been achieved for electrochemical roughening in the presence of KCl or NaCl as the supporting electrolyte. The poor solubility of AgCl that is formed in the oxidation step prevents diffusion of the Ag^+ cations into the solution, which would otherwise slow down reduction. Increasing the current appears to increase the roughness also on smaller length scales, which in turn leads to an initially higher surface enhancement but also to a faster decay of the SER intensities during the experiments.

Electrochemical cells used for SER experiments are laboratory-built and consequently there are a large number of different designs that have been proposed. Essential components of the cells are the three electrodes including the SER-active working electrode, the Pt counter electrode, and the reference electrode (e.g., Ag/AgCl), which are connected to a potentiostat for controlling the potential at the working electrode. Two examples of the working electrode design that are particularly useful for studying biological system are shown in Fig. 4.10. In one

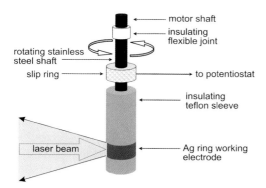

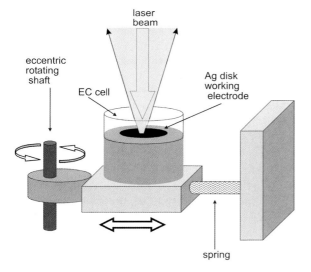

Fig. 4.10 Designs of electrochemical cells for SER spectroscopy of biological systems. Top: rotating ring working electrode mounted on a stainless-steel shaft that also serves as an electrical conductor. The connection to the potentiostat is established via a slip ring. The shaft is driven via a flexible insulating joint with the motor (Murgida and Hildebrandt 2001). Bottom: eccentrically rotating electrochemical (EC) cell including a disk working electrode (Bonifacio et al. 2004).

example, the working electrode is shaped as a ring that is mounted on a rotating stainless-steel high precision shaft while establishing a good electrical contact. This shaft is attached to the motor via an insulating tube joint (Murgida and Hildebrandt 2001). The electrical contact from the shaft to the potentiostat is ensured via a slip ring. The rotation of the shaft with frequencies between 1 and 25 Hz guarantees the continuous movement of the sample through the exciting laser and has been found to be essential for avoiding unwanted photoinduced (degradation) processes of the adsorbed biomolecules.

In the second device, the working electrode is a stationary metal disk that is directly connected to the potentiostat. Here, the entire cell, which is eccentrically mounted on a rotor, is moved with respect to the laser focus (Bonifacio et al. 2004). This device is advantageous for experiments under strictly anaerobic conditions as continuous purging of the cell with purified Ar during the SER experiments is more efficient at removing oxygen than in cells with rotating electrodes. In addition, the cell can be better adapted to short working distances of microscope objectives in confocal measurements. In contrast to the disk-electrode cell, the rotating-ring electrode cell can be used for both 90° and backscattering geometries using two and one optical windows of the cell, respectively. In the rotating-ring electrode device, the risk of photoinduced sample damage can be further lowered by mounting the cell on a translational stage that periodically moves up and down (in the z-direction) such that the laser beam passes across the entire height of the working electrode surface.

To construct electrochemical cells for SER studies, similar constraints have to be considered as for IR-spectroelectrochemical cells (see Section 4.2.1.3): the reference electrode has to be placed in the closest possible proximity to the working electrode, and the arrangement of the working and counter electrodes should be chosen such that the distribution of the electric field lines is largely homogeneous. Unlike IR-spectroelectrochemical cells, however, long diffusion distances do not represent a problem as the SER spectroscopy probes surface confined processes. Further options, such as cooling and purging devices, may be included according to the requirements of the experiments.

4.3.3
Metal Films Deposited on ATR Elements

In SEIRA experiments operated in the ATR configuration, the ATR crystal is an integral component of the electrochemical cell (Ataka and Heberle 2003). In the example shown in Fig. 4.11, a Si single-reflection ATR hemispheric prism is used. On top of the ATR element, a metal (Au) film is deposited via evaporation or chemical reduction of the corresponding metal salt ($KAuCl_4$). The conditions are chosen such that a complete coverage is achieved with an estimated roughness that is comparable to that of roughened electrodes.

Unlike conventional ATR-IR spectroscopy (see Section 4.2.1.2), multiple reflections cannot improve the signal as the amplitude of the evanescent wave is effectively damped by the metal coating. In addition, there is also another funda-

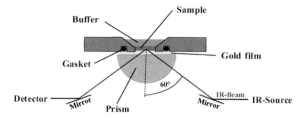

Fig. 4.11 Principle of a device used for SEIRA spectroscopy. The gold island film is deposited on the hemispheric prism of silica. The prism is operated in the single-reflection ATR mode. Only plane mirrors are needed to direct and collect the IR beam. Sample is applied to the gold film and buffer is overlayed.

mental difference between ATR-IR and ATR-SEIRA spectroscopy. Because the enhanced absorption is a near-field effect that decays exponentially with the distance from the metal film [see Section 2.3; Eq. (2.80)], the sampling range of ATR-SEIRA spectroscopy is limited to less than 10 nm. When biological material is adsorbed onto the surface, it is the first monolayer that is therefore predominantly probed, which, depending on the type of system, may have a thickness of a few nanometres (e.g., membranes). In ATR-IR spectroscopy the effective penetration depth d_e of the radiation field into the optically less dense medium is ca. 100 times larger. Thus, the ATR-IR signals that are measured are not restricted to a monolayer, and moreover, due to the lack of surface enhancement, relatively thick films (ca. 1000 nm) of the molecules to be studied are required for a satisfactory S/N ratio. Such thick films are usually obtained by drying the sample onto the ATR surface before it is brought in contact with the aqueous solution. In this way, a stack of hundreds of layers (e.g., membrane sheets) is obtained, albeit not necessarily in an ordered manner. Also in this respect, the ATR-SEIRA technique is advantageous as ordered monolayers can be obtained by various procedures (see Section 4.4.6).

4.3.4
Metal/Electrolyte Interfaces

Insertion of metal into an aqueous electrolyte-containing solution creates a charged interface due to the accumulation of positive and negative charges in the solid and the solution phase (Bockris et al. 2000). Depending on the type of metal, the electron density in the metal is either reduced or increased and consequently oppositely charged ions are concentrated on the solution side. The formation of the charged interface does not require the application of an external potential within an electrochemical cell and thus also occurs for metal particles (colloids) suspended in aqueous solutions or for a metal electrode at an open circuit.

The charge distribution in the solution phase is described by the concept of the electrical double layer. Firstly, a compact layer of specifically adsorbed ions and highly oriented water molecules is formed. The properties of this layer, also denoted as the inner Helmholtz layer or Stern layer, strongly depend on the type of metal and the electrolyte in the solution. Ag in contact with an aqueous solution carries a net positive charge, which then leads to the chemisorption of anions. The free energy gain in formation of the chemisorptive bonds must override the partial loss of the hydration shell. Thus, chloride, sulfate, or phosphate show a high tendency for specific adsorption whereas perchlorate and fluoride are considered as non-binding anions. The specific binding of anions usually overcompensate the positive charge of the metal resulting in a sharp potential drop within the compact double layer, i.e., over a distance of less than 10 Å. Adjacent to the compact double layer, the diffuse double layer or outer Helmholtz layer is formed with cations and anions that are considered to diffuse freely.

As can be seen from Fig. 4.12, there is a rather complex variation of the potential distribution in the vicinity of the metal surface. This distribution can be approximately described within a simple electrostatic model considering the charge densities σ_i localised in the various interfaces (Smith and White 1992). Applying Gauss' law, the charge density on the metal σ_M is related to the potential difference between the metal (ϕ_M) and the inner Helmholtz plane (ϕ_C)

$$\sigma_M = \frac{\varepsilon_0 \varepsilon_C}{d_C} (\phi_M - \phi_C) \tag{4.16}$$

where ε_0 is the permittivity, and ε_C and d_C are the dielectric constant and the thickness of the inner Helmholtz layer, respectively. The charge density in the solution σ_S is derived from Gouy–Chapman theory and related to the potential difference between the inner Helmholtz plane and the bulk solution ϕ_S ($d \to \infty$)

$$\sigma_S = -\varepsilon_0 \varepsilon_S \kappa \frac{2kT}{e} \sinh\left[\frac{e}{2kT}(\phi_C - \phi_S)\right] \tag{4.17}$$

Here ε_S is the dielectric constant of the solution, T is the temperature, and e and k are the elementary charge and Boltzmann constant, respectively. The quantity κ is the Debye length and describes the thickness of the diffuse double layer, given by the drop in the potential from ϕ_C to $(1/e) \cdot \phi_C$. The Debye length decreases with increasing ionic strength. To fulfill overall charge neutrality, σ_M, σ_S, and the charge density at the inner Helmholtz plane σ_C must sum to zero

$$\sigma_M + \sigma_C + \sigma_S = 0 \tag{4.18}$$

The interfacial potentials and thus the charge densities are altered upon applying an external potential, which raises or lowers ϕ_M. When ϕ_M approaches ϕ_S, the charge density on the metal is steadily lowered and eventually becomes zero. This is the potential of zero charge, E_{pzc}, which is ca. -0.7 V and $+0.2$ V (versus

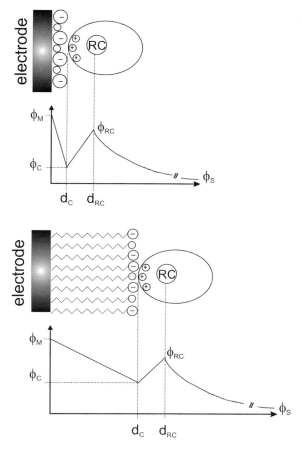

Fig. 4.12 Schematic representation of the potential distribution in the electrical double layer of a metal electrode coated with anions (top) and an SAM of carboxyl-terminated thiols (bottom) including an electrostatically adsorbed redox protein. Here "RC" denotes the reaction centre of the protein.

NHE) for Ag and Au, respectively. Above and below this potential, the inner Helmholtz layer is constituted predominantly by anions and cations, respectively.

In typical SER or SEIRA experiments, the solutions that are in contact with the metal contain electrolytes with ions that can specifically adsorb to the surface. This is also true for colloidal suspensions in which specifically adsorbing ions (e.g., citrate) are introduced during the preparation procedure or subsequently added to induce particle aggregation. Consequently, larger molecules to be studied by SER/SEIRA spectroscopy do not usually bind directly to the metal surface but to the layer of specifically adsorbed ions. The only exceptions refer to those molecules that are capable of replacing the chemisorbed ions. Among these are molecules that carry a thiol function, as both Ag and Au undergo firm chemical bonds with the sulfur.

For all other molecules that cannot penetrate the inner Helmholtz layer, adsorption behavior can be qualitatively predicted on the basis of the charge distribution in the electrical double layer. When the electrode potential is set above the potential of zero charge, the electrode is covered with specifically adsorbed anions and thus preferentially positively charged molecules are adsorbed. Upon approaching E_{pcz}, the charge density decreases and also neutral or even non-polar molecules can bind, whereas below E_{pcz} binding of negatively charged molecules is preferred. For Ag hydrosols the actual potential of the citrate-covered metal colloids that can be considered as microelectrodes is ca. +0.1 V and thus far above E_{pcz}, such that cationic molecules are preferentially adsorbed (Henglein and Lilie 1981).

Thus, a positively charged protein such as cytochrome c electrostatically binds to the chemisorbed anions with a highly negative free adsorption enthalpy ΔG_{ads} of -52 kJ M^{-1} (Hildebrandt and Stockburger 1986) and the adsorption behaviour can be described according to

$$\frac{k}{k_{max} - k} f = \frac{c_{dis}}{c_{water}} \exp\left(-\frac{\Delta G_{ads}}{RT}\right) \tag{4.19}$$

where k is the number of occupied adsorption sites and k_{max} refers to the maximum number of adsorption sites. The concentrations of the dissolved protein and water are denoted by c_{dis} and c_{water} and f represents a term to take into account interactions between adsorbed molecules. For a Langmuir adsorption isotherm, f is equal to 1, whereas for the Frumkin isotherm, which describes the adsorption behaviour of cytochrome c on Ag colloids, f is given by

$$f = \exp\left(-2a\frac{k}{k_m}\right) \tag{4.20}$$

Equations (4.19 and 4.20) are useful to estimate the protein concentration c_0 required for SERR experiments, as is illustrated in the case of cytochrome c. Combining Eqs. (4.19 and 4.20) and using the degree of saturation of the adsorption site $\theta = k/k_{max}$ one obtains

$$c_0 = \frac{\theta}{1-\theta} \cdot c_{water} \cdot \exp(-2a\theta) \cdot \exp\left(\frac{\Delta G_{ads}}{RT}\right) + c_{ads} \tag{4.21}$$

where c_{ads} is the concentration of the adsorbed proteins. For this quantity we obtain

$$c_{ads} = \frac{\theta \cdot k_{max}}{N_A \cdot V_{cell}} = \frac{\theta}{N_A \cdot V_{cell}} \cdot \frac{A_{surf}}{A_{mol}} \tag{4.22}$$

with N_A and V_{cell} denoting the Avogadro number and the solution volume in the cuvette or in the electrochemical cell. For the surface area A_{surf} of an electrochemical roughened electrode the geometrical area (e.g., 1.0 cm^2) has to be multiplied by the roughness factor, which is typically 1.5 to 4. The space required by the adsorbed protein is more difficult to assess but it appears to be ca. four times larger than the geometrical area, as judged from three-dimensional structure data. This increased spaced requirement implies that only ca. 25% of the available surface area can be at best occupied by proteins. With $V = 5$ mL, $A_{surf} = 1.5$ cm^2, $A_{mol} = 50$ nm^2, and $a = 1.3$, a value of the total concentration of cytochrome c of ca. $4 \cdot 10^{-8}$ mol L^{-1} is required to achieve a degree of adsorption of $\theta = 0.9$.

The accessible potential range in SER experiments is limited, in particular for Ag. The upper limit is given by the oxidation potential of the metal, which is ca. +0.65 V in sulfate-containing solutions but ca. 0.4 V more negative in the presence of chloride. For Au, the oxidation potential is distinctly more positive (ca. +1.4 V).

Slow metal oxidation destroys the SER-active surface structure, and, moreover, metal cations formed at potentials close to the oxidation potential bear the risk of chemical attacks of covalent bonds in the macromolecule, such as disulfide bonds in proteins. The concentration of cations exponentially increases when the electrode potential approaches the redox potential, such that the electrode potential should be kept ca. 200 mV below the oxidation potential. The lower potential limit is given by the onset of the reduction of hydrogen ions, which, depending on the solution, may occur at potentials below -1.0 V. However, even at much more positive potentials (ca. -0.4 V), oxygen reduction starts and may produce reactive intermediates that are capable of attacking biomolecules.

4.3.5
Adsorption-induced Structural Changes of Biopolymers

Adsorption of biomolecules either directly to the metal or to the inner Helmholtz layer may cause irreversible structural changes. One origin of such unwanted effects is the enormous electric field strengths in the electrical double layer. The electric field strength E_F, experienced by molecules adsorbed to the layer of specifically adsorbed ions, may be estimated according to Eq. (4.16). This is given by the potential drop across the compact double layer (Fig. 4.12)

$$E_F = \frac{\phi_M - \phi_C}{d_C} = \frac{\sigma_M}{\varepsilon_0 \varepsilon_C} \tag{4.23}$$

Neglecting overcompensation of the charge on the metal by the specifically adsorbed ions, one can assume $\sigma_M \approx -\sigma_C$. As an example, we can consider an Ag electrode in contact with a sulfate containing solution. At ca. +0.2 V, σ_C is ca. -0.25 C m^{-2} and with $\varepsilon_C = 6$ an electric field strength of $4.7 \cdot 10^9$ V m^{-1} is calculated. Such field strengths are sufficient to perturb, for instance, acid–base equilibria in biopolymers, which in turn can initiate far-reaching structural changes.

The electric field strength increases with increasing $|E - E_{pzc}|$. The electric field vector points in opposite directions on Au and Ag in the respective accessible potential ranges. Approaching E_{pzc}, the electric field strength decreases accompanied by a decrease in the amount of adsorbed anions (cations) on the electrode surface. This, however, also causes a decrease in the adsorption constant for positively (negatively) charged biomolecules concomitant with a loss of SER intensity. For proteins an additional consequence of the low surface charge at $E \approx E_{pzc}$ is the increased tendency of nonpolar and hydrophobic amino acid residues to interact with the uncharged metal surface. As a result, hydrophobic amino acids, which are originally buried in the interior of the protein, interact with the surface and, hence, cause the unfolding of the polypeptide chain.

In this respect, Au is the more appropriate metal for studying biological molecules as, due to the more positive oxidation potential, a larger potential range is accessible without running the risk of creating aggressive Au cations. Furthermore, the positive potential of zero charge implies that in the potential range of interest electric field strengths are smaller than for the Ag.

4.3.6
Biocompatible Surface Coatings

Irreversible structural changes of proteins due to the electric field or the attack of metal ions can largely be avoided by covering the metal surface with biocompatible coatings. These coatings not only prevent direct contact with the surface but also interact with the biomolecules in a manner that is more closely related to their natural environment. The most versatile and simple coatings are based on the self-assembly of amphiphiles carrying a thiol function. The alkanethiol derivatives bind to Ag or Au via formation of covalent metal–sulfur bond and can constitute densely packed monolayers (self-assembled monolayers, SAMs) due to the hydrophobic interactions between the aliphatic chains (Ulman 1996). Adsorption of the biomolecules to be studied then occurs on the surface of the SAM, which may be appropriately functionalised to allow for electrostatic, hydrophobic, or covalent binding (Murgida and Hildebrandt 2005). SAMs made of carboxyl-terminated thiols have been widely used to immobilise positively charged proteins via electrostatic interactions. The charge density on the SAM surface is much lower than in the inner Helmholtz layer constituted by specifically adsorbed anions. This is due to the fact that the pK_A of the carboxylate group in the layer is distinctly higher than that of the molecule in solution. This pK_A shift, which results from the electrostatic repulsions of adjacent carboxylate functions, depends on the electrode potential, the ionic strength, and the distance from the electrode. At an ionic strength of 0.1 mol L^{-1} and an electrode potential of 0 V, the pK_A of the carboxylate head groups of SAM containing thiols with 10 methylene groups is between 8 and 9, and thus ca. 5 units higher than for the molecules in solution. As a result, only a small fraction of the SAM head groups are in fact deprotonated at pH 7.0, which on the one hand is sufficient for electrostatic binding of basic proteins but, on the other, is associated with electric field strengths that are distinctly lower compared with chloride-covered Ag electrodes. The situation is different when the carboxylate head group is replaced by a phosphonate or sulfonate function that carries one negative charge in a wide pH range and thus exhibits a relatively high charge density corresponding to a very strong local electric field. Besides the SAM head group, the length of the alkane chain also offers the possibility of adjusting the local electric field, which depends on the interfacial potential drop across the SAM and thus decreases with increasing distance from the electrode (Fig. 4.12). The same considerations hold for amino-terminated alkanethiols that can be used to bind acidic proteins.

The electrostatic binding of proteins may also represent a first step toward covalent cross-linking. After addition of 1-ethyl-3-[3-(dimethylamino) propyl] carbodiimide hydrochloride (EDC) to the electrostatically pre-aligned protein–SAM complex, a covalent bridge is formed between the adjacent amino and carboxylate functions. Alternative strategies for covalent cross-linking are based on free thiol groups of surface-exposed cysteine residues that can be linked to a cysteamine monolayer via N-succinimidyl-3-maleimidopropionate. This approach offers the advantage that different orientations of the immobilised protein can be achieved given that cysteines are introduced at selected positions on the protein surface via genetic engineering.

A particular variant of covalent attachment can be employed for proteins that carry a cofactor. Here the cofactor can be directly wired to an appropriately functionalised surface either through coordinative or covalent bonds (Murgida and Hildebrandt 2005).

Hydrophobic coatings of metal surfaces using methyl-terminated alkanethiols are associated with very low electric fields. This immobilisation strategy is applicable when soluble proteins possess hydrophobic patches on the surface but it bears the risks of partial unfolding as buried hydrophobic amino acids may be pulled out of the interior of the protein.

Tailor-made functionalisation of metal surfaces for immobilising proteins and enzymes is a rapidly growing field due to the potential impact in biotechnology and bioelectronics. The various approaches that have been developed are not restricted to the self-assembly of amphiphiles but also include Langmuir–Blodget layers, polyelectrolyte or peptide coatings. These developments may provide a welcome "spin-off" to be used for spectroscopic studies in fundamental research.

However, not all of these biocompatible coatings are suitable for SER/SERR and SEIRA spectroscopy as the surface enhancement decays with the distance from the electrode. As illustrated in Fig. 2.13, for a surface roughness of ca. 20 nm the enhancement of the SEIRA and SER signal at a distance of 3.5 nm is decreased by a factor of ca. 10 and 2.5, respectively. This loss of intensity may be acceptable because the enhancement may still be sufficient for studying proteins bound onto or integrated into complete phospholipid bilayers that are attached to the metal surface. In this respect, the concept of protein-tethered membranes on solid supports is particularly intriguing. It is based on the coating of electrode surfaces with dithiobis-(N-succinimidyl propionate) and N_α,N_α-bis(carboxymethyl)-L-lysine, such that after addition of Ni or Zn ions a complex

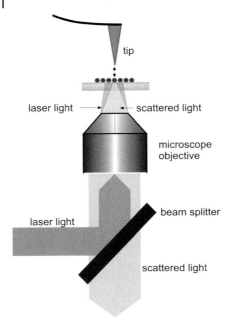

Fig. 4.13 Design for a tip-enhanced Raman experiment using a confocal illumination and scattering geometry. The laser beam (dark grey) is focused onto an array of molecules but only those that are in the near-field of the tip give rise to an enhanced Raman scattering (light grey), which is collected by the microscope objective (cf. Rasmussen and Deckert 2005).

is formed with the histidine tag of a solubilised membrane protein. Once attached to the functionalised surface, the detergent can be substituted by lipids to form an extended lipid bilayer around the immobilised proteins. An alternative, less elegant but simpler, approach is based on the direct adsorption of the solubilised enzyme as the detergent molecules may provide a sort of "protecting" surface coating.

4.3.7
Tip-enhanced Raman Scattering

Direct contact of the biomolecule with the metal surface is avoided in tip-enhanced Raman scattering as the metallic tip is kept a few Ångströms above the sample. Although this technique is still at its infancy and has been applied to biological systems only in a few cases, it may show a high potential in the future because it combines the advantages of SER/SERR spectroscopy and scanning probe microscopy (Rasmussen and Deckert 2005). However, there are still substantial experimental problems to be solved. Specifically, an optimum solution has to be found to reconcile the requirements for efficient enhancement of the incident and scattered radiation and for maximising collection of the Raman scattered photons. It is mainly the component of the oscillating electric field that is parallel to the tip that is enhanced, so the incident laser beam should be parallel to the sample, which is of course a highly unfavourable orientation to excite the molecules and to collect the scattered light. Conversely, a poorer enhancement but an optimum illumination and transfer of the scattered light to the spectrometer is achieved in a confocal arrangement with 180° backscattering geometry, as is possible by using for instance an inverted microscope (Fig. 4.13). For other configurations the reader is referred to the literature (Rasmussen and Deckert 2005).

A second practical problem refers to the preparation of the tip, typically made of Au for the spectroscopic applications. The morphology of the tip and specifically its sharpness are the crucial determinants of the plasmons resonance frequency and thus for the frequency at which maximum enhancement can be achieved. These properties may differ from tip preparation to tip preparation, such that for a given excitation wavelength individual tips may provide a good enhancement whereas others do not. In addition to these experimental difficulties, there is also a significant problem that cannot be avoided. For a tip in close proximity to the sample, the enhancement of the Raman scattering will be largely restricted to an area of less than 5 nm^2, such that the enhancement factor must be ca. 10^6 to achieve at least the same intensity as for a normal unenhanced spectrum probing a spot of typically 5 µm^2 of a monolayer.

Regardless of these obstacles, the intense efforts are justified as the combination of atomic-scale microscopy and molecular spectroscopy will open up new horizons in analysing complex biological systems at different levels of the structural hierarchy.

4.4
Time-resolved Vibrational Spectroscopic Techniques

Concomitant with the exciting developments in vibrational spectroscopies in the 1980s, specific emphasis has been laid on time-resolved methods to probe the dynamics of molecular processes. It is interesting to note that many of the methodological achievements have been made using biological systems as test cases (e.g., bacteriorhodopsin). Conversely, both time-resolved IR and Raman techniques have contributed substantially to the understanding of the dynamics of biomolecules, specifically proteins and enzymes.

As with all time-resolved spectroscopic approaches, the essential element of the set-up is a device to synchronise either the probe or the detection event with the initialisation of the process to be studied (Fig. 4.14). The method by which synchronisation is accomplished depends on the nature of the process. The largest time range can be covered for photoinduced reactions using an additional photolysis laser beam that triggers the reaction. For these processes the limit of the time resolution can be extended to the pico- and femtosecond range (Ondrias and

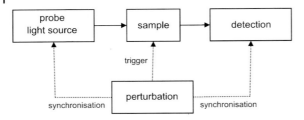

Fig. 4.14 Block diagram for the set-up of a time-resolved spectroscopic experiment. The perturbation that triggers the process to be studied is synchronised either with the probe light source or with the detection system.

Simpson 1996; McCamant et al. 2005). Processes that are induced by addition of a reactant can be studied down to a time-resolution limited by the time required for mixing the reaction partners, which at best may be achieved within tens of microseconds (Cherepanov and de Vries 2004; Takahashi et al. 1995). Limits for studying relaxation processes induced by a rapid change of the temperature or the electrode potential lie in the nano- and microsecond range (Murgida and Hildebrandt, 2001; Gruebele et al. 1998).

In most time-resolved IR spectroscopic techniques (step-scan, rapid-scan), time-resolution is achieved via the detection system, which requires implementation of additional hardware components to the interferometer. For this reason, the principles of these techniques are described in Section 3.1.4. On the other hand, time-resolved Raman spectroscopic techniques can, in most instances, be employed with the standard spectrometer and detection systems used for stationary experiments. Alternatively, time-resolution is frequently achieved by gating the Raman probe beam or by flowing samples. These approaches are largely restricted to RR rather than non-resonance Raman spectroscopy as higher demands are imposed on the sensitivity of the spectroscopic detection. Although the major part of this section refers to time-resolved RR spectroscopic techniques, some of the methodological approaches for initiating and probe biomolecular processes can also be combined with time-resolved IR spectroscopic detection, such as rapid mixing techniques.

4.4.1
Pump–Probe Resonance Raman Experiments

In pump–probe experiments, which are restricted to proteins containing photosensitive co-factors, a photolysis and a probe laser are required, which hit the sample in a well-defined temporal correlation. The photolysis beam initiates the reaction that can either be a photochemical process of the system under consideration or the photoinduced release of a reactant that subsequently initiates a chemical reaction. In these experiments, the probe beam is chosen to be in resonance with the cofactor such that it exclusively probes the vibrational spectrum of the cofactor.

Pump–probe experiments can be carried out with cw- and pulsed-lasers depending on the desired time-resolution. Whereas with cw-laser experiments the time-resolution is largely restricted to the micro- and millisecond range and, only under special conditions, can reach the long nanosecond time scale, the time-resolution of pulsed-laser experiments can be extended down to the short picosecond and even to the femtosecond time scale.

4.4.1.1 Continuous-wave Excitation

In cw-pump–probe experiments, time-resolution is usually achieved by the flow of the sample first passing the pump and subsequently the probe laser beam. The design of such an experiment is depicted schematically in Fig. 4.15 and is applicable for studying a photoinduced process that starts with a photochemical reaction and then runs through a sequence of thermal relaxation steps. Typical examples are photoinduced processes of photoreceptors, such as bacteriorhodopsin (Althaus et al. 1995). The first event is the irradiation of a volume element of the sample by the pump beam, which initiates the reaction sequence. The energy flux for the pump laser and the residence time of the sample in the laser beam (Δt_p) constitute the photoconversion parameter, which determines the fraction of the molecules undergoing the photoinduced reaction sequence [Eq. (4.1)]. This photolysed volume element then passes the probe beam for the RR spectra to be measured. The time required for the volume element to flow from the

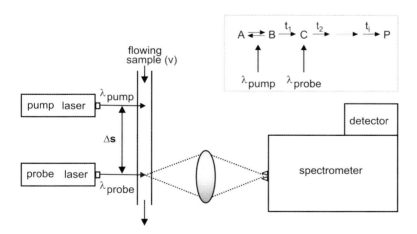

Fig. 4.15 Pump–probe flow system for time-resolved RR spectroscopy with cw-laser excitation. The delay time between the pump and the probe event is controlled by the spatial separation Δs of the pump and probe beams. The approach may be applied to probe the intermediates formed during a photoinduced reaction sequence such as that shown in the inset of the figure. Then the delay time may be adjusted according to the lifetimes t_i of the individual species.

pump to the probe beam is the delay time, which depends on the flow rate v and the spatial separation of both laser beams Δs

$$\delta = \frac{v}{\Delta s} \tag{4.24}$$

To probe the RR spectra of the intermediates a high degree of photoconversion is of course advantageous and, therefore, the photochemical rate constant at the pump wavelength $l_0(\lambda_p)$ [Eq. (4.2)] and Δt_p should be set as high as possible. Natural upper limits for these quantities are given by the photochemical and kinetic properties of the photoactive system to be studied. Specifically, care has to be taken to avoid secondary photoreactions that may occur if, during Δt_p, intermediates are formed that under the action of the photolysis beam can undergo a photoinduced branching reaction. On the other hand, the probability of sequential (or even simultaneous) two-photon absorption processes or photoinduced damage of the molecules is rather low in cw-pump–probe experiments with a flowing sample device.

In contrast to the photolysis beam, the energy flux and the residence time of the sample in the probe beam (λ_r) should be reduced as much as possible to establish small photoconversion parameters $[\Delta t_r \cdot l_0(\lambda_r) < 0.05]$. Higher photoconversion parameters would cause unwanted perturbations to the kinetics to be studied. The residence time of a flowing sample in the probe laser beam is given by

$$\Delta t_r = \frac{2 \cdot r_r}{v} \tag{4.25}$$

where r_r is the radius of the probe laser beam. Combining Eq. (4.25) with the expression for $l_0(\lambda_r)$ (Eq. 4.2) one obtains

$$\Delta t_r \cdot l_0(\lambda_r) = 9.62 \cdot 10^{-10} \left(\frac{\gamma \cdot \varepsilon_r \cdot P_r \cdot \lambda_r}{v \cdot r_r} \right) \tag{4.26}$$

and the corresponding expression holds for the photolysis beam. The flow rate v is the same in the expressions for the pump and the probe beams and also the quantum yield γ can be considered to be constant and wavelength independent in a first approximation.

To achieve the desired values for the photoconversion parameters of the pump and probe beam of ca. 0.7 and 0.05, respectively, one, therefore, has to vary the wavelength, focus, and power of the pump and probe laser beams. However, these variables are not freely adjustable. To ensure that only photolysed sample is irradiated by the probe beam, r_p has to be set ca. four-times larger than r_r. Furthermore, the variability of the pump and probe wavelengths is limited. Whereas λ_p is selected to provide efficient excitation conditions for the parent state, corresponding to a maximum value for ε_p, λ_r is usually chosen to achieve optimum resonance enhancement to probe the intermediate of interest. Consequently, ε_r is high for this intermediate but possibly low for the parent state.

As an illustrative example we will consider a photoactive system with an extinction coefficient of 30 000 L mol^{-1} cm^{-1} at 514.5 nm (λ_p) and a quantum yield of 0.5. For a flow rate of 10 m s^{-1}, a radius of the photolysis beam focus of 100 μm, and a laser power of 0.1 W, the photoconversion parameter $\Delta t_p \cdot l_0(\lambda_p)$ is calculated to be ca. 0.75, corresponding to a degree of photoconversion of ca. 50% (if a photochemical back reaction is neglected). With a probe beam at 413 nm that is in resonance with the absorption of an intermediate ($\varepsilon_r = 30\,000$ L mol^{-1} cm^{-1}) a value for photoconversion parameter of ca. 0.075 is obtained using $r_r = 20$ μm and $P_r = 0.0025$ W. Under these conditions, photoconversion of the intermediate is less than 10% assuming a quantum yield of 0.5. The degree of photoconversion of the unphotolysed parent state in the probe beam is even lower due to the less favourable excitation conditions. These conditions are acceptable as the perturbation of the photoinduced reaction sequence by the probe beam remains relatively small.

Now we will consider the relationship between δ and Δt, which ultimately refers to the time-resolution of the experiment and thus the individual reaction steps that can be resolved. The reaction sequence is initiated when the sample is in the pump beam for the residence time Δt_p. The probe beam hits the sample after a delay time δ and a spectrum is taken during the residence time of the sample in the probe beam Δt_r. Variation of δ then affords RR spectra with different contributions of the various parent and intermediate states. These spectra can be quantitatively analysed in terms of spectral contributions of the individual states and thus allow the analysis of the kinetics of the processes. A principal restriction, however, arises from the fact that the RR experiment can only distinguish between states that differ by the cofactor structure. Reaction steps in which the cofactor structure remains unchanged but involve protein structural changes are RR-spectroscopically silent and can only be detected by transient IR spectroscopic techniques.

The individual RR experiments do not refer to a "point" on the time axis but to an interval, which, hence, introduces an intrinsic uncertainty for kinetic analyses. Furthermore, the time-resolution is not only controlled by the delay time between the pump and probe event but also by Δt_r. For these reasons it is important to choose Δt_r as small as possible, which requires a tight focussing of the probe beam (small r_r) and a high flow rate v. One may consider a beam geometry at which the pump and probe foci just touch each other as the lower limit of the delay time according to

$$\frac{r_p + r_r}{v} = \delta \qquad (4.27)$$

where r_p is the radius of the pump beam focus.

Two approaches have been established to provide a flowing sample device, a capillary flow system and a rotating cell. A capillary flow system driven by a peri-

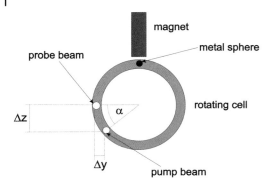

Fig. 4.16 Top view of a rotating cell in a pump–probe RR experiment. The time required for a sample volume element passing from the pump to the probe beam can be evaluated from the rotational frequency and the diameter of the cell, and the spatial displacement of the laser beams defined by Δy, Δz, and α (Naumann et al. 2006). A metal sphere may be kept at a fixed position downstream in the rotating cell via a magnet for efficient mixing of photolysed and unphotolysed sample.

staltic pump offers the advantage that a large reservoir can be used to afford flow rates up to 20 m s^{-1}, which according to Eq. (4.27) corresponds to an upper limit of the delay time of 5 µs assuming pump and probe beam foci of 80 and 20 µm, respectively. However, the actual limit may be even lower as such high flow rates are associated with extreme shear forces that may cause denaturation of the protein.

In contrast to the capillary flow system, which is also applicable to irreversible photoinduced processes, the rotating cell arrangement is restricted to reversible photoinduced reaction sequences (photocycles). In a rotating cell, the "flow rate" of the sample is given by the rotational frequency of the cell v_0, which usually cannot exceed 70 s^{-1} (Fig. 4.16) (Althaus et al. 1995). For a cell with a radius r_c of 1 cm, the same pump and probe beam radii as in the example above would then correspond to a delay time of ca. 9 µs according to

$$\frac{r_p + r_r}{2 \cdot \pi \cdot r_c \cdot v_0} = \delta \tag{4.28}$$

However, there is an additional constraint that may demand distinctly slower flow rates. Neglecting lateral diffusion of the protein, the same volume element that had been photolysed re-enters the laser beams after one rotation of the cell. To avoid quasi-photostationary mixtures with poorly defined distributions among the various states of the protein, the "fresh sample" condition must be fulfilled, implying that the rotational period must be sufficiently long to ensure the full recovery of the original parent state before the sample is photolysed again. This condition is usually fulfilled if the rotational period is ca. four-times longer than the slowest step in the reaction cycle. Then the degree of recovery is more than 98% and accumulation of long-lived intermediates during the subsequent pump–probe sequences is negligible. If the rate-limiting step for the recovery of the parent state occurs with a relaxation time of ca. 3.5 ms, a rotational frequency of 70 Hz is acceptable. The extension to a four-times larger recovery time is possible if the solution is thoroughly mixed prior to re-entering the photolysis beam. This can be achieved by fixing a "mixing metal sphere" via an external magnet downstream of the laser beams in the rotating cuvette (Fig. 4.16).

The restrictions imposed by the "fresh sample" condition are less severe for a capillary flow system as it can be linked to a sufficiently large sample reservoir allowing for the recovery of the system. Both approaches imply a trade-off between time-resolution, recovery time, and the amount of sample available.

A promising alternative is based on the use of electro-optical shutters to gate the photolysis and the probe beam (Fig. 4.17) (Naumann et al. 2006). This approach offers the advantage of covering a large dynamic range of time-resolution of six decades or even more without increasing the sample volume above 0.5 mL, the

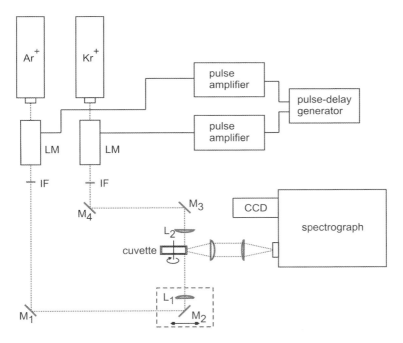

Fig. 4.17 Set-up for time-resolved pump–probe RR experiments with gated cw-excitation (Ar^+ and Kr^+ lasers). Gating is achieved by means of laser intensity modulators (LM) that are synchronised via a pulse-delay generator (sequence of pulses is shown in Fig. 4.18). The pump and probe beams are collinearly focussed into the sample contained in a rotating cuvette. Further optical components include interference filters (IF), mirrors (M_1, M_2, M_3, M_4) and lenses (L_1, L_2).

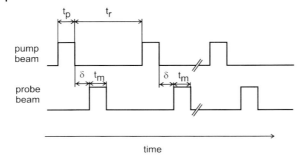

Fig. 4.18 Sequences of pump and probe events in time-resolved two-colour RR experiments using gated cw-excitation as depicted in Fig. 4.17. The pump and probe events are denoted by t_p and t_m whereas t_r refers to the recovery time of the sample (Naumann et al. 2006).

typical content of a rotating cuvette. In this instance, the residence times for the sample in the pump and probe beams, t_p and t_m, and the period of time allocated to the recovery of the sample, t_r, is controlled by gating the laser beams via Pockels cells that are triggered and synchronised by a pulse delay generator (Fig. 4.18). Particularly short time-resolutions down to the long nanosecond time range can then be achieved by focussing the pump and probe beams at the same spot and the sequence of correlated pump and probe pulses is repeated until a rotation of the cell is completed. Then both laser beams are blocked for the full recovery of the sample. It is clear that due to the gating the number of probing photons is reduced, resulting in a correspondingly increased measuring time.

The same alignment of the pump and probe beams is also used for measurements at long time-resolutions. However, now the lengths of the pump and the measurement pulses are set equal to the rotational period of the cell, such that a ring-shaped volume element is first illuminated and subsequently probed. The upper limit of the delay time depends on the diffusion coefficient of the protein D and must be chosen to be distinctly shorter than $(r_p)^2/(2D)$ to avoid mixing of photolysed and unphotolysed sample via lateral diffusion prior to the probe event.

4.4.1.2 Pulsed-laser Excitation

Pump–probe experiments with pulsed-laser excitation follow similar principles as those with gated cw-excitation. In both examples the photochemical conversion is given by Eqs. (4.1 and 4.2) where Δt is defined by the duration of the pulse. However, synchronisation of the pump and probe pulses can be achieved in different ways (Ondrias and Simpson 1996).

If two independent lasers are used, the pump and probe pulses must be triggered through an external digital delay generator as described above (Naumann et al. 2006). This device is typically employed in experiments using short nanosecond pulses that can be provided by Nd:YAG or excimer lasers. In particular, Nd:YAG lasers represent a relatively cheap solution and are easy to handle. Because wavelengths in the visible or UV spectral region are usually required, the second, third, or fourth harmonic of the 1064-nm line are employed, for which pulse energies of between 1 and 10 mJ are readily obtained. These systems yield 7–10 ns pulses and repetition rates of 10–20 Hz. The pulse energies are high enough to produce Raman-shifted lines or to pump dye or Ti:sapphire lasers such that a wide range of pump and probe wavelengths are accessible. This allows the design of two-colour experiments adjusted to the properties of the photoactive system to be studied, i.e., with pump and probe wavelengths in resonance with the absorption bands of the parent state and the photoinduced product, respectively. Achieving a high degree of photoconversion is usually not difficult as it can be judged from a simple estimation. For a pulse energy of the pump beam of 0.1 mJ, a pulse width of 10 ns, $\lambda_p = 532$ nm with a beam radius of 100 µm, an extinction coefficient of 30 000 L mol^{-1} cm^{-1}, and a quantum yield for the photochemical process of 0.3, Eq. (4.2) yields $I_0 = 2.3 \cdot 10^{-9}$ s^{-1} and Eq. (4.1) a photoconversion parameter of 23, which ensures the maximum possible degree of photoconversion. To avoid photoinduced processes in the probe beam, however, it is desirable to attenuate the pulse energy by ca. two orders of magnitude, which, for a repetition rate of 10 Hz, corresponds to 10 µJ s^{-1}. Such a low energy flux is not sufficient to obtain RR spectra illustrating the dilemma of pulsed RR experiments, i.e., to find a balance between an acceptable S/N ratio and avoiding unwanted photoinduced processes. Specifically, when applying the technique to biological systems, flowing sample devices are mandatory. Owing to the typically low repetition rate, the condition that a volume element of the sample is not irradiated by two consecutive pulses is easily fulfilled.

In the picosecond time range, pump–probe experiments are based on a single excitation source, which is usually a mode-locked Ar ion or (frequency-doubled) Nd:YAG laser. The output of this laser is divided by a beam splitter into appropriate portions to provide separate pulse trains that are either used directly for photolysing and probing the sample, or for generating appropriate pump and probe wavelengths through one or two dye or Ti:sapphire lasers. Thus, pump and probe pulses are, *per se*, synchronised as they originate from the same source. To achieve a well-defined temporal delay between the pump and probe pulses, delay lines are introduced into the set-up (Fig. 4.19). They are designed such that a probe pulse of a conjugate pump–probe pulse pair has to travel over a longer distance Δs than the pump pulse before it irradiates the sample. Then the delay time δ between pump and probe pulses is given by

$$\delta = \frac{\Delta s}{c} \tag{4.29}$$

where c is the speed of light. Thus, a delay time of 10 ps corresponds to a spatial difference of 0.3 cm, which illustrates the high demands of ps pump–probe experiments on the optical set-up. Mode-locked lasers operate with repetition rates in the MHz range and the pulse energy is much lower than for ns-pulsed lasers. Thus, the risk of unwanted photoinduced processes is significantly reduced but

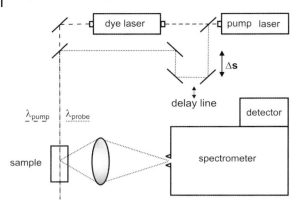

Fig. 4.19 Pump–probe set-up for time-resolved RR spectroscopy with pulsed-laser excitation. By means of a delay line the travelling time of the probe pulses (dotted line) from the laser to the sample is increased by 2Δs relative to that of the pump pulses (dashed line). Pump and probe pulses are focussed collinearly in the sample.

even at high flow rates repetitive irradiation of the sample by consecutive laser pulses is unavoidable due to the high repetition rate of the laser. Assuming a beam radius of 100 μm, the movement of the sample out of the beam requires ca. 20 μs, given a flow rate of 10 m s^{-1}. Within a typical repetition rate of more than 50 MHz, this volume element is irradiated by ca. 1000 pulses. Although pulses energies are lower, the consecutive pulses may lead to secondary photoprocesses. In addition to these difficulties, one has to keep in mind that narrow frequency bandwidths of the probe laser, as close as possible to the transform limit, are essential for RR applications in ps experiments.

Because of this constraint, the natural limit of classical pump–probe RR experiments is at a few ps. To gain information about ultrafast processes in the femtosecond range, nonlinear techniques have to be employed. A recently developed approach utilises the stimulated Raman effect for molecules in the electronically excited state (McCamant et al. 2005). Here the photoreceptor is excited with a sub-50-fs pulse of sufficient energy to produce a high population of the excited state. Subsequently, two pulses are combined with different spectral and temporal widths: a "long" ps pulse with a narrow bandwidth and a wavelength in resonance with the second excited state, and a broad bandwidth near-infrared 15-fs pulse that includes those spectral components appropriate for stimulated Raman emission, which then occurs with a temporal resolution much better than the picosecond Raman pump pulse. Thus, it is the spectral bandwidth of this picosecond pulse that determines the spectral resolution. The set-up of this approach is demanding and clearly not a routine approach for the study of biological systems. However, it represents one of the few methods to gain insight into ultrafast structural changes of chromophores following photoexcitation. The potential of this technique has been demonstrated in studying the sub-picosecond events in the visual receptor rhodopsin (Kukura et al. 2005). The high sensitivity allows a complete spectrum to be obtained of an organic solvent with a single picosecond pump, femtosecond probe–pulse pair (McCamant et al. 2003).

4.4.1.3 Photoinduced Processes with Caged Compounds

Pump–probe vibrational spectroscopy refers to biological photoreceptors and in this sense it is restricted to proteins carrying an intrinsic chromophore that, upon light absorption, undergoes molecular changes. *A priori*, other processes which consist of or are initiated by bimolecular reactions between an exogenous compound and, for example, a protein cannot be probed in this way. However, in some instances this limitation can be overcome by using so-called caged compounds. Here, the compound that starts the molecular process is protected by a chromophoric group, usually absorbing in the near-UV. Addition of this caged compound to the protein in the dark, therefore, does induce the reaction. When, however, the protecting chromophore is irradiated by an intense laser flash, a photochemical process destroys the "cage" and the reactive compound (e.g., a proton or an ATP molecule) is released into the solution, such that the reaction with the protein starts. Thus, the system is made artificially light-sensitive and the pump–probe technique can be applied.

There are two processes that may limit the time-resolution: the photolysis process itself and the diffusion of the released compound to the protein. The fastest start of a reaction is achieved if the caged compound is attached to the protein in close proximity to the catalytic centre. If the distance is sufficiently small, diffusion can be neglected and the time-resolution is largely controlled by the photochemical decomposition of the cage. Under these conditions, time-resolution down to the short microsecond time scale can be achieved.

There are several possibilities for encaging reactants. Previously, 2-nitrobenzyl derivatives have been employed (for a review see McCray and Trentham 1989). Owing to the conversion of the chromophoric cage into a by-product (2-nitrosobenzaldehyde) that is harmful to biological system, 1-(2-nitrophenyl)ethyl compounds have been used as a substitute (Kaplan et al. 1978). Both compounds have in common that the decaging process is relatively slow (1–10 ms). The recently developed *p*-hydroxyphenacyl cages circumvent both problems as the photolytic process is very fast (<1 μs) and the by-product has no known effect on biological systems (Park and Givens 1997). Caged compounds have become increasingly important, particularly in time-resolved IR spectroscopy, as will be demonstrated on the basis of selected examples in Section 6.4.

4.4.2
Rapid Mixing Techniques

The conventional approach for studying bimolecular reactions in solution is based on rapid mixing of the reactants. Rapid mixing techniques are well-established approaches and are frequently used to investigate the kinetics of enzymatic reactions, usually by monitoring characteristic UV–vis absorption

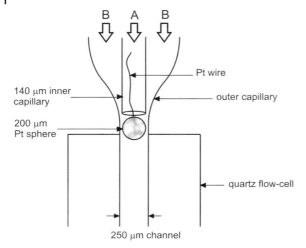

Fig. 4.20 Chamber for rapid mixing of two solutions A and B flowing with high velocity through an inner (A) and an outer capillary (B). Turbulent mixing is achieved via a Pt sphere. The mixed solution then enters the quartz flow cell for spectroscopic investigations (according to Takahashi et al. 1995).

changes. Special variants to this approach can also be combined with Raman and IR detection.

The central component of these techniques is a mixing chamber in which two solutions are rapidly mixed. Mixing chambers that are commercially available, usually as part of a complete set-up, are characterised by mixing times of a few milliseconds. Laboratory-built mixing chambers can be designed for shorter mixing times of tens of microseconds (Cherepanov and deVries 2004; Takahashi et al. 1995). Most of these are based on two concentrically arranged capillary tubes (Fig. 4.20). Both solutions are pressed synchronously, with an initial rate of ca. 10 m s^{-1}, into the two capillaries by pneumatically driven syringes. The solution passing through the inner capillary (140 µm diameter) hits a platinum bead that causes the flow streaming to merge with solution flowing through the outer capillary. In this way, highly efficient and fast mixing is achieved before the resultant solution leaves the mixing chamber for the observation cuvette. This device has been successfully employed for time-resolved RR spectroscopy (see Chapter 7). For rapid mixing/stopped flow experiments, a third syringe, mounted at the exit of the cuvette, is synchronised with the two mixing syringes to stop the sample flow abruptly for the spectroscopic monitoring of the kinetics.

For IR spectroscopic detection, specific restrictions, related to the strong absorption of water, have to be considered when designing the device. Firstly, highly concentrated protein solutions have to be used, which are very viscous. Thus, it is extremely difficult to press such solutions through capillaries and, moreover, through the observation cuvette, which, as a second requirement, should be very thin with an optical pathlength of only a few micrometers. These difficulties can at least be overcome partly by using ^2H_2O as the solvent. In this way, a cuvette thickness of 25–50 µm can be tolerated and, therefore, the concentration can be reduced by a factor of 5–10 as compared with an H_2O solution. For these conditions, a rapid-mixing device has been constructed (White et al. 1995) and further improved in several laboratories [e.g. (Fig. 4.21), Georg 1999]. With such a device, the practical dead-time is ca. 10 ms. The main difference as compared with the device used in UV–vis spectroscopy is that no stop-syringe is needed. Owing to the highly viscous solution and the small diameter of the cuvette the back-pressure is already sufficiently high to stop the flow of the sample. The mixed so-

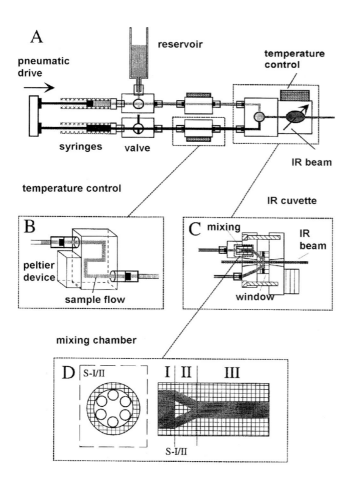

Fig. 4.21 Rapid mixing device for time-resolved IR spectroscopy. A, Complete set-up, syringes operated by pneumatic drive; B, temperature control, both branches can be controlled separately; C, IR cuvette with mixing chamber; D, mixing chamber (Georg 1999).

lution is transferred into the cuvette by a hole in one window and a second hole serves as the exit for the spill. The mixing chamber is mounted directly above the filling hole. As a large pressure is built up in the cuvette, the thickness of the windows has to be around 5 mm, i.e., larger as compared with a usual IR cuvette (e.g., a sandwich cuvette). Because of the high pressure, the two windows have to be clamped tightly together in order to avoid changes to the window spacing while the pressure is built up. Nevertheless, residual distortions contribute to the dead time of the device.

This mixing event determines the starting point of the reaction to be studied and also defines the limit for the time-resolution of the spectroscopic experiment. The reaction is then monitored either in the "stopped flow" mode, which is more favourable for IR detection (*vide supra*), or in the "rapid flow" mode, which is used in Raman spectroscopy (see Section 4.4.2.1). In addition, the mixing chamber can be linked with the "rapid freeze–quenching" technique, which has also been combined with Raman spectroscopy.

4.4.2.1 Rapid Flow

In the rapid flow or continuous flow mode, the solution leaves the mixing chamber and enters an optical flow cell. The probe beam is then positioned with a spatial separation Δs downstream from the mixing chamber corresponding to a delay time δ_s of

$$\delta_s = \frac{\Delta s}{v} \tag{4.30}$$

where v is the flow rate of the solution leaving the mixing chamber. The limit of the time resolution t_{lim} is then given by

$$t_{\text{lim}} = t_{\text{mix}} + \delta_{s,\text{min}} + \frac{1}{2} t_{\text{meas}} \tag{4.31}$$

where t_{mix} is the mixing time and $\delta_{s,\text{min}}$ is the minimum delay time corresponding to the closest possible approach of the probe beam with respect to the mixing chamber. The quantity t_{meas} depends on the diameter of the probe beam and is determined in analogy to Eq. (4.30). For a probe beam diameter of 100 µm and a flow rate of 40 m s^{-1}, t_{meas} is 2.5 µs and thus much shorter than t_{mix} (>15 µs) and $\delta_{s,\text{min}}$, which due to unavoidable technical constraints is likely to be ca. 30 µs. A serious disadvantage of the continuous flow technique is the high demand on the sample volume. Assuming a minimum acquisition time of 1 s, the total sample volume required for measuring a spectrum at a single kinetic data point is ca. 8 mL. Moreover, owing to the small optical pathlengths of the optical flow cell (ca. 250 µs), relatively high concentrations of the biomolecules are needed, which may be higher by a factor of 10 than in conventional Raman or RR experiments. Thus, it is not surprising that the rapid flow technique is restricted to samples that are available in large amounts. It has been employed in combination with RR detection and has also been implemented for time-resolved IR spectroscopy using a microfabricated cell with diffusive mixing of two very thin laminar flow streams (Kauffmann et al. 2001).

4.4.2.2 Rapid Freeze–Quench

The sample restrictions do not hold for the rapid freeze–quenching technique (Cherepanov and de Vries 2004; Oellerich et al. 2000). Here the solution leaves the mixer and, after a delay line of variable length (see Section 4.2.1), it is injected into cold liquid pentane at ca. $-120\,°C$ (Fig. 4.22). At such a low temperature, thermal reactions are usually blocked. Varying the length of delay line (i.e., δ_s) it is then possible to freeze the sample at various times along the reaction coordinate. The frozen snow-like sample is collected from the pentane bath and deposited into an appropriate sample holder that can be inserted into a cryostat for the Raman or RR spectroscopic measurements.

The technique offers the additional advantage that the same samples can subsequently be studied by other spectroscopic techniques, such as RR and electron paramagnetic resonance (EPR) spectroscopy, without any restriction on the measuring time. However, optical spectroscopies that have to be operated in the transmission mode are not applicable. The limit of the time-resolution in conventional set-ups

$$t_{\lim} = t_{\mathrm{mix}} + \delta_{s,\min} + t_{\mathrm{cool}} \qquad (4.32)$$

is controlled by the cooling time t_{cool} which is ca. 3–4 ms. Consequently, rapid quenching experiments are carried out with commercially available mixers that have mixing times in the short millisecond range. For specially designed set-ups

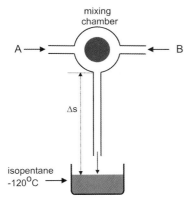

Fig. 4.22 Device for rapid freeze–quenching including a mixing chamber in which the solutions A and B are rapidly mixed. The combined solution is pressed through a delay line into liquid isopentane at $-120\,°C$.

that provide time-resolutions in the microsecond time scale the reader is referred to the literature (Cherepanov and de Vries 2004).

4.4.3
Relaxation Methods

Thermal reactions that are faster than the mixing time can be studied by means of relaxation techniques (Bernasconi 1976). In this approach, the equilibrated system constitutes the starting point. A rapid change of a parameter that perturbs the equilibrium is applied and the subsequent restoration of the equilibrium is monitored by time-resolved spectroscopic techniques. Perturbation of the equilibrium can be achieved by jump-like changes to the temperature, the pressure, the electric field, the electrode potential, or any other variable that controls the equilibrium. Let us consider a simple monomolecular reaction between two states A and B, e.g., an acid/base equilibrium. The time-dependent response of A is then given by

$$[A]_t = [A]_{t=0} \exp[-(k_f + k_b)t] \tag{4.33}$$

where $[A]_t$ is the concentration of A at time t after the perturbation event, $[A]_{t=0}$ is the initial equilibrium concentration of A before the perturbation, and k_f and k_b are the rate constants for the forward and back reactions, respectively. As these rate constants are related to each other through the equilibrium constant K_{eq}

$$K_{eq} = \frac{k_f}{k_b} \tag{4.34}$$

both rate constants are readily determined from a semi-logarithmic plot of the concentration changes of A monitored after the equilibrium perturbation. For monomolecular reactions, this expression holds independently of the magnitude of the equilibrium perturbation, whereas for bimolecular reactions the perturbation should be kept small to allow linearisation of the kinetic equations.

The relaxation method has been adapted to study potential-dependent processes of redox proteins immobilised on Ag or Au electrodes by SERR spectroscopy (Murgida and Hildebrandt 2001, 2005). In this approach, a rapid potential jump from the initial potential E_i to the final potential E_f is employed to perturb the equilibrium of the immobilised species (Fig. 4.23). To monitor the relaxation to the new equilibrium corresponding to E_f, the SERR spectrum is probed during the time interval Δt subsequent to a delay time δ with respect to the potential jump. This spectrum then reflects the actual composition of the species at δ' ($\cong \delta + \Delta t/2$) and so the systematic variation of δ allows the temporal evolution of the new equilibrium to be followed.

The experimental challenge of this approach is associated with the time scale of the processes to be probed. If they occur on the millisecond or microsecond time scale, a single shot spectrum with an accumulation time of Δt does not provide an

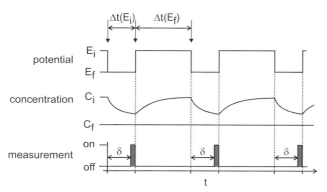

Fig. 4.23 Correlation of potential jump sequence and measuring intervals Δt (shaded areas). Following a potential jump from E_i to E_f the potential dependent equilibrium is perturbed. The relaxation process corresponds to a change in the concentration from c_i to c_f of the species involved. This process is probed after a delay time δ during the measuring interval Δt. Subsequently, the potential is reset to the original value for $\Delta t(E_i)$ until the initial equilibria are restored.

S/N ratio that is sufficient for a quantitative analysis of the spectrum as Δt should be distinctly smaller than the relaxation time. Hence, the electrode potential is reset to the initial value immediately after the measuring interval to restore the original equilibrium at E_i. The sequence of potential jumps and measuring intervals is then repeated n times until a satisfactory spectrum is obtained (Fig. 4.23). The most elegant technical solution is based on the detection by a confocal spectrometer (spectrograph) with a low-noise CCD detector and electro-optically gated cw-excitation (see Section 4.4.1.1). If the dark current is low enough, the total signal accumulation ($n \cdot \Delta t$) occurs on the chip without deterioration of the S/N ratio during the dark time, which is usually longer than Δt by more than an order of magnitude. The sequence of the individual events, i.e., potential jumps and gating of the laser beam, are synchronised by a pulse-delay generator. This approach is applicable for processes over a wide time scale and it is only limited by the response time of the electrochemical cell, which can be kept as low as a few microseconds. The crucial prerequisite, however, is that the processes under investigation are fully reversible. This is also true for monitoring the potential-jump induced processes by step-scan SEIRA spectroscopy (Murgida and Hildebrandt 2005).

In principle, the rapid perturbations of equilibria can also be achieved by sudden changes in other variables such as pressure, temperature, or the electric field. For biological systems in solutions, temperature jump experiments have been used to probe relaxation kinetics, such as the unfolding processes of proteins. Two approaches have been employed with reference to two different time regimes. On the one hand, electrically induced heating of the solution by more than 10° can readily be achieved within milliseconds, leading to relatively large concentration changes that can safely be detected by vibrational spectroscopy (Backmann et al. 1995). Shorter time scales require laser-induced temperature jumps that can be generated by laser pulses of high energy in the IR region (e.g., 1.9 µm) (Grübele et al. 1998; see Section 5.6). The wavelength is chosen to match vibrational overtones of the solvent (H_2O, ^2H_2O). The laser-induced temperature-jump method has been widely employed in combination with UV–vis absorption detection mainly for studying protein unfolding processes. More recently, it has also been extended to the experimentally more demanding detection by vibrational spectroscopies (Colley et al. 2000).

4.4.4
Spatially Resolved Vibrational Spectroscopy

In confocal Raman spectroscopy, the incident laser beam is tightly focussed by a microscope objective such that the Raman scattered light is collected from a small volume of the sample. As in light microscopy, the spatial resolution d is limited by the Rayleigh criterion

$$d = \frac{0.61}{n \cdot \sin \alpha} \lambda \tag{4.35}$$

where $n \cdot \sin \alpha$ is the numerical aperture and λ is the excitation wavelength. As numerical apertures of 1.0 can readily be achieved, Raman spectra can be measured with spatial resolutions of ca.1 µm. Upon moving the sample through the laser focus in step widths that correspond to the optical resolution, Raman images of heterogeneous samples can be obtained thereby simultaneously providing molecular structure (chemical composition) and microscopic structure information (Colarusso et al. 2000). Raman microscopy, which is a standard feature of modern confocal Raman spectrometers, may be particularly useful for studying complex biological systems such as cells or membranes.

Higher spatial resolutions that go beyond the Rayleigh limit [Eq. (4.30)] are accessible by scanning optical near-field microscopy (SNOM). In this technique, the sample is irradiated through an optical fibre with an aperture that is smaller than the wavelength. When, in addition, the distance of the aperture to the sample is also smaller than the wavelength, the light interacts with the sample prior to diffraction. The response of the illuminated spot, which in the near-field corresponds to the size of the aperture, can then be detected in the reflection mode through the same fibre. In analogy to scanning probe microscopies, the fibre tip is scanned across the sample such that sequential nanometre-sized spots of the sample are probed. This technique has been widely used by monitoring the fluorescence of the illuminated molecules but has also been adapted to the detection of IR absorption and Raman scattering (Rasmussen and Deckert 2005; Futamata and Bruckbauer 2001). However, the price for the increased spatial resolution is the decreased sensitivity. As the sample is probed only in the near-field, the number of molecules contributing to the Raman scattering is drastically reduced. In addition, the number of probe photons passing through the capillary is reduced by several orders of magnitude. Thus, the application of SNOM vibrational spectroscopy in life sciences is as yet limited to a few examples, although there is no doubt about the potential of this technique for studying processes in cells. This is also true for tip-enhanced Raman spectroscopy (see Section 4.3.7), which can be considered as apertureless SNOM.

4.5
Analysis of Spectra

Vibrational spectra of biological systems are usually fairly complicated. Specifically, biopolymers but also large cofactors exhibit a high density of vibrational modes per wavenumber increment, such that individual bands may overlap and lead to broad asymmetric features in the spectra. The complexity of the spectra is further increased if more than one molecular species is present. In addition, non-structured background may interfere with the vibrational bands. Thus, a reliable and reproducible procedure is required for the analysis of the spectra.

The starting point for the analysis of spectra is the measured spectra that are accumulated up to a particular S/N ratio. If required, the quality of the spectra should be improved by prolonged accumulation times and/or combining individual spectra measured sequentially. The addition of separately measured spectra is, however, only justified if time-dependent changes during the series of measurements can be ruled out. This condition can readily be checked by mutual subtraction of the individual spectra that should yield only noise but no difference signals. Prior to further data treatment, the spectra may be corrected for the instrumental response, such as different sensitivities of CCD elements and the wavelength-dependence of the grating reflectivity and detector sensitivity. However, it is not advisable to employ smoothing procedures because they result in a loss of information and *appear to* but do not produce higher quality spectra.

The second step refers to the subtraction of the background, which, for instance, may result from fluorescence in the case of Raman spectroscopy. In spectra covering only small spectral regions (<300 cm^{-1}) this background can be frequently considered as linear but in general a polynomial subtraction is necessary. Such a subtraction, however, is associated with some arbitrariness and may lead to errors in the relative band intensities and even in the frequencies, specifically of bands at the edges of the spectrum that is recorded. This uncertainty may cause problems when comparing spectra of series of measurements from different samples. In this instance, an alternative approach for background subtraction can be employed using singular value decomposition (SVD) methods (*vide infra*).

In some situations, bands of the solvent may contribute to the spectrum. For biological samples, these contributions are often very low as, unlike with IR, Raman bands of water, the usual solvent, are rather weak in the region below 2000 cm^{-1}. Raman bands of the buffer have to be considered if it includes oxoanions such as phosphate or sulfate because these anions give rise to relatively strong Raman bands below 1000 cm^{-1}. For absolute FTIR spectra, subtraction of the strong bands of water is more critical. However, in most applications, FTIR spectroscopy is operated in the difference mode, such that background solvent absorption bands are cancelled out, given that the water content in the sample is unchanged during the experiment.

Once "pure" spectra of the biological systems are obtained, various methods can be employed for extracting further information. For structurally and chemically pure samples, a band fitting procedure can substantially support the identification of the spectral parameters (frequencies, relative intensities, and bandwidths) of the individual vibrational modes. Such band fitting programs are commercially available and often implemented in the data acquisition software of the spectrometers. In general, band fitting requires a physically meaningful input by the operator. As a rule of thumb, the fitting program can only reliably determine band components in overlapping structures if visual inspection has already identified shoulders or asymmetries in the band envelope. Determining the second derivatives of the spectra can support the determination of the number of band components involved. Furthermore, band fitting routines seek for the mathematically best solution, which is not necessarily a physically meaningful one. For instance, bandwidths of fitted peaks larger than 20 cm^{-1} indicate the overlap of several closely spaced bands as typical bandwidths are between 10 and 15 cm^{-1}.

In general, individual bands exhibit a Lorentzian band shape if the spectral resolution is smaller than the intrinsic bandwidth. Otherwise the band profile is modulated by the instrumental function. The perturbations of the band profiles is more significant in FT Raman and FTIR spectra due to the limited integration in the Fourier transformation and the folding with the apodisation function (see Chapter 3, Box 3A). Here, the resultant band profiles are at best simulated by mixed Lorentzian–Gaussian functions.

The analysis of spectra that include contributions from various chemically or structurally different species is more demanding and conventional band fitting frequently leads to ambiguous results. In those instances where some of the species that are involved are known and their spectra can be obtained in a pure form, the component analysis is applicable (Döpner et al. 1996). In this method a series of experimental spectra is simulated by the superposition of complete spectra of the known components and a number of individual bands constituting initial ("guess") component spectra of the unknown species. In an iterative fitting procedure, the spectral parameters of these latter bands are refined to obtain a set of component spectra that allows a satisfactory global fit to all experimental spectra. This procedure is reliable provided the experimental spectra set is large and the number of unknown species is small. If all component spectra are known or eventually confidently determined, the quantitative analysis of the experimental procedure is very precise as the only fitting parameters are the amplitudes of the individual components.

In contrast to the component analysis, which is based on self-written routines, various commercially available software packages offer multivariate analysis methods such as singular value decomposition (SVD) (Henry and Hofrichter 1992; Hendler and Shrager 1994). These techniques are purely mathematical procedures and thus reduce the "intuition" factor that may cause ambiguities in fitting procedures. The formalism does not consider individual bands or component spectra but takes into account that at each of the m wavenumber increments of the experimental spectrum the intensity originates from all species albeit with different and *a priori* unknown contributions. If n experimental spectra with different compositions are available, all data can be arranged in an $n \times m$ matrix A, such that each column contains the information about the contribution of all species to the respective experimental spectra. The rows represent the contributions of each species to the intensity at the respective wavenumber increment. The objective of SVD algorithms is now to decompose this matrix A in terms of a matrix U containing spectral components that are sufficient to describe all experimental spectra, and a matrix V including the information on how these spectral components vary in the series of measurements.

$$A = USV^T \tag{4.36}$$

The matrix S is a square and diagonal matrix including the weights for the individual spectral components and aids to distinguish between spectral components (large weights) and noise (small weights). In principle, SVD allows the determination of the number of components involved albeit not necessarily the individual spectra. An additional side product of the SVD analysis is that the background also represents a spectral component. Thus, employing SVD analysis prior to the raw spectra may provide a reliable method for background subtraction (*vide supra*).

References

Althaus, T., Eisfeld, W., Lohrmann, R., Stockburger, M., **1995**, "Application of Raman spectroscopy to retinal proteins", *Isr. J. Chem.* **35**, 227–251.

Ataka, K., Heberle, J., **2003**, "Electrochemically induced surface-enhanced infrared difference absorption (SEIDA) spectroscopy of a protein monolayer", *J. Am. Chem. Soc.* **125**, 4986–4987.

Backmann, J., Fabian, H., Naumann, D., **1995**, "Temperature-jump induced refolding of ribonuclease A: a time-resolved FTIR spectroscopic study", *FEBS Lett.* **364**, 175–178.

Baymann, F., Moss, D. A., Mäntele, W., **1991**, "An electrochemical assay for the characterisation of redox proteins from biological electron transfer chains", *Anal. Biochem.* **199**, 269–274.

Benecky, M., Li, T. Y., Schmidt, J., Frerman, F., Watters, K. L., McFarland, J., **1979**, "Resonance Raman study of flavins and the flavoprotein fatty acyl coenzyme A dehydrogenase", *Biochemistry* **18**, 3471–3476.

Bernasconi, C. F., **1976**, "*Relaxation Kinetics*", Academic Press, New York.

Bockris, J. O'M., Reddy, A. K. N., Gamboa-Aldeco, M., **2000**, "*Modern Electrochemistry 2A*", Kluwer Academic, New York.

Bonifacio, A., Millo, D., Gooijer, C., Boegschoten, R., van der Zwan, G., **2004**, "Linearly moving low-volume spectro-electrochemical cell for microliter-scale surface-enhanced resonance Raman spectroscopy of heme proteins", *Anal. Chem.* **76**, 1529–1531.

Braiman, M., Mathies, R. A., **1982**, "Resonance Raman spectra of bacteriorhodopsin's primary photoproduct – evidence for a distorted 13-*cis* retinal chromophore", *Proc. Natl. Acad. USA* **79**, 403–407.

Cherepanov, A. V., de Vries, S., **2004**, "Microsecond freeze-hyperquenching: development of a new ultrafast micromixing and sampling technology and application to enzyme catalysis", *Biochim. Biophys. Acta* **1656**, 1–31.

Colarusso, P., Kidder, L. H., Levin, I. W., Lewis, E. N., **2000**, "Raman and infrared microspectrosocpy", in *Encyclopedia of Spectroscopy and Spectrometry*, Lindon, J. C., Tranter, G. E., Holmes, J. L., (Eds.), Academic Press, New York, pp. 1945–1954.

Colley, C. S., Clark, I. P., Griffiths-Jones, S. R., George, M. W., Searle, M. S., **2000**, "Steady state and time-resolved IR spectroscopy of the native and unfolded states of bovine ubiquitin: protein stability and temperature-jump kinetic measurements of protein folding at low pH", *Chem. Commun.* 1493–1494.

Cotton, T. M., Schultz, S. G., Vanduyne, R. P., **1980**, "Surface-enhanced resonance Raman scattering from cytochrome *c* and myoglobin adsorbed on a silver electrode", *J. Am. Chem. Soc.* **102**, 7960–7962.

Döpner, S., Hildebrandt, P., Mauk, A. G., Lenk, H., Stempfle, W., **1996**, "Analysis of vibrational spectra of multicomponent systems. Application to a resonance Raman spectroscopic study of cytochrome *c*", *Spectrochim. Acta, Part A* **52**, 573–584.

Eng, J. F., Czernuszewicz, R. S., Spiro, T. G., **1985**, "Raman difference spectroscopy via backscattering from a spinning tube and from a low-temperature tuning fork", *J. Raman Spectrosc.* **16**, 432–437.

Futamata, M., Bruckbauer, A., **2001**, "ATR-SNOM-Raman spectroscopy", *Chem. Phys. Lett.* **341**, 425–430.

Georg, H., **1999**, "*Untersuchungen zum Faltungsmechanismus der Ribonuklease A mittels Erzeugung eines Temperatursprungs durch schnelles Mischen*", Thesis, Faculty of Chemistry and Pharmacy, Albert-Ludwigs-Universität Freiburg.

Goormaghtigh, E., Raussens, V., Ruysschaert, J.-M., **1999**, "Attenuated total reflection infrared spectroscopy of proteins and lipids in biological membranes", *Biochim. Biophys. Acta* **1422**, 105–185.

Grübele, M., Sabelko, J., Ballew, R., Ervin, J., **1998**, "Laser temperature jump induced protein refolding", *Acc. Chem. Res.* **31**, 699–707.

Harrick, N. J., **1987**, "*Internal Reflection Spectroscopy*", Harrick Scientific Corporation, Ossining, NY, USA.

Heinze, J. **1984**, "*Cyclovoltametrie, die 'Spektroskopie' des Elektrochemikers*", Verlag Chemie GmbH, Weinheim.

Hendler, R. W., Shrager, R. I., **1994**, "Deconvolutions based on singular value decomposition and the pseudoinverse: a guide for beginners", *J. Biochem. Biophys. Methods* **28**, 1–33.

Henglein, A., Lilie, J., **1981**, "Storage of electrons in aqueous solution – the rates of chemical charging and discharging the colloidal silver microelectrode", *J. Am. Chem. Soc.* **103**, 1059–1066.

Henry, E. R., Hofrichter, J., **1992**, "Singular value decomposition: application to analysis of experimental data", *Methods Enzymol.* **210**, 129–193.

Hildebrandt, P., Stockburger, M., **1984**, "Surface enhanced resonance Raman spectroscopy of Rhodamine 6G adsorbed on colloidal silver", *J. Phys. Chem.* **88**, 5935–5944.

Hildebrandt, P., Stockburger, M., **1986**, "Surface enhanced resonance Raman spectroscopy of cytochrome *c* at room and low temperatures", *J. Phys. Chem.* **90**, 6017–6024.

Kaplan, J. H., Forbush, B., Hoffmann, J. F., **1978**, "Rapid photolytic release of adenosine 5′-triphosphate from a protected analogue: utilization by the Na:K pump of human red blood cell ghosts", *Biochemistry* **17**, 1929–1935.

Kauffmann, E., Darnton, N. C., Austin, R. H., Batt, C., Gerwert, K., **2001**, "Lifetime of intermediates in the β-sheet to α-helix transition of β-lactoglobulin by using a diffusional IR mixer", *Proc. Natl. Acad. Sci. USA* **98**, 6646–6649.

Kerker, M., Wang, D. S., Chew, S., **1980**, "Surface enhanced Raman scattering (SERS) by molecules adsorbed on spherical particles", *Appl. Opt.* **19**, 4159–4273.

Kiefer, W., **1973**, "Raman difference spectroscopy with rotating cell", *Appl. Spectrosc.* **27**, 253–257.

Kneipp, K., Moskovits, M., Kneipp, H. (Eds.), **2006**, "Surface-enhanced Raman scattering:

physics and applications", *Top. Appl. Phys.* **103**, Springer, Berlin.

Kukura, P., McCamant, D. W., Yoon, S., Wandschneider, D. B., Mathies, R. A., **2005**, "Structural observation of the primary isomerization in vision with femtosecond-stimulated Raman", *Science* **310**, 1006–1009.

Matousek, P., Towrie, M., Ma, C., Kwok, W. M., Phillips, D., Toner, W. T., Parker, A. W., **2001**, "Fluorescence suppression in resonance Raman spectroscopy using a high-performance picosecond Kerr gate", *J. Raman Spectrosc.* **32**, 983–988.

McCamant, D. W., Kukura, P., Mathies, R. A., **2003**, "Femtosecond broadband stimulated Raman: a new approach for high-performance vibrational spectroscopy", *Appl. Spectrosc.* **57**, 1317–1323.

McCamant, D. W., Kukura, P., Mathies, R. A., **2005**, "Femtosecond stimulated Raman study of excited-state evolution in bacteriorhodopsin", *J. Phys. Chem. B*, **109**, 10449–10457.

McCray, J. A., Trentham, D. R., **1989**, "Properties and uses of caged compounds", *Annu. Rev. Biophys. Biophys. Chem.* **18**, 239–270.

Murgida, D. H., Hildebrandt, P., **2001**, "Proton coupled electron transfer in cytochrome c", *J. Am. Chem. Soc.* **123**, 4062–4068.

Murgida, D. H., Hildebrandt, P., **2005**, "Redox and redox-coupled processes of heme proteins and enzymes at electrochemical interfaces", *Phys. Chem. Chem. Phys.* **7**, 3773–3784.

Naumann, H., Engelhard, M., Murgida, D. H., Hildebrandt, P., **2006**, "Time-resolved resonance Raman spectroscopy of sensory Rhodopsin II from *Natronobacterium pharaonis*", *J. Raman Spectrosc.* **37**, 436–441.

Oellerich, S., Bill, E., Hildebrandt, P., **2000**, "Freeze-quench resonance Raman and electron paramagnetic resonance spectroscopy for studying enzyme kinetics. Application to azide binding to myoglobin", *Appl. Spectrosc.* **54**, 1480–1484.

Ondrias, M. R., Simpson, M. C., **1996**, "Time-resolved resonance Raman spectroscopy", in *Modern Techniques in Raman Spectroscopy*, Laserna, J. J., (Ed.), Wiley, New York, pp. 343–386.

Park, C.-H., Givens, R. S., **1997**, "New photoactivated protecting groups. 6. p-hydroxyphenacyl: a phototrigger for chemical and biochemical probes", *J. Am. Chem. Soc.* **119**, 2453–2463.

Rasmussen, A., Deckert, V., **2005**, "New dimension in nano-imaging: breaking through the diffraction limit with scanning near-field optical microscopy", *Anal. Bioanal. Chem.* **381**, 165–172.

Rich, P. R., Breton, J., **2002**, "Attenuated total reflection Fourier transform infrared studies of redox changes in bovine cytochrome c oxidase: resolution of the redox Fourier transform infrared difference spectrum of heme a_3", *Biochemistry* **41**, 967–973.

Schneider, G., Diller, R., Stockburger, M., **1989**, "Photochemical quantum yield of bacteriorhodopsin from resonance Raman scattering as a probe for photolysis", *Chem. Phys.* **131**, 17–29.

Shreve, A. P., Cherepy, N. J., Mathies, R. A., **1992**, "Effective rejection of fluorescence interference in Raman spectroscopy using a shifted excitation difference technique", *Appl. Spectrosc.* **46**, 707–711.

Smith, C. P., White, H. S., **1992**, "Theory of the interfacial potential distribution and reversible voltammetric response of electrodes coated with electroactive molecular films", *Anal. Chem.* **64**, 2398–2405.

Takahashi, S., Ching, Y.-C., Wang, J., Rousseau, D. L., **1995**, "Microsecond generation of oxygen-bound cytochrome c oxidase by rapid solution mixing", *J. Biol. Chem.* **270**, 8405–8407.

Turro, N. J., **1991**, "*Modern Molecular Photochemistry*", University Science Books, New York.

Ulman, A., **1996**, "Formation and structure of self-assembled monolayers", *Chem. Rev.* **96**, 1533–1554.

Vogel, R., Siebert, F., **2003**, "New insights from FTIR spectroscopy into molecular properties and activation mechanisms of the visual pigment Rhodopsin", *Biopolymers (Biospectrosc.)* **72**, 133–148.

White, A. J., Drabble, K., Wharton, C. W., **1995**, "A stopped-flow apparatus for infrared spectroscopy of aqueous solutions", *Biochem. J.* **306**, 843–849.

5
Structural Studies

5.1
Basic Considerations

To date structural studies using vibrational spectroscopy have mainly been restricted to the determination of the secondary structure, which is given by the geometry of the peptide backbone of the protein. This backbone is formed by the peptide bond linking two amino acids, and the two adjacent C–C and N–C bonds. The peptide bond itself is basically fixed geometrically, as the C–N bond has partial double bond character and, therefore, the peptide group, i.e., OC–NH, is essentially planar. It usually exhibits a *trans* geometry, i.e., the oxygen and hydrogen are in *trans* positions in the peptide bond. In Fig. 5.1 a section of a backbone structure is shown. Because the peptide bond geometry is fixed, variations in the

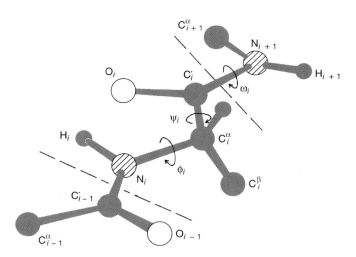

Fig. 5.1 Short stretch of a peptide chain indicating the peptide bonds and the dihedral angles Φ and Ψ. ω is the angle of the peptide bond, which is 180° or 0° for the *trans*- and *cis*-peptide geometries.

Vibrational Spectroscopy in Life Science. Friedrich Siebert and Peter Hildebrandt
Copyright © 2008 WILEY-VCH Verlag GmbH & Co. KGaA, Weinheim
ISBN: 978-3-527-40506-0

geometry can only be achieved by variations in the dihedral angles ϕ and ψ. Taking into account the steric hindrance as a constraint, basic structures can be derived: the right-handed α-helix, the β-sheet structure, and the unordered or random coil structure. These are the secondary structural elements, which can be addressed by vibrational spectroscopy. There are several additional secondary structural components, which are, in some instances, also amenable to the particular techniques of vibrational spectroscopy, and we will discuss some of these techniques later. An excellent reference to the general architecture and properties of proteins is the book by Creighton (1993).

Which vibrational modes of a protein/peptide will be influenced by the secondary structure? For this discussion, it first has to be emphasised that the backbone consists of a chain of identical repeat units linked together. Thus, the vibrational modes of the chain will initially be determined by the modes of a single peptide unit. Vibrations sensitive to the secondary structure are shown on the left side of Fig. 5.2. The amide I mode essentially consists of the C=O stretching vibration with a small admixture of the NH bending. The amide II mode consists of a combination of the NH bending and C–N stretching and, similarly, this is also true for the amide III mode, albeit with a different sign in the combination of the coordinates. In addition to these three modes, there is the NH stretching vibration (amide A), which is highly localised. It is particularly sensitive to the local environment, such as hydrogen bonding, but its diagnostic value for secondary struc-

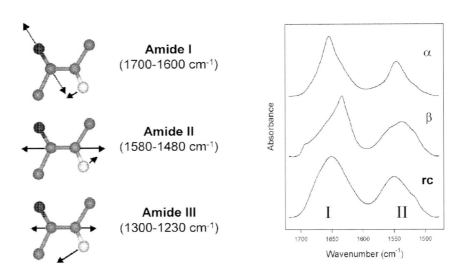

Fig. 5.2 IR spectra of proteins with predominantly one type of secondary structure. Proteins have been dissolved in H_2O. On the left side, the molecular motions of the main amide vibrations, i.e., amide I, amide II, and amide III, are indicated. α: hemoglobin, β: concanavalin A, coil: casein, all in H_2O in transmission, 6 micron path (courtesy of T. A. Keiderling).

ture determination is rather low. There are also low-frequency modes, which are rarely analysed. The amide A, I, and II modes are strong in the IR, whereas amide III exhibits only a weak intensity. For Raman studies, amide III has been used successfully but amide II is very weak.

What properties of the secondary structural elements could influence these modes? This topic has been subject to extensive research during recent years, using both theoretical and experimental methods. It is easy to see that hydrogen bonding with the NH group affects the amide A, II and III modes, and hydrogen bond interactions with the C=O group also have some influence on the amide I mode. However, hydrogen bonding alone cannot account for the sensitivity of these modes towards the secondary structure of the peptide bond. This issue will be discussed later in more detail. At this point, we will use this sensitivity of the amide modes in a more heuristic approach.

For the analysis of the amide I band by IR spectroscopy we will start with the spectra of proteins with defined secondary structure predominantly of one type. Hemoglobin is largely α-helical, concanavalin has mainly β-sheet, and casein is random coil. The IR spectra of these proteins are shown in Fig. 5.2 (right panel). The β-sheet structure has fairly distinct features, whereas the α-helical and random coil structures mainly differ in band position and band half-width. A common property of the amide I band is that it is intrinsically rather broad and it is not immediately evident that the amide I band of a protein, which includes different secondary structural elements, can be analysed correspondingly.

There is another difficulty for the study of proteins. The experiments have to be performed in water, and as outlined in Section 4.1.1, water has a strong absorption band in the amide I region. To some extent this can be circumvented by taking measurements in ^2H_2O. For peptides, the peptide groups are exposed to the solvent and the amide hydrogen is exchanged by deuterium. The amide I band has a small admixture of the NH bending coordinate. Because the intrinsic N^2H bending frequency is shifted to below 1000 cm^{-1}, the NH bending and C=O stretching coordinates are decoupled, resulting in a small downshift of the mode that is now called amide I'. In the study of proteins the difficulty is that not all amide protons are equally exchangeable, depending on the location within the protein. Some of them, which are highly protected, cannot be exchanged even after incubation of the protein in ^2H_2O for months. Thus, in most instances incubation of a protein in ^2H_2O will result in a mixture of amide I and amide I' modes. Although the exchange pattern and kinetics have been used in specific cases for structural investigations (Goormaghtigh et al. 1999), this inhomogeneity may complicate the spectral analysis. Particularly for smaller proteins, there is the possibility of unfolding in a ^2H_2O environment (see Sections 5.2 and 5.6), thereby enabling full exchange of the amide hydrogens. Whenever possible, this strategy should be used.

In Fig. 5.3 we show the amide I' band of three peptides having α-helical, β-sheet, and random coil structures. Whereas the spectrum of the β-sheet peptide is very similar to that in Fig. 5.2, the relative band positions in the spectra of random coil and α-helical peptides are reversed. Thus, in addition to the secondary

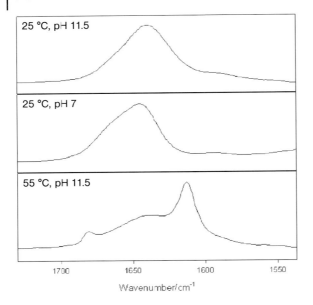

Fig. 5.3 FTIR spectra of polylysine, which adopts an α-helical (top), random coil (middle), or β-sheet secondary structure (bottom) depending on the temperature and pH, as indicated on the figure. Peptides have been dissolved in ^2H_2O, therefore, the amide II band, which has a large contribution from NH bending, is missing. Transmission spectra, path: 100 micron, conc. ca. 10 mg/mL (courtesy of T. A. Keiderling).

structure other factors influence amide I′ band positions. When discussing the theory of the amide I modes, we will see that hydrogen bonding to solvent molecules, here ^2H_2O, shifts the amide I′ modes downwards, but in a different manner for α-helical and random coil structures (Starzyk et al. 2005; Kubelka et al. 2005).

5.2
Practical Approaches

As measurements in ^2H_2O are problematic for larger proteins, procedures have been developed to subtract the spectrum of the aqueous buffer. One simple method uses the fact that in general proteins do not absorb between 1900 and 1750 cm^{-1}, whereas the absorbance of water shows a minimum around 1850 cm^{-1} and increases both at higher and lower wavenumbers. Thus, the spectrum of the aqueous buffer may be subtracted interactively from the spectrum of the protein solution in such a way that a flat baseline is obtained in this spectral range (Dong et al. 1990). In a more elaborate way, the subtraction can be performed more objectively using an algorithm that will also provide a flat baseline (Powell et al. 1986). Whereas these methods work adequately for the bulk solvent, they cannot correct for water molecules interacting with the protein. This is a general problem for the IR spectroscopy of compounds dissolved in a solvent that itself absorbs in the IR. The number of solvent molecules affected by the interaction with the solute is equal to or larger than the total number of solute molecules. For proteins dissolved in water, one has to relate the number of strongly interacting water molecules to the number of peptide bonds (see below). Thus, there is an inherent problem in deriving "pure" IR spectra of proteins dissolved in water. Nevertheless, the subtraction of bulk water represents a first approximation.

In Fig. 5.4 the amide I and amide I' spectra of the small protein ribonuclease T_1 are shown. The spectra have been measured under different conditions, i.e., in H_2O, ^2H_2O after 1-h exposure, and ^2H_2O after complete exchange by unfolding and refolding of the protein (1st, 3rd, and 5th spectrum, respectively) (Fabian et al. 1993). The solvent contribution has been subtracted according to the "flat baseline" method discussed above. As anticipated, the bands are very broad. There are some shoulders present, indicating different secondary structural elements.

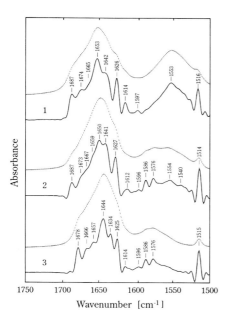

Fig. 5.4 Spectra of ribonuclease T_1 and their self-deconvolution: 1, measured in H_2O; 2, after 1-h exposure to ^2H_2O, and 3, after denaturation and renaturation in ^2H_2O. The progressing $^1H/^2H$ exchange is most easily seen with the disappearance of the amide II mode around 1553 cm^{-1} (taken from Fabian et al. 1993, with permission).

In order to better identify the individual spectral components in the otherwise often featureless broad amide I (or amide I′) band envelope, its shape is analysed by two basic and related techniques, second-derivative formation and self-deconvolution. Second-derivative formation of the band contour clearly detects variations in the slope. Self-deconvolution is an artificial resolution enhancement. Both mathematical procedures allow the detection of spectral contributions in a broad absorption band. For more details the reader is referred to the original literature (Surewizc and Mantsch 1988). The application of self-deconvolution to the three spectra is shown by the 2nd, 4th, and 6th spectra. It is interesting to note that the spectra after 1 h of exposure to ^2H_2O and after full exchange differ considerably, emphasising the problem of incomplete $H/^2H$ exchange discussed above. The resulting band positions obtained by resolution enhancement (or by second-derivative formation) can subsequently be used to describe the measured absorption envelope by individual band components. This simulation usually works well (see Section 4.5). Using empirical rules, the intensities of the individual bands are then interpreted in terms of percentages of the secondary structural elements, sometimes even including γ-turns (Dong et al. 1990; Surewicz and Mantsch 1988). As the empirical rules are based on the spectra of proteins with known structure, this relatively simple procedure has some success, although it lacks a theoretical justification. Among others, it does not directly take into account the different band shapes for the three main secondary structural elements shown in Fig. 5.3. Thus, we will not describe these procedures for quantitative secondary structure determination in any further detail.

A completely different method uses the amide I bands of proteins with known structures as a database. The idea is that each element of a certain secondary structure is characterised by a well-defined amide I band. Thus, the first goal is to derive these spectra from the database. For this reason, the structural data have to be expressed in percentages of the various secondary structural elements, including β- and γ-turns. The determination of these basic spectra represents a typical task for chemometrics. Factor analysis and singular value decomposition are, among others, corresponding mathematical methods (e.g. Malinowski 1991) (see section 4.5). The complete set of basic spectra is often called a basis set. As a first test for the success of the methods, the spectra of the database will be fitted by the spectra of the basis set, and the corresponding relative contributions of secondary structure elements should be reproduced. As a further test, spectra of proteins with known structure but not contained in the database are fitted by the spectra of the basis set, and the results are compared with the experimental data. If the performance of these tests is satisfactory, it should be possible to analyse the amide I band envelope of proteins of unknown structure with respect to the secondary composition.

There have been several investigations using these methods, also in combination with UV circular dichroism (CD). The inclusion of UV CD was motivated in particular by the similarity of the amide I IR spectra of an α-helix and of the random coil structure (see Fig. 5.3), whereas the UV CD spectra differ considerably. In more recent publications, in which older literature is also cited, these methodologies are critically discussed (Wi et al. 1998; Oberg et al. 2003). It is emphasised that the choice of the database that is used to derive the basis is critical, and should comprise as many different structures as possible. In addition, in critical examples, it is pointed out that IR and CD analysis should be applied separately. In a more recent publication it is suggested that there are three distinct positions in the spectra that contain all the non-redundant information which can be related to the secondary structure content (Goormaghtigh et al. 2006). These workers also emphasise that the inclusion of other positions of the spectra may even cause a poorer structural prediction.

Now we have to ask how the bound water molecules, for which the spectral contribution is difficult or even impossible to quantify (*vide supra*), affect these evaluations? It is interesting to note that the spectra of the database may be either measured from a concentrated protein solution or a hydrated film deposited on an ATR crystal. In the former, the contribution of bulk water is subtracted, whereas for the film samples no correction is applied. Thus, in both instances bound water molecules contribute to the spectra. As the different methods of amide I band analysis are fairly successful, one might speculate that the contributions of bound water molecules to the spectra do not depend very much on the structure of the protein. In addition, bound water molecules give rise to a considerably lower absorption in the amide I range than all the peptide C=O oscillators of the protein. Because the number of bound water molecules is smaller than the number of peptide bonds, the contribution of bound water molecules can, in the main, be neglected for this type of analysis.

It should be emphasised that the databases documented so far have been established for soluble proteins. Thus, only the secondary structure composition of soluble proteins can be evaluated using the statistical methods. For membrane proteins sufficient structural data are not available to set up a corresponding database. Furthermore, reproducible experimental conditions are more difficult to define because protein-containing membranes tend to orient and this will cause dichroic effects, influencing the absorption strength of the various secondary structural elements, depending on the orientation. Concentrated detergent solutions can also be studied, but it is difficult to assess the effect the detergent exerts on the spectra. However, large fractions of α-helical or β-structure content can also be determined for membrane proteins, using the empirical rules mentioned above.

5.3
Studies on the Origin of the Sensitivity of Amide I Bands to Secondary Structure

Up to now, the analysis of the amide I band has been based either on empirical rules or on statistical evaluations. We will now describe experiments, together with their theoretical interpretation, which shed more light on this issue. As discussed above, the peculiarity of a protein consists in the backbone chain being represented by a series of chemically identical units, the peptide groups, which are arranged in a defined structure. With respect to the amide I band, one has to consider the row of oscillators that consists mainly of the C=O stretching vibrations. These oscillators are basically in resonance. It is well known from excitonic systems, which also consist of closely spaced identical units (in these instances, chromophores), that for the correct description of the absorption properties the dipolar coupling of the individual optical transition moments plays a major role. Correspondingly, for the series of molecular oscillators such as the peptide groups, the dipolar coupling of the IR transition moments is important. Of course, the coupling depends on the mutual orientation of the transition moments and on the distance of the oscillators. Therefore, this coupling brings structural information into the amide I band of a peptide chain. An early and extensive study using computational methods provided a good introduction to the general problem of normal mode analysis of peptides and proteins (Krimm and Bandekar 1986). The calculations were based on an empirical force field (see section 2.1.3). As already suggested in earlier studies cited in this work, the transition dipole moment coupling (TDC) was incorporated by adding a force field term to the atoms involved in the amide I normal mode derived from the dipolar interaction energy. The significance is most obvious for the anti-parallel β-sheet structure. There are two infrared active modes B_1 and B_2, which are also termed "parallel" and "perpendicular" because of their polarisations being parallel or perpendicular to the chain axis. The two modes are located at 1695 and 1630 cm^{-1}. Neglecting TDC, these modes are calculated to be at 1673 and 1665 cm^{-1}, corresponding to a much smaller splitting, mainly caused by the upshift of the B_2 mode. The main effect of TDC is the strong downshift of the B_2 mode and, to a lesser extent, the upshift of the B_1 mode.

The importance of TDC for the amide I absorbance was emphasised even more in model calculations for real proteins (Torii and Tasumi 1992). As it appears to be very difficult to apply the empirical force field method to real proteins and to incorporate TDC, a much simpler approach was chosen. Instead of using the complete chain of amino acids, these workers just used the amide I oscillators, not connected to each other, but placed at the positions determined from the crystal structure. Thus, the coupling of the amide I oscillators through valence and hydrogen bonds are considered not to be important. In order to describe the geometrical factors correctly, the location of the amide I oscillator in the peptide bond is deduced from the calculated normal mode of polyglycine or polyalanine, using an empirical force field (Krimm and Bandekar 1986). The diagonal force constants are assumed to correspond to the frequency of an isolated oscillator located at 1650 cm^{-1}. The off-diagonal force constants are derived from the dipolar coupling of the transition moments, i.e.,

$$F_{jk} = \frac{0.1}{\varepsilon} \frac{\delta\hat{\mu}_j \cdot \delta\hat{\mu}_k - 3(\delta\hat{\mu}_j \cdot \hat{n}_{jk})(\delta\hat{\mu}_k \cdot \hat{n}_{jk})}{R_{jk}^3} \tag{5.1}$$

Here, ε is the dielectric constant, $\delta\hat{\mu}_l$ is the transition moment of the lth peptide group, R_{jk} is the length of the line connecting peptide groups j and k, and $\hat{n}_{jk}$ is the unit vector along this line. The magnitude of the transition moment has been estimated from the integral amide I absorption band of peptides and proteins. The envelope of the amide I absorption of this simplified system is obtained by the polarisation of each normal mode. This polarisation is obtained by the vector sum of the polarisation of each peptide group in this normal mode. Finally, the square of the polarisation is multiplied by the square of the magnitude of the transition moment.

Surprisingly, this simple model can describe the absorption spectra fairly well, as demonstrated for several examples of proteins with known structure. This approach demonstrates that the main sensitivity of the amide I band to the secondary structure is caused by TDC. One important result has to be emphasised. The IR spectrum of myoglobin, a small soluble protein consisting of short α-helical stretches connected by stretches of irregular structure, also shows, beside the prominent absorption band around 1655 cm^{-1} characteristic of an α-helix, bands between 1630 and 1640 cm^{-1}, a region typical of β-sheet structure. The calculations confirm that for an infinitely long helix the splitting between the main IR active A mode and the weaker E_1 mode is small [less than 5 cm^{-1} (Krimm and Bandekar 1986]. For short helices, as for myoglobin, the splitting becomes larger, causing the low-frequency amide I absorption. Again, the size of the splitting is determined by TDC. However, it should be mentioned that some of the C=O oscillators had to be adjusted in order to improve the agreement with the experimental spectra. Although there appears to be some systematic rules involved, a physical explanation for this adjustment could not be given.

With the simple model described here, irregular proteins without symmetry properties can be treated. This feature of the method is the main advantage as compared with earlier studies, which relied on the well-defined structure of one type (extended α-helix or extended β-sheet) or on very short stretches such as β- and γ-turns (Krimm and Bandekar 1986). For those calculations TDC was incorporated as a type of perturbation of the general empirical force field.

Further support for the role of TDC has recently been obtained by recognising that this coupling will be strongly affected if one or more of the oscillators can be detuned by isotopic labelling. Here, ^{13}C-labelling of the C=O group offers a convenient technique. The single-oscillator frequency of the C=O stretching is downshifted from ca. 1650 to ca. 1610 cm^{-1} upon ^{13}C/^{12}C substitution. As the calculations treated so far relied on the resonance condition in TDC, which is no longer valid for the isotopomers, more elaborate calculations are necessary in order to compare experiment and theory (Decatur and Antoni 1999; Brauner et al. 2000; Silva et al. 2000). In a pioneering study, the effect of ^{13}C-labelling on the amide I band of a hydrophobic peptide has been addressed both experimentally and theoretically (Brauner et al. 2000). The underlying model is similar to the previous one. However, in addition to TDC, coupling of the oscillators via single bonds in the peptide chain and via hydrogen bonding have also been considered. In order to describe the label effects, the corresponding diagonal term is also altered. Furthermore, for more closely spaced oscillators it is not the dipole approximation but a more accurate formalism that is used to account for the effect of a dipole on the neighbouring one.

Figure 5.5 (A) shows the measured and calculated IR spectra of the peptide Lys$_2$LeuAla*(LeuAla)$_5$, where the asterisk indicates the position of the ^{13}C=O label in the sequence. In the spectrum of the unlabelled peptide (not shown), the low-frequency band around 1610 cm^{-1} is missing, implying that labelling causes this rather large band. As compared with the main band around 1630 cm^{-1}, the downshift is smaller than expected for a single C=O oscillator. In addition, the area under the shifted band is much larger than (1/13)th of the area of the main band of the (unlabelled) peptide that contains 13 amide bonds. The band at 1695 cm^{-1} is caused by the so-called longitudinal mode, which is polarised parallel to the peptide chain of the β-sheet. Its contribution is much weaker and can be neglected for the subsequent analysis. The reduced isotopic shift and the large intensity of the 1610-cm^{-1} band indicate that the labelled oscillator couples to unlabelled ones. This conclusion is confirmed by the vibrational analysis. In Fig. 5.5 (B) the normal mode composition of the band around 1610 cm^{-1} is shown (upper panel, a). The main amplitude is located at the labelled oscillator (black dot), but the mode has considerable contributions from two oscillators of the neighbouring chain. A detailed analysis of the effect of the different coupling mechanisms shows that hydrogen bonding and TDC is the main source of coupling between the labelled and the unlabelled oscillator, whereas coupling between two unlabelled oscillators occurs via a through-bond mechanism.

The large intensity of the shifted band is explained by this coupling to the unlabelled oscillators, which "borrows" intensity from the main band. Conversely, calculations also demonstrate that the intensity ratio for the main band of the singly-labelled and the unlabelled peptide is smaller than 12/13. In addition, the amplitude of the oscillators is reduced. Both factors cause a considerable reduction in the intensity of the main band. It should be mentioned that the model used recently for the calculations could be successfully applied to calculate the envelope of the amide I band of globular proteins (Brauner et al. 2005).

These studies have been extended by more refined calculations (Kubelka and Keiderling 2001). Here, short (basically two amino acids) strands of β-sheet involving two strands were treated by density functional theory (DFT) (see Section 2.1.4). The calculated force field and atomic polar tensors were then transferred to the larger structures (longer and more strands). It has been shown previously that this method allows an adequate description of the amide I' vibrational circular dichroism. It is clearly demonstrated that a better agreement with the experiment is obtained by including inter-chain coupling of up to five strands in the calculations. Here, it is important to mention that in DFT calculations TDC is already contained in the calculated force field, and it is difficult to separate this coupling from the other contributions. For the larger structures, the additional TDC had to be introduced on the basis of the transition moments obtained from the DFT calculations. In later work on two-strands peptides, different computational models and the influence of the strand were studied (Bour and Keiderling 2004). A detailed presentation of these results is, however, beyond the scope of this tutorial.

These investigations are important, for example, to analyse the IR spectra of amyloid fibrils, which are composed of a β-sheet structure. The structure of fibrils is difficult to determine by other techniques, and it appears that important in-

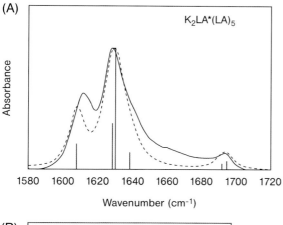

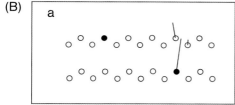

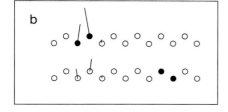

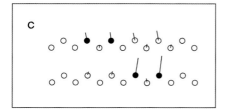

Fig. 5.5 (A) FTIR spectrum of the peptide Lys$_2$LeuAla*(LeuAla)$_5$ in 2-chloroethanol. The asterisk denotes the ^{13}C labelling of this amino acid at the C=O group (solid line); the computed spectrum (dashed line) was obtained as described in the text (taken from Brauner et al. 2000, with permission). (B) Symbolic description of the normal mode composition of the shifted band described in (A) (panel a). It shows that in this normal mode (1610 cm^{-1}) the labelled oscillator (black dot) has a large amplitude, but the normal mode also borrows intensity from the unlabelled oscillators of the neighbouring strand (open circles) lines show amplitude of the oscillators. Panels b and c show different label patterns not discussed here (taken from Brauner et al. 2000, with permission).

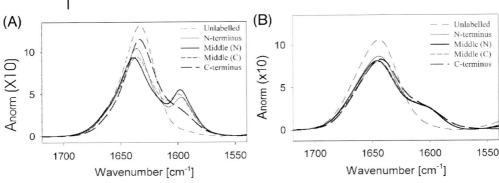

Fig. 5.6 (A) Amide I′ FTIR spectra of unlabelled and labelled peptides Ac-Ala$_4$-Lys-Ala$_4$-Lys-Ala$_4$-Lys-Ala$_4$-Tyr-NH$_2$: unlabelled, dotted/dashed line. The different labelled peptides include ^{13}C=O labels in the first segment including amino acids 1–4 (thick solid line), the second segment (5–8; thin solid line), the third segment (9–12; thin dashed line), and the fourth segment (13–16; thick dashed line). Measurements were performed at 5 °C. At this temperature the peptides adopt α-helical conformation. (B) Amide I′ FTIR spectra of the peptides shown in (A), but measurements performed at 50 °C with the peptides adopting a random coil structure (Silva et al. 2000).

sight into the structure and architecture can be obtained from IR spectra, especially if specific isotopic labelling of the peptides is included (Silva et al. 2003; Hiramatsu and Kitagawa 2005).

^{13}C-labelling of helical peptides causes more subtle effects. The pronounced "intensity borrowing" described above for β-sheet structures does not take place. Fairly straightforward behaviour is observed if four consecutive residues are labelled and this labelled segment is displaced along the peptide chain (Silva et al. 2000). Figure 5.6 (A) shows the spectra of the unlabelled and labelled peptide Ac-Ala$_4$-Lys-Ala$_4$-Lys-Ala$_4$-Lys-Ala$_4$-Tyr-NH$_2$ where "Ac" indicates the acetylation of the C-terminus. Measurements have been performed in ^{2}H$_2$O. Therefore, the amide I′ band is observed (see Section 5.1). For the labelled peptides four consecutive alanines were always labelled, including the first segment with the positions 1–4, the second (5–8), the third (9–12), and the fourth (13–16). Labelling leads to a band around 1595 cm^{-1}. Whereas for the first three spectra a clearly shifted band is observed, labelling of the fourth segment at the C-terminus only leads to a broad shoulder. This result implies that this part of the peptide is considerably less ordered, a conclusion supported by measurements at 50 °C. At this temperature the peptide has lost the helical structure [Fig. 5.6 (B)] (see Section 5.6) and all the spectra of the four differently labelled peptides are similar, exhibiting the broad shifted band. When the band positions of the shifted (^{13}C=O) band are plotted against temperature, a broad sigmoid behaviour is observed, characteristic of a broad phase transition describing the unfolding of the peptide. Comparing the sigmoid behaviour of the four labelled peptides, it could be shown that the transitions are shifted to lower temperatures for peptides labelled at the C- and N-termini (first and fourth segment) as compared with peptides labelled in the central segments (second and third). This finding indicates that the unfolding is not uniform along the peptide chain. Investigations on the effect of labelling of the helix have been extended by additional label patterns. The measurements have been accompanied by calculations using the transferred DFT method as described above (Huang et al. 2004).

These studies demonstrate that IR spectroscopy not only provides global structural information, but may also contribute to the elucidation of local structures when combined with specific isotopic labelling of the peptide. It should be mentioned that IR vibrational CD represents another powerful method to study secondary structure of proteins and peptides. As in UV CD, it is based on the coupling of transition dipole moments, here of the amide I (and also amide II) oscillators. Detailed descriptions of the technique are given in publications from the group working with Keiderling (Kubelka et al. 2005; Huang et al. 2004; Kubelka and Keiderling 2001, and references cited therein).

5.4
Direct Measurement of the Interaction of the Amide I Oscillators

The great success of modern NMR spectroscopy in structure determination is based on the possibility to determine the interactions between nuclear spins. For these approaches coherences are generated between spin states that are modified by the interactions between the spins. A simple example for the generation of such a coherence is the application of a $\pi/2$ radio-frequency pulse, flipping the magnetisation from the z-direction into the x,y-plane. In fact, there is an equivalence between a spin-1/2 system and an optically driven two-level system, such as a vibrational transition (see Zhuang et al. 2005; Mukamel and Zhuang 2005, and references cited therein). Therefore, one might ask whether it is possible to generate vibrational coherences in analogy to spin coherences? The essential difference between the two systems is the relaxation time: whereas for magnetisation it is in the order of milliseconds and longer (depending on the system), for a vibrational excitation of a molecule in the condensed phase it is in the order of picoseconds and less. This implies that, in order to perform experiments in the IR that mimic those in NMR spectroscopy, extremely short IR pulses have to be applied. Such an approach has become possible with the advent of techniques enabling the production of short IR pulses.

One of the first applications to study the interaction of amide I oscillators was published in 1998 (Hamm et al. 1998). In this study, the physical framework for the interpretation of the data was also developed. As the methodology requires a deeper insight into the formalism to treat the nonlinear interaction of radiation with matter, the description of this formalism is beyond the scope of this tutorial. We will instead use a more intuitive way to outline the principles of such experiments, following the description given by Bredenbeck (Bredenbeck 2005).

In general, the techniques are termed two-dimensional IR spectroscopy (2D-IR). In the simplest version, IR pump/IR probe measurements are performed

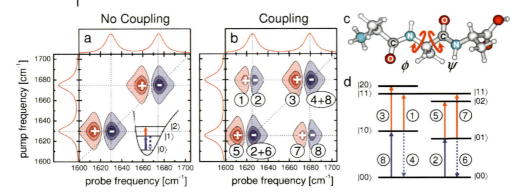

Fig. 5.7 Schematic representation of 2D spectra of two amide I oscillators [peptide shown in (c)]; (a) uncoupled oscillators; (b) coupled oscillators; (d) transitions if the coupling is turned on (courtesy of J. Bredenbeck, PhD Thesis, 2005).

using the equipment shown in Fig. 3.5 (a). As described in Section 3.1.5, the pump pulse passes through a computer-controlled Fabry–Pérot filter, converting the broad IR pulse into a narrow band pulse, which is scanned over the bandwidth of the original pulse. The probe pulse still has the original broad bandwidth of approximately 100 cm^{-1}. Both pulses are directed onto the sample, but a delay between them can be adjusted. After the sample, the probe pulse passes through a monochromator and is imaged onto an array detector, providing the spectral information in one axis of the 2D experiment.

The basic idea of a 2D IR experiment is conveniently illustrated in a system having two oscillators (e.g., amide I) with slightly different frequencies. This situation is depicted in Fig. 5.7 (c), showing a peptide of three amino acids with two peptide groups (two amide I oscillators). If the pump pulse is in resonance with one oscillator, it brings it into the first excited state. This excitation causes a bleach for the corresponding $0 \to 1$ transition. In addition, the probe beam induces stimulated emission resulting in the $1 \to 0$ transition. Bleaching and stimulated emission leads to an absorption decrease according to

$$\Delta A = -\lg[I(\tau)/I_0] \tag{5.2}$$

where $I(\tau)$ and I_0 are the intensities seen by the array detector after the delay τ and before the arrival of the pump pulse, respectively. The absorbance increases and decreases in Figs. 5.7 (a), (b), and (d) are given by the red and blue colours, respectively.

The probe pulse induces the excited state absorption corresponding to the $1 \to 2$ transition. This process occurs at a slightly lower frequency due to the anharmonicity of the amide I oscillator. Thus, at this frequency the probe beam is attenuated compared with the situation prior to the arrival of the pump pulse, and an absorbance increase is observed. It can be shown that, in the absence of anharmonicity, the excited state absorbance, bleaching, and stimulated emission would just cancel each other out and no absorbance changes would be observed. If the second oscillator is included, and the pump pulse is scanned over the accessible spectral range a two-dimensional plot of the spectral changes is obtained as shown in Fig. 5.7 (a). The two axes correspond to the probe and the pump frequencies.

If the interaction between the two oscillators is now turned on, the 2D plot becomes more interesting. The interactions can be rationalised by the energy scheme representing the two oscillators [Fig. 5.7 (d)]. Because of the interactions, the frequency of the transition of one oscillator depends on the state of the other oscillator. Thus, the transition $(0, 0) \rightarrow (1, 0)$ [denoted as "8" in Fig. 5.7 (b) and (d)] is different from the transition $(0, 1) \rightarrow (1, 1)$ (denoted as "7"). Using femtosecond pump and probe pulses, the interactions cause the so-called cross peaks in the lower right corner, one positive (7) and one negative (8). If there is no interaction, the two cross peaks will coincide and cancel each other out. Exchanging the role of the two oscillators, the two cross peaks in the upper left corner are obtained. If there is a model for the molecular interaction, structural information can, in principle, be obtained by analyzing the 2D spectrum. It is clear that variation of the delay time of the probe pulse allows extraction of information on the lifetime of the vibrations. More elaborate experiments, in fact, monitor the spectral changes in the 2D plot by varying the delay time. In this way, one may obtain information on homogeneous and inhomogeneous line widths and thus on the dynamics of the system (Bredenbeck and Hamm 2003).

During the past few years, a growing number of publications have appeared, demonstrating the impact of these techniques on the understanding of the molecular interactions within peptides and also of peptides with their environment. These studies have been accompanied by an increasing number of theoretical investigations. As one would expect from the foregoing discussion, the use of specific isotopic labelling of the C=O oscillator also appears to be a promising tool for these techniques. In a recent publication the interaction between two trans-membrane helices (helix dimer) has been addressed on the basis of these techniques (Fang et al. 2006). This study is especially important as the analysis confirms results on helix dimers in micelles obtained by NMR spectroscopy.

Because the underlying theory for the interpretation of 2D spectroscopy is fairly complicated, it is not possible to discuss original publications here in a comprehensive manner but we will simply refer to a few instructive publications (Kolano et al. 2006; Fang et al. 2004; Ganim and Tokmakoff 2006).

5.5
UV-resonance Raman Studies Using the Amide III Mode

So far, we have only described IR techniques that are suitable for structural studies of proteins and peptides. Although the amide I mode is Raman active, the intensity appears to be too low to allow for the sufficiently precise measurements required for band analyses. The peptide bond has electronic transitions between 190 and 220 nm. Thus, it is expected that some of the peptide modes will couple to these electronic transitions, and thus gain resonance enhancement upon excitation in the deep UV. However, only with the development of nonlinear optics has this wavelength range become accessible for RR spectroscopic investigations. As described in Section 3.2.1, there are several methods available for generating high intensity laser lines in this spectral range. Among them are fifth-order anti-Stokes shifting of the third harmonic of a pulsed Nd:YAG laser in a high-pressure H_2-gas Raman cell (204 nm), fourth harmonic generation of the output of a Ti:sapphire laser (194 nm), or the intracavity doubling of the 413-nm line of a Kr ion laser. In these UV RR experiments, special care has to be taken to avoid damage of the samples by the UV laser light (see Section 4.2.2.1). Specifically, the continuous flow of the sample through the exciting laser line is a prerequisite.

The UV RR spectra show a preferential enhancement of the amide II and amide III modes, whereas the intensity gain for the amide I mode is much smaller. The enhancement of the amide II mode is remarkable, because in nonresonant Raman scattering the amide II mode is almost inactive. The amide III mode has been described as a mode involving the C–N stretching and the NH bending, similar to the amide II, but with an opposite sign of the coupling. However, in pioneering studies by Asher and coworkers, it was shown that the coupling with the $C_\alpha H$ bending coordinate is of great importance for distinguishing between different secondary structure elements. This relationship has been demonstrated by the UV RR spectra of helical peptides and proteins in the folded form and unfolded random coil states (Asher et al. 2001) [Fig. 5.8 (A) and (B)].

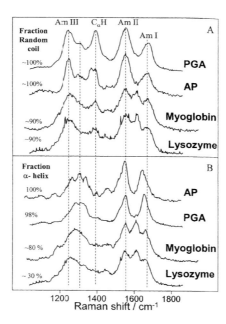

Fig. 5.8 UV RR spectra of peptides and proteins. Panel (A) random coil conformation; panel (B), native folded conformation as indicated. PGA, polyglutamic acid; AP, alanyl-based peptide $A_5(ARA)_3A$ (taken from Asher et al. 2001, with permission).

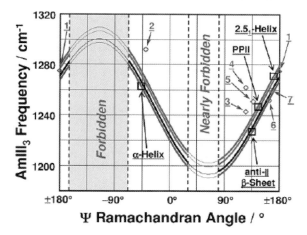

Fig. 5.9 Correlation between the dihedral angle Ψ and the amide III frequency, theoretical dependence and measured frequencies. Measured frequencies 1–7 are from peptide crystals; values for the α-helix, anti-parallel β-sheet and 2.5_1 helix, and polyproline II helix are indicated. Calculations are based on different models, in particular with respect to solvent exposure. For details see Mikhonon et al. (2006).

In the random coil states, the $C_\alpha H$ bending is clearly visible at 1400 cm^{-1}, whereas it almost disappears in the folded states. It has been further demonstrated that, unlike the amide I and amide II modes, the amide III mode is relatively localised. Among others, one explanation could be that because of the low intrinsic IR transition moment, coupling with transition dipole moments of adjacent amide bonds is small. Instead, the amide III mode of each peptide bond could sense the individual coupling with the neighbouring $C_\alpha H$ bending coordinate. This coupling depends sensitively on the dihedral angle Ψ (Mikhonin et al. 2006), whereas the dependence on the angle Φ is considerably weaker. This is demonstrated in Fig. 5.9, where experimental and theoretical amide III frequencies are shown for systems in different hydrogen-bond interactions. The methodology is still under development and new applications especially in the field of protein folding studies (see 5.6) appear to be particularly promising.

5.6
Protein Folding and Unfolding Studies Using Vibrational Spectroscopy

Exploring the mechanism and pathways of protein folding and unfolding reactions represents an important field of modern biophysics. Good overviews of this field are provided by the book *Proteins* by T. E. Creighton (1992), and by the book *Protein Folding* edited by T. E. Creighton (1993). As with other structurally sensitive methods, vibrational spectroscopies can also be employed to monitor protein folding and unfolding processes. In this section, we will present a selection of approaches that have been used in vibrational spectroscopy.

There are various procedures to induce unfolding, including the addition of denaturants such as urea or guanidinium chloride, or changes of pH, pressure, and temperature. Denaturants are not very practical in combination with vibrational spectroscopy as they give rise to additional and partly relatively strong bands that may obscure the protein bands of interest. Variations of pH, on the other hand, are difficult to establish in a IR cuvette. It should be mentioned that pressure variations can also be applied, but specially designed cuvettes are necessary, which can resist the high pressures required for unfolding (Panick et al. 1998). Thus, the most convenient method to initiate folding or unfolding is based on temperature changes. However, proteins unfolded at high temperatures have the tendency to aggregate irreversibly, making the analysis of the unfolding process difficult. Therefore, special pH conditions and/or additions of denaturants are often combined, inducing the unfolding at somewhat lower temperature.

As a typical example of a protein, the temperature-induced protein unfolding is shown for RNase T1 in Fig. 5.10 (Fabian et al. 1993). In order to facilitate the analysis, measurements are often performed in ^2H_2O. Because of the reduced

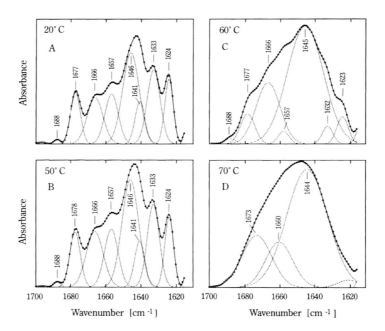

Fig. 5.10 Unfolding of ribonuclease T_1 by temperature. Spectra are measured at the temperatures indicated, protein dissolved in ^2H_2O. Self-deconvoluted spectra are shown. Most of the fine structure is lost at 70 °C, i.e., the protein is unfolded. Nevertheless, as the spectrum still shows some residual structure, some of the residual structured bends may still be present. The protein can be completely refolded by lowering the temperature (Fabian et al. 1993).

absorption of the solvent, cuvettes with larger spacings and lower protein concentrations can be employed, reducing the danger of aggregation. Furthermore, the lower absorption of ^2H_2O results in smaller temperature-dependent contributions of the solvent to the temperature-dependent protein spectra, and therefore smaller corrections have to be applied. The spectra that are shown have been measured at temperatures of 20, 50, 60, and 70 °C at $p^2H = 7$. Prior to the measurements, $H/^2H$ exchange was accomplished as shown in Fig. 5.4 (see Section 5.2). The spectra have been self-deconvoluted and subjected to a band fitting analysis as described in Section 5.2. Although, as outlined previously, this procedure does not necessarily provide exact information on the secondary structure composition, it can be used to assign the main features. From the crystal structure it is known that there is one α-helix and several β-sheet segments involved in differing hydrogen bond interactions. The remainder of the protein is composed of unordered structures. Accordingly, the two bands at 1633 and 1624 cm^{-1} are assigned to amide I modes of the β-sheet segments, whereas the band at 1646 cm^{-1} represents the amide I of the α-helix. All the fine structure in the amide I band region is preserved up to a temperature of 50 °C, whereas at higher temperatures it starts to vanish until it is almost completely lost at 70 °C. The spectral changes have been found to be completely reversible, indicating that no aggregation occurs in the unfolded state and that the protein fully refolds upon cooling to ambient temperature. As the spectrum measured at 70 °C indicates some residual structure, it might be that this unfolded state still has some structured bends. The analysis using smaller temperature intervals shows that the temperature-dependence of the spectra is consistent with a two-state model for the equilibrium between the folded and unfolded states, i.e., the spectra measured between 50 and 70 °C can be described by a superposition of the spectra of the folded and unfolded states. Such a result is in fact expected for small single domain proteins (Creighton 1992, 1993). Finally, we would like to mention a recent study in which 2D methods (see Section 5.4) have been used to analyse the unfolding of the related protein RNase A (Chung et al. 2004).

To probe the kinetics of the unfolding/folding reactions, time-resolved techniques have to be applied. However, here the method of triggering the reaction requires special attention. The temperature increase that induces the unfolding processes must be applied as a rapid temperature jump within a time interval that is shorter than the desired time-resolution. This requirement can be fulfilled by various approaches.

In section 4.4.2 the rapid mixing device for time-resolved IR spectroscopy was introduced. When a cold protein solution is rapidly mixed with a hot buffer and this mixture is driven into the preheated IR cuvette, unfolding processes longer than approximately 10 ms can be monitored using the rapid-scan FTIR technique. As an example, we will describe folding/unfolding studies of the ribonuclease RNase A. As in the static measurements, ^2H_2O was used as the solvent. This is particularly important using the rapid-mixing chambers, as the lower protein concentration and the larger spacing of the windows allows a faster mixing and a shorter transfer of the solution into the cuvette. As compared with RNase

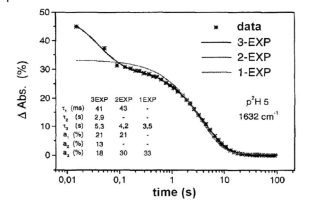

Fig. 5.11 Kinetics for the refolding of ribonuclease A, induced by a rapid-mixing temperature jump from 81 to 51 °C, monitored at 1632 cm^{-1}. Spectra have been measured with the rapid-scan FTIR technique. Also shown are fits to a sum of one, two, and three exponentials. The insets give the time constants and amplitudes (H. Georg, PhD Thesis, 1999).

T1 (104 amino acids), RNase A is slightly larger (124 amino acids). The secondary structure compositions are very similar for both proteins, but the fold is quite different. For example, RNase A has three short α-helices instead of the single but longer helix in RNase T1. Between pH 3 and 5, the protein can be reversibly unfolded. The rapid mixing device has the advantage that temperature jumps can be produced in both directions, and therefore, folding kinetics can also be followed (Georg 1999). The time-resolved rapid-scan FTIR spectra had a time-resolution of approximately 10 ms.

In Fig. 5.11 the refolding kinetics are monitored at a single wavenumber (1632 cm^{-1}) characteristic of the re-formation of the β-sheet structure (the reference spectrum is that of the folded protein). The intensity traces at this wavenumber have been extracted from the time-resolved spectra, measured for a temperature jump from 81 to 51 °C at p^{2}H 5. Evidently, the process cannot be described by a single time constant, and the analysis shows that two time constants are required. The time-resolved spectra have been fitted over the measured spectral range between 1800 and 1300 cm^{-1} by a sum of exponentials (cf. Section 6.2). The results are shown in Fig. 5.12 for a temperature jump from 69 to 39 °C at p^{2}H 3. The lower temperature jump has been chosen to slow down the reaction. In Fig. 5.12 (A) the two amplitude spectra (ms and s kinetics) are shown and compared with the static difference spectrum comprising the same temperature change.

In principle, the sum of the two amplitude spectra should reproduce the static difference spectrum. However, if the two amplitudes are summed together, a considerable part is still missing, which indicates that even faster processes take place that cannot be resolved with this technique. However, the two amplitude spectra are somewhat different (relative size of the band at 1632 and 1667 cm^{-1}),

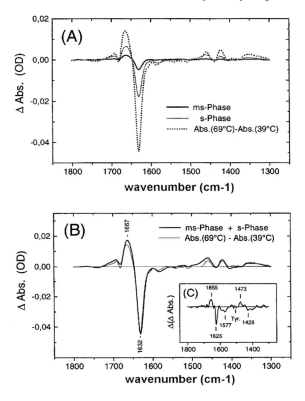

Fig. 5.12 Ribonuclease A refolding measured with the rapid-mixing temperature jump from 69 to 39 °C. The time-resolved spectra have been obtained with the rapid-scan technique. The spectra have been fitted to a sum of two exponentials. (A) Amplitude spectra of the fit to two exponentials. (B) Comparison of the amplitude spectra with the static difference spectrum obtained with the same temperature jump. For a better comparison, the spectra have been normalised. The insert shows the difference between the static difference spectrum and the sum of the two amplitude spectra, demonstrating that a fast component has escaped the measurement (H. Georg, PhD Thesis, 1999).

indicating that the process is more complicated than that which is described by a simple two-state transition. This becomes even more evident if the missing part, i.e., the faster component not covered by this technique, is analysed. In Fig 5.12 (B), the sum of the amplitude spectra is compared with the static difference spectrum, using the 1632 cm^{-1} band for normalisation. Again, it is clear that there are pronounced deviations. The insert shows the difference. There is a fast disappearance of a band at 1655 cm^{-1} and a fast appearance of a band at 1625 cm^{-1}. In a first analysis, this can be interpreted by a fast disappearance of a random coil structure and a fast appearance of a special β-sheet structure, although, as outlined earlier, additional techniques such as isotopic labelling are required to test this interpretation.

Studying processes at shorter times first requires faster temperature jumps. For this purpose, the aqueous solvent is directly heated by a short (ns) near-IR pulse. Such a pulse can be generated by Raman shifting (Stokes) the output of a Nd:YAG laser using a high pressure H_2 Raman cell (*vide supra*). This approach provides an output at 1.9 µm, with a pulse width of about 3 ns (Wang et al. 2004). In a cuvette with a spacing of 50 µm a temperature pulse of up to 10 °C can be generated. For monitoring the spectral changes with high time resolution, a tunable IR laser (diode laser) and a fast photovoltaic MCT detector (see Section 3.1.1.2) often represent the basic equipment. Although this temperature-jump technique is only suited to induce unfolding, information on the folding dynamics can also be obtained within the framework of relaxation kinetics (see Section 4.4). For the most simple situation, the transition between two states, the observed rate constant k_{obs} derived from the analysis of the time-dependent spectral changes is given by

$$k_{obs} = k_f + k_u \tag{5.3}$$

where k_u, and k_f are the rate constants for unfolding and folding step, respectively. These rate constants are related to the equilibrium constant K_{eq} by

$$K_{eq} = \frac{k_u}{k_f} \tag{5.4}$$

In order to determine the unidirectional rate constants, a complete thermodynamic analysis of K_{eq} has to be performed. Using this methodology, fast folding and unfolding processes in the nanosecond time range could be observed for peptides having well-defined secondary structure in the folded state. For alanine-based peptides this is demonstrated in Figs. 5.13 and 5.14. In Fig. 5.13, the static temperature-induced IR difference spectra for the peptide known as SPE_4 [Ac-YGSPEAAA(KAAAA)$_4$-NH$_2$] are shown. In Fig. 5.14 the relaxation kinetics for a temperature jump from 1 to 11 °C are displayed for this peptide in addition to those for the shorter peptide SPE_2. For a detailed analysis of the data the reader is referred to the original publication.

UV resonance Raman spectroscopy can be used for time-resolved studies of folding/unfolding kinetics monitoring the time-dependent changes of the amide III band (Balakrishnan et al. 2006). Also in this instance, a temperature jump is applied by a near-IR pulse to heat the aqueous buffer, and the amide III intensity changes are monitored with 194-nm excitation, revealing fast unfolding processes on the nanosecond-time scale.

The few selected examples described above demonstrate the great potential of vibrational spectroscopies in elucidating folding/unfolding processes of proteins. The combination with isotopic labelling of model peptides appears to be particularly promising as they may allow for a time-resolved description of global and localised structural changes.

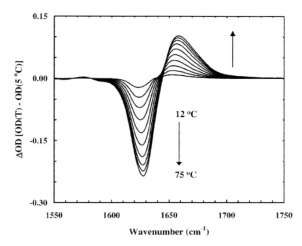

Fig. 5.13 Static temperature-induced folding/unfolding difference spectra of the peptide SPE$_4$ [Ac-YGSPEAAA(KAAAA)$_4$-NH$_2$]. Temperatures are as indicated. The spectra clearly indicate the helix-coil transition (Wang et al. 2004).

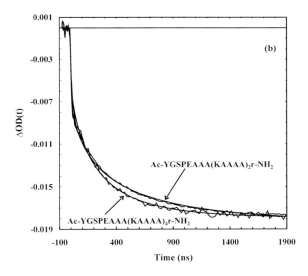

Fig. 5.14 Relaxation kinetics of the SPE$_2$ and SPE$_4$ peptides induced by a temperature jump from 1 to 11 °C, demonstrating the fast unfolding within several 100 ns. They also demonstrate the length dependence of the kinetics (Wang et al. 2004).

References

Asher, S. A., Ianoul, A., Mix, G., Boyden, M. N., Karnoup, A., Diem, M., Schweitzer-Stenner, R., **2001**, "Dihedral Ψ angle dependence of the amide III vibration: a uniquely sensitive UV resonance Raman secondary structural probe", *J. Am. Chem. Soc.* **123**, 11775–11781.

Balakrishnan, G., Hu, Y., Case, M. A., and Spiro, T. G., **2006**, "Microsecond melting of a folding intermediate in a coiled-coil peptide, monitored by T-jump/UV Raman spectroscopy", *J. Phys. Chem. B* **110**, 19877–19883.

Bour, P., Keiderling, T. A., **2004**, "Ab initio modeling of amide I coupling in antiparallel β-sheets and the effect of ^{13}C isotopic labeling on infrared spectra", *J. Phys. Chem. B* **109**, 5348–5357.

Brauner, J. W., Dugan, C., Mendelsohn, R., **2000**, "^{13}C Isotope labeling of hydrophobic peptides. Origin of the anomalous intensity distribution in the infrared amide I spectral region of β-sheet structures", *J. Am. Chem. Soc.* **122**, 677–683.

Brauner, J. W., Flach, C. R., Mendelsohn, R., **2005**, "A quantitative reconstruction of the amide I contour in the IR spectra of globular proteins: from structure to spectrum", *J. Am. Chem. Soc.* **127**, 100–109.

Bredenbeck, J. **2005**, "*Transient 2D-IR Spectroscopy: Towards Ultrafast Structural Dynamics of Peptides and Proteins*", Thesis, University of Zurich, http://www.dissertationen.unizh.ch/2005/bredenbeck/

Bredenbeck, J., Hamm, P., **2003**, "Peptide structure determination by two-dimensional infrared spectroscopy in the presence of homogeneous and inhomogeneous broadening", *J. Chem. Phys.* **119**, 1569–1578.

Chung, H. S., Khalil, M., Tokmakoff, A., **2004**, "Nonlinear infrared spectroscopy of protein conformational change during thermal unfolding", *J. Phys. Chem. B* **108**, 15332–15342.

Creighton, T. E. (Ed.), **1992** "*Protein Folding*", W. H. Freeman and Company, New York.

Creighton, T. E., **1993**, "*Proteins: Structure and Molecular Properties*", W. H. Freeman and Company, New York.

Decatur, S. M., Antonic, J., **1999**, "Isotope-edited infrared spectroscopy of helical peptides", *J. Am. Chem. Soc.* **121**, 11914–11915.

Dong, A., Huang, P., Caughey, W. S., **1990**, "Protein secondary structures in water from second derivative amide I infrared spectra", *Biochemistry* **29**, 3303–3308.

Fabian, H., Schultz, C., Naumann, D., Landt, O., Hahn, U., Saenger, W., **1993**, "Secondary structure and temperature-induced unfolding and refolding of ribonuclease T_1 in aqueous solution: a Fourier transform infrared spectroscopic study", *J. Mol. Biol.* **232**, 967–981.

Fang, C., Wang, J., Kim, Y. S., Charnley, A. K., Barber-Armstrong, W., Smith III, A. B., Decatur, S. M., Hochstrasser, R. M., **2004**, "Two-dimensional infrared spectroscopy of isotopomers of an alanine rich α-helix", *J. Phys. Chem. B* **108**, 10415–10427.

Fang, C., Senes, A., Cristian, L., DeGrado, W. F., Hochstrasser, R. M., **2006**, "Amide vibrations are delocalized across the hydrophobic interface of a transmembrane helix dimer", *Proc. Natl. Acad. Sci. USA* **103**, 16740–16745.

Ganim, Z., Tokmakoff, A., **2006**, "Spectral signatures of heterogeneous protein ensembles revealed by MD simulations and 2DIR spectra", *Biophys. J.* **91**, 2636–2646.

Georg, H., **1999**, "*Untersuchungen zum Faltungsmechanismus der Ribonuklease A mittels Erzeugung eines Temperatursprungs durch schnelles Mischen*", Thesis, Faculty of Chemistry and Pharmacy, Albert-Ludwigs-Universität Freiburg.

Goormaghtigh, E., Raussens, V., Ruysschaert, J.-M., **1999**, "Attenuated total reflection infrared spectroscopy of proteins and lipids in biological membranes", *Biochim. Biophys. Acta* **1422**, 105–185.

Goormaghtigh, E., Ruysschaert, J.-M., Raussens, V., **2006**, "Evaluation of the information content in infrared spectra of protein secondary structure determination", *Biophys. J.* **90**, 2946–2957.

Hamm, P., Lim, M. H., Hochstrasser, R. M., **1998**, "Structure of the amide I band of peptides measured by femtosecond nonlinear-infrared spectroscopy", *J. Phys. Chem. B* **102**, 6123–6138.

Hiramatsu, H., Kitagawa, T., **2005**, "FT-IR approaches on amyloid fibril structure", *Biochim. Biophys. Acta* **1753**, 100–107.

Huang, R., Kubelka, J., Barber-Armstrong, W., Silva, R. A. G. D., Decatur, S. M., Keiderling, T. A., **2004**, "Nature of vibrational coupling in helical peptides: an isotopic labeling study", *J. Am. Chem. Soc.* **126**, 2346–2354.

Kolano, C., Helbing, J., Kozinski, M., Sander, W., Hamm, P., **2006**, "Watching hydrogen-bond dynamics in a β-turn by transient two-dimensional infrared spectroscopy", *Nature* **444**, 469–472.

Krimm, S., Bandekar, J., **1986**, "Vibrational spectroscopy and conformation of peptides, polypeptides, and proteins", *Prot. Chem.* **38**, 181–364.

Kubelka, J., Keiderling, T. A., **2001**, "The anomalous infrared amide I intensity distribution in ^{13}C isotopically labeled peptide β-sheets comes from extended, multiple-stranded structures. An ab initio study", *J. Am. Chem. Soc.* **123**, 6142–6150.

Kubelka, J., Huang, R., Keiderling, T. A., **2005**, "Solvent effects on IR and VCD spectra of helical peptides: DFT-based static spectral simulations with explicit water", *J. Phys. Chem. B* **109**, 8231–8243.

Malinowski, E. R., **1991**, "*Factor Analysis in Chemistry*", Wiley Interscience, New York.

Mikhonin, A. V., Bykov, S. V., Myshakina, N. S., Asher, S. A., **2006**, "Peptide secondary structure folding reaction coordinate. Correlation between UV Raman amide III frequency, Ψ Ramachandran angle, and hydrogen bonding", *J. Phys. Chem. B* **110**, 1928–1943.

Mukamel, S., Zhuang, W., **2005**, "Coherent femtosecond multidimensional probes of molecular vibrations", *Proc. Natl. Acad. Sci. USA* **102**, 13717–13718.

Oberg, K. A., Russchaert, J.-M., Goormaghtigh, E., **2003**, "Rationally selected basis proteins: a new approach to selecting proteins for spectroscopic secondary structure analysis", *Prot. Sci.* **12**, 2015–2031.

Panick, G., Malessa, R., Winter, R., Rapp, G., Frye, K. J., Royer, C. A., **1998**, "Structural characterization of the pressure-denatured state and unfolding/refolding kinetics of staphylococcal nuclease by synchrotron small-angle X-ray scattering and Fourier-transform infrared spectroscopy", *J. Mol. Biol.* **275**, 389–402.

Powell, J. R., Wasacz, F. M., Jakobson, R. J., **1986**, "An algorithm for the reproducible spectral subtraction of water from the FTIR spectra of proteins in dilute solutions and adsorbed monolayers", *Appl. Sectrosc.* **40**, 339–344.

Silva, R. A. G. D., Kubelka, J., Bour, P., Decatur, S. M., Keiderling, T. A., **2000**, "Site-specific conformational determination in thermal unfolding studies of helical peptides using vibrational circular dichroism with isotopic substitution", *Proc. Natl. Acad. Sci. USA* **97**, 8318–8323.

Silva, R. A. G. D., Barber-Armstrong, W., Decatur, S. M., **2003**, "The organization of a β-sheet formed by a prion peptide in solution: an isotope-edited FTIR study", *J. Am. Chem. Soc.* **125**, 13674–13675.

Starzyk, A., Barber-Armstrong, W., Sridharan, M., Decatur, S. M., **2005**, "Spectroscopic evidence for backbone desolvation of helical peptides by 2,2,2-trifluoroethanol: an isotope-edited FTIR study", *Biochemistry* **44**, 369–376.

Surewicz, W. K., Mantsch, H. H., **1988**, "New insight into protein secondary structure from resolution-enhanced infrared spectra", *Biochim. Biophys. Acta* **952**, 115–130.

Torii, H., Tasumi, M., **1992**, "Model calculations on the amide-I infrared bands of globular proteins", *J. Phys. Chem.* **96**, 3379–3387.

Wang, T., Zhu, Y., Getahun, Z., Du, D., Huang, C.-Y., DeGrado, W. F., Gai, F., **2004**, "Length dependence helix-coil transition kinetics of nine alanine-based peptides", *J. Phys. Chem. B* **108**, 15301–15310.

Wi, S., Pancoska, P., Keiderling, T. A., **1998**, "Prediction of protein secondary structures using factor analysis on Fourier transform infrared spectra: effect of Fourier self-deconvolution of the amide I and amide II bands", *Biospectroscopy* **4**, 93–106.

Zhuang, W., Abramavicius, D., Mukamel, S., **2005**, "Dissecting coherent vibrational spectra of small proteins into secondary structural elements by sensitive analysis", *Proc. Natl. Acad. Sci. USA* **102**, 7443–7448.

6
Retinal Proteins and Photoinduced Processes

The development of vibrational spectroscopic techniques for analysing photo-induced processes in life sciences has contributed substantially to the understanding of the molecular functioning of biological photoreceptors. For many representatives of this class of proteins, specifically retinal proteins, the current knowledge of their molecular mechanism is, to a large extent, based on the results obtained by IR and RR spectroscopy. Furthermore, many important methodological developments have been made using retinal proteins as test cases. This chapter will, therefore, focus on the vibrational spectroscopic studies on the function of retinal proteins, describing the essential progress that has been made on the basis of various dedicated vibrational spectroscopic approaches. We will further illustrate that these techniques are not necessarily restricted to photoreceptors but may also be applied to photo-inactive proteins and enzymes when the biological process to be studied can be initiated by an exogenous photochemical trigger.

Retinal proteins represent a class of photo-active proteins that have the cofactor retinal (vitamin A aldehyde) bound to a specific lysine via a Schiff base (Fig. 6.1). In most instances, the Schiff base is protonated, shifting the absorption maximum from 360 nm to the red. A broad range extending from approximately 430 to 600 nm has been found for the various proteins. The specific absorption maximum controls the wavelength-dependent functional response of the retinal protein upon light irradiation as reflected by the so-called action spectrum that follows the absorption spectrum of the retinal protein. It is a characteristic of this chromophore that it can adopt a variety of isomers, differing in the geometry of the double (*trans–cis*) and single (*s-cis–s-anti*) bonds. Several of these isomers occur naturally either in the dark (non-illuminated) or photoactivated states of these proteins. In addition, the Schiff base bond can adopt a *syn-* or *anti-*geometry [see Ottolenghi (1980) for physico-chemical aspects of retinal proteins]. Most retinal proteins are membrane proteins constituted of seven transmembrane helices, and the chromophore is always bound to a lysine on helix 7. They encompass all visual pigments of animals (see below for the visual pigment rhodopsin), other vertebrate photoreceptors such as melanopsin (Doyle et al. 2006), and pinopsin (Holthues et al. 2005). Furthermore, there are also retinal proteins that are photoreceptors in certain bacteria, algae, and fungi. Other retinal proteins function as light-driven ion pumps (proton, anions, see below for bacteriorhodopsin and

Vibrational Spectroscopy in Life Science. Friedrich Siebert and Peter Hildebrandt
Copyright © 2008 WILEY-VCH Verlag GmbH & Co. KGaA, Weinheim
ISBN: 978-3-527-40506-0

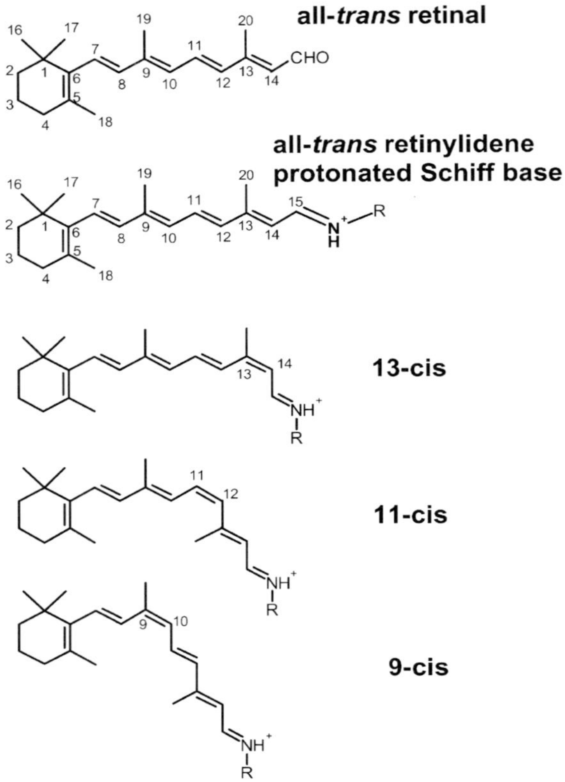

Fig. 6.1 Retinal isomers and retinylidene Schiff base (C=N linkage). The nitrogen of the Schiff base can be protonated.

halorhodopsin) and, in this connection, they are used for light energy conversion. Whereas for many years these ion pumps have been regarded from a biological stand point as rather exotic systems, since the discovery of proteorhodopsin as a putative proton pump in the marine environment, an important bioenergetic role has been attributed to this system (see Spudich et al. 2000, Jung and Spudich 2004 for the various microbiol rhodopsins). Recently, light-switchable channels have been identified as retinal proteins (Nagel et al. 2003). Because of their broad range of functions, and as photoreception, especially in the animal kingdom, has a specific role in sensory physiology, retinal proteins have been of great interest in the past. The relatively simple architecture, and the well-characterised physico-chemical properties of the chromophore and its Schiff base have made them an ideal research target for biophysicists and also for vibrational spectroscopists. The astonishing properties of these proteins have also stimulated theoreticians to study the initial light-induced reaction and the spectroscopic properties of the chromophore in the binding pocket of the protein. In addition, numerous important mechanistic aspects have been discovered using vibrational spectroscopy, thus we regard them as an ideal system for a tutorial. We will focus on the visual pigment rhodopsin and on the light-driven proton pump bacteriorhodopsin. Several useful reviews have appeared on these two proteins. Their crystal structures have been determined, which is a fortunate situation for the application of vibrational spectroscopy. For the demonstration of special techniques of vibrational spectroscopy, a few applications to some other retinal proteins will be included.

6.1
Rhodopsin

Rhodopsin is the visual pigment of vertebrate rod photoreceptor cells (Okada et al. 2001, Sakmar et al. 2002). It is located in the stack of disk membranes in the outer segment of the rod cell. The retinal has the 11-*cis* geometry (Fig. 6.1), and the initial action of light is the isomerisation to the all-*trans* form, occurring in the electronic excited state. The geometrical changes of the chromophore induce changes of the protein that are important for the signalling function of rhodopsin. In the active form generated by light, the cytoplasmatic surface changes its structure allowing the interaction with the so-called heterotrimeric G protein, which is thus activated. Rhodopsin belongs to the large class of G protein-coupled receptors (see biochemistry tutorials for details of the signal transduction by G protein-coupled receptors and by rhodopsin).

The focus here is on the molecular events in rhodopsin initiated through the chromophore by the absorption of light. It is worth noting that vibrational spectroscopy has contributed considerably to our understanding of these processes. Important mechanistic questions have been addressed: (a) what are the molecular processes of the chromophore and what are their time dependence?, (b) what are the molecular processes of the protein together with their time dependence?, and (c) what are the interactions between the chromophore and protein that are responsible for the induction of the protein changes? From their sensitivity and selectivity, resonance Raman (RR) spectroscopy is especially suited to monitoring chromophore changes, whereas infrared difference spectroscopy is sensitive for changes of the protein and of the chromophore. As molecular interactions, e.g., hydrogen bonding, influence molecular vibrations of groups involved in this interaction, both methods can monitor such interactions and their changes.

The photoinduced reaction pathway of rhodopsin is shown in Fig. 6.2. The active form metarhodopsin II (MII), which is capable of activating the G protein, is generated within milliseconds. As indicated, there is a large blue-shift of the absorption maximum to 380 nm, indicating that the Schiff base is unprotonated or that the Schiff base linkage is cleaved and free retinal is generated. All the earlier, still inactive intermediates have absorption maxima above 440 nm, indicating that the Schiff base is protonated. Metarhodopsin I (MI) and MII are in a pH and temperature dependent equilibrium, i.e., if this equilibrium is formed by illumination of rhodopsin, it can be shifted by more acid pH to the active form. The

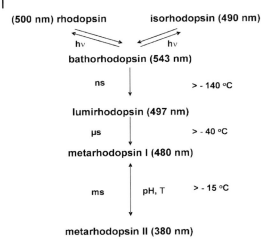

Fig. 6.2 Light-induced reaction pathway of rhodopsin. The absorption maxima of the intermediates in addition to the stabilising temperature and the approximate reaction times are given.

pK_a at 10 °C is 7. The titration indicates that one proton is taken up in the transition. The initial action of light is the isomerisation of the chromophore from 11-cis to all-trans. It is thought that in photorhodopsin (a precursor of batahorhodopsin that cannot be stabilised at low temperature), which is already formed within 400 fs, the chromophore geometry is essentially all-trans, and, as we will show, vibrational spectroscopy has contributed to the proof of this assumption. Thus, there are two important molecular events in the activation process, i.e., the light-induced chromophore isomerisation and the later uptake of a proton causing the transition from the still inactive MI to the active MII. In addition to 11-cis retinal, the protein can also bind the 9-cis isomer. This so-called isorhodopsin state is also inactive, and the absorption of light causes the 9-cis → all-trans isomerisation. The photoreaction proceeds via the same bathorhodopsin intermediate. A peculiarity of many retinal proteins, and also of rhodopsin, is the possibility to extract the chromophore from the protein by illumination in the presence of hydroxylamine. Under these conditions, the Schiff base linkage is hydrolysed and retinal oxime is formed, the chromophore no longer being in the binding pocket of the protein. With retinal oxime, the oxygen at C_{15} is replaced by hydrogen and a hydroxyl group. In most instances, the retinal-free protein can be regenerated by the addition of the chromophore, e.g., of 11-cis or 9-cis retinal. This shows that the chromophore binds spontaneously to the lysine, forming the protonated Schiff base. In this way, rhodopsin or isorhodopsin can be formed. This technique will be extensively employed in the study of retinal proteins, regenerating the protein with isotopically labelled or chemically modified chromophores.

As in many photobiological systems it is possible to stabilise intermediates of the photoinduced reaction at low temperature. In Fig. 6.2, temperatures are given at which the intermediates can be stabilised after illumination of rhodopsin. Although it might be that these intermediates deviate structurally somewhat from those characterised by time-resolved investigations, Cryogenic stabilisation nevertheless represents a convenient technique to study their properties. It is indicated that rhodopsin, bathorhodopsin and isorhodopsin can be converted into each other, i.e., illumination of bathorhodopsin produces rhodopsin and isorhodopsin. The same also holds for the other intermediates up to MI. In low-temperature studies of the intermediates they accumulate upon illumination, and, having an absorption maximum not so different from that of the parent state, they can themselves undergo a light reaction. Thus, the state produced by illumination may contain not only the desired intermediate but also photoproducts from its secondary photoreaction. The situation is usually less severe if flash illumination is used with flash durations of 20 ns and shorter. Such flashes can be obtained from excimer/dye-laser or Nd:YAG laser based systems. Here, one has only to consider the three states rhodopsin, bathorhodopsin, and isorhodopsin. Furthermore, by limiting the light energy in order to keep the primary photoproduct yield in the linear range the generation of secondary photoproducts becomes, in most instances, negligible.

6.1.1
Resonance Raman Studies of Rhodopsin

In RR studies of rhodopsin, one of the problems is its light sensitivity, i.e., the monitoring laser beam itself evokes the photoreaction. In the chapter on sampling methods for RR spectroscopy (Chapter 4) several procedures are described to deal with this problem. What they all have in common is that through a rapid-flow device fresh sample solution is brought into the laser beam under conditions of a low photoconversion parameter [see Eqs. (4.1 and 4.2)], such that the degree of photolysed protein in the sampling volume is kept as low as possible. Furthermore, as rhodopsin does not undergo a cyclic photoreaction, i.e., the system does not reversibly return to the dark state after illumination, the illuminated sample has to be discarded. As described in Section 4.4.1, by using a second laser for pumping the photoreaction and spatially displacing the monitoring laser, spectra of intermediates can be obtained. Such flow systems, however, usually require large amounts of rhodopsin. This can be circumvented by performing experiments at around 80 K, where bathorhodopsin is stabilised. Illumination of the sample with the probing beam will lead to a photostationary state containing rhodopsin, isorhodopsin, and bathorhodopsin. Additional illumination with red light (>570 nm) will shift this equilibrium towards isorhodopsin, whereas additional illumination with blue light ($440 < \lambda < 470$ nm) shifts the equilibrium mainly to the bathorhodopsin side. By using the RR spectrum of rhodopsin from the rapid-flow experiments, the low-temperature spectra of bathorhodopsin and isorhodopsin can be obtained by subtraction of the respective unwanted contributions from the photostationary states. Such experiments will be now described.

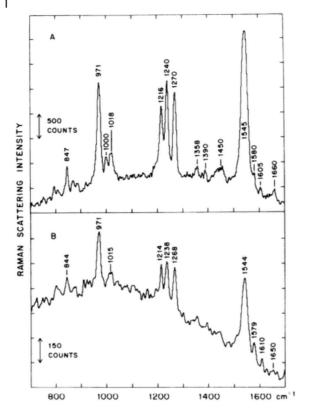

Fig. 6.3 Rapid-flow RR spectra of solubilised rhodopsin (A) and rhodopsin in sonicated disc membranes (B) (Mathies et al. 1976).

In Fig. 6.3 (A) the rapid-flow RR spectrum of rhodopsin is shown (Mathies et al. 1976). The probe laser beam was the 600 nm output from a dye laser pumped by an argon laser. Rhodopsin was solubilised and purified over a hydroxylapatite column. Because of the large resonance enhancement of the Raman scattering, the bands can all be assigned to chromophore vibrations, with no contribution from the protein. Spectrum (B) shows rhodopsin in sonicated disc membranes. With exception of the larger noise in (B), the spectra are identical. This demonstrates that the detergent does not influence the chromophore in rhodopsin, i.e., the binding pocket is not distorted. In order to obtain information on the chromophore configuration, rapid-flow RR spectra of the model compounds have been measured (Fig. 6.4), i.e., the chloride salts of the n-butylamine protonated Schiff base of 11-*cis* (a), 9-*cis* (b), 13-*cis* (c), and all-*trans* (d) isomers of retinal in ethanol. As these model compounds are also light sensitive, the rapid-flow technique had to be applied (Mathies et al. 1977). As we will see, in all RR spectra of retinals and

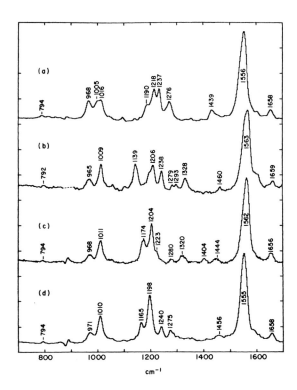

Fig. 6.4 Rapid-flow RR spectra of isomers of protonated retinylidene Schiff bases: (a) 11-*cis*, (b) 9-*cis*, (c) 13-*cis*, (d) all-*trans* (Mathies et al. 1977).

their protonated and unprotonated Schiff bases, a band between 1510 and 1570 cm^{-1} has the highest intensity. It can be assigned to the in-phase combination of all C=C stretching vibrations of the polyene, and it is therefore also termed the ethylenic stretching mode. It can be shown that its position correlates linearly with the wavelength of the visible absorption maximum, a longer wavelength is characterised by a lower frequency. An explanation can be found in a review on retinals and retinal proteins (Ottolenghi 1980).

It is obvious that the spectral range between 1100 and 1300 cm^{-1}, the so-called fingerprint region, is sensitive for the retinal geometry. A comparison of the RR spectrum of rhodopsin with that of the 11-*cis* model compound shows great similarities. This further demonstrates that the spectrum of rhodopsin only contains bands caused by the chromophore. The agreement of the fingerprint region can be seen more clearly in Fig. 6.5, where the 11-*cis* model compound and rhodopsin are directly compared (Mathies et al. 1987). In this figure, the bands are also labelled according to their normal mode composition, based on specific ^{13}C label-

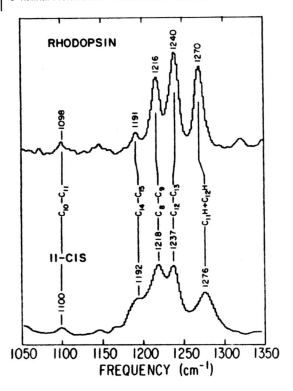

Fig. 6.5 Comparison of the fingerprint region of the RR spectra of rhodopsin and of 11-*cis* retinylidene protonated Schiff base (Mathies et al. 1987).

ling of the chromophore. Similar experiments have also been performed for isorhodopsin and the 9-*cis* model compound, and again, good agreement between the spectra has been obtained (Mathies et al. 1987). Isorhodopsin has been obtained as described above from opsin and 9-*cis* retinal. However, before discussing spectral changes caused by isotopic labelling, we will first demonstrate the feasibility of low-temperature experiments, where the photoequilibria mentioned above are exploited.

6.1.2
Resonance Raman Spectra of Bathorhodopsin

The key experiments to obtaining spectra of this early intermediate are shown in Fig. 6.6 (Eyring and Mathies 1979). Experiments had been performed on solubilised rhodopsin at $-160\,°C$, sufficiently low to stabilise bathorhodopsin. The sample was probed by the 585 nm output from a dye laser, which was pumped by the all-line output of an argon laser. If the sample is only irradiated by the probe

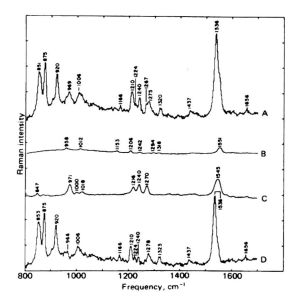

Fig. 6.6 Low-temperature RR experiments of solubilised rhodopsin at −160 °C, probe beam 585 nm from an argon laser: (A) additional all-line illumination from the argon laser containing wavelengths at 514, 496, 488, and 476 nm, (B) probe-beam only, spectrum represents essentially isorhodopsin, (C) rapid-flow RR spectrum of rhodopsin, (D) spectrum of bathorhodopsin, extracted from (A) (Eyring and Mathies 1979).

beam, the equilibrium is largely on the side of isorhodopsin (490 nm), as both rhodopsin (500 nm) and bathorhodopsin (543 nm) absorb at longer wavelengths. If in addition the sample is irradiated with the all-line output of the argon laser, containing mainly wavelengths at 514, 496, 488, and 476 nm, the equilibrium now contains contributions from all three species. The composition of the irradiation sample can be determined by warming up the solution to room temperature. Bathorhodopsin decays to MII, which, in the presence of hydroxylamine, reacts to chromophore-free opsin and retinal oxime. Retinal oxime, absorbing at 360 nm, can be quantified easily, and therefore, the amount of bathorhodopsin can be determined. The remaining absorbance around 500 nm is due to rhodopsin and isorhodopsin. From the corresponding spectra of the pure states, the respective contributions in the mixture can be derived. For irradiation with the probe-beam only, the following composition was obtained: 0% bathorhodopsin, less than 5% rhodopsin, more than 95% isorhodopsin, i.e, the spectrum essentially represents isorhodopsin [Fig. 6.6 (B)]. With the additional all-line illumination [Fig. 6.6 (A)], the composition is now 45% bathorhodopsin, 39% rhodopsin, and 16% isorhodopsin. Using the rapid-flow spectrum of rhodopsin depicted in Fig. 6.3 and repeated in Fig. 6.6 (C), and additionally the spectrum of isorhodopsin, their contributions to spectrum (A) can be subtracted and a pure bathorhodopsin spectrum is obtained (D). As expected, the ethylenic mode is at lower frequencies (1536 cm^{-1}) as the absorption maximum is red-shifted (543 nm). Furthermore, this spectrum shows fairly unusual bands of high intensity between 800 and 950 cm^{-1}.

Now we are ready discuss some of the bands in greater detail. Of special interest is the elucidation of the alterations of the chromophore in this early intermediate. As it has been suggested that the isomerisation has already taken place in the electronic excited state, one would expect an all-*trans* configuration in bathorhodopsin. We have shown above that the fingerprint region is sensitive to the chromophore configuration. Although some similarities with spectrum (d) of Fig. 6.4 (all-*trans* protonated Schiff base model compound) are apparent, a final decision cannot be made on the basis of this simple inspection. However, before analysing this problem, we will first discuss the strong low-frequency modes.

The region between 700 and 1000 cm^{-1} is characteristic for hydrogen-out-of-plane (HOOP) bending vibrations in polyenes, i.e., the hydrogens move perpendicularly to the plane of the retinal. If there are two hydrogens across a double bond, the two vibration couple to an in-phase and an out-of-phase mode. The higher frequency mode is around 960 cm^{-1}, and the lower frequency mode around 750 cm^{-1}. If the hydrogens are in the *trans* position, the higher frequency mode represents the in-phase coupling, and this mode is IR active, but Raman inactive, and the opposite holds for the mode at lower frequency. If the hydrogens are in the *cis* position, the situation is reversed. If there is no second hydrogen across the double bond, the isolated HOOP mode is around 850 cm^{-1}, and it is both IR and Raman active. In the RR spectrum of bathorhodopsin there are three bands in this spectral region: at 853, 875, and 920 cm^{-1}. It is tempting to assign the former two to isolated HOOP modes, whereas the position of the last is unusual. In order to assign these bands to specific modes, rhodopsin has been regenerated with retinals deuterated at specific positions using the regeneration technique described above, and corresponding low-temperature RR spectra have been measured. The results are shown in Fig. 6.7 (a and b). The spectra are labelled with the specific deuteration, and contributions from rhodopsin and isorhodopsin are marked by crosshatching of the bands. These contributions have been determined separately in the same way as for the unlabelled species (Eyring et al. 1982).

Deuteration at position 11 clearly abolishes the band at 921 cm^{-1}, and a new band appears at 743 cm^{-1}. Thus, these bands must contain contributions from 11-H and 11-^{2}H, respectively. Deuteration at position 12 narrows the band at 853 cm^{-1} and shifts it to 842 cm^{-1}. A new band appears at 699 cm^{-1}. Thus, the band at 853 cm^{-1} contains two modes. Double deuteration essentially causes the sum of the two single-label effects. This is fairly unusual, as it shows that the two hydrogens are not coupled, as one would expect, because they are across a double bond. The band at 921 cm^{-1} must be assigned to the 11-HOOP mode, whereas one of the contributions to the band at 853 cm^{-1} is due to the 12-HOOP mode. The other contribution is assigned to the 14-HOOP mode, as the spectrum of

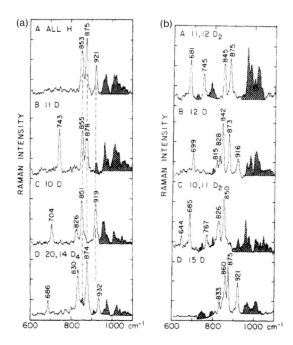

Fig. 6.7 Resonance Raman spectra of bathorhodopsin in the HOOP region. The pigment contains specifically deuterated retinals as indicated (Eyring et al. 1982).

14,20-^{2}H$_4$ demonstrates. Labelling at position 10 abolishes the band at 875 cm^{-1}, which therefore is assigned to the 10-HOOP mode. The shifted band appears at 704 cm^{-1}. Subsequently, the two contributions to the band at 853 cm^{-1} were better resolved (Palings et al. 1989). Thus, in the spectrum of bathorhodopsin only single, essentially uncoupled HOOP modes of high intensity are present. Whereas for positions 10 and 14 this is expected from the structure (a methyl group is across the double bond), the results for positions 11 and 12 are unexpected, as here the two hydrogens are across a double bond and coupling would have been expected. Therefore, two observations in the spectrum of bathorhodopsin have to be explained: (1) the high intensity of the HOOP modes, and (2) the decoupling of the 11,12-HOOP modes.

As discussed in Section 2.2.3, the RR intensity of a normal mode of a molecule correlates with the geometrical change in the resonant electronically excited state along this coordinate. For the high RR activity of the ethylenic mode, the explanation is straightforward as the first electronic transition is associated with a decrease in the alternating bond pattern, such that the bond lengths of the double bonds and single bonds increase and decrease, respectively. Because the ethylenic mode is dominated by the C=C stretching coordinates, a relative large excited-state displacement results and the RR intensity is very high. For the HOOP modes the explanation is not as simple. It has been suggested that the high intensity is caused by distortions of the chromophore in bathorhodopsin around single and double bonds at the respective hydrogens (Eyring et al. 1982). However, it was not so clear which changes in the electronic excited state cause the large intensities. Later, theoretical studies indicated that if the distortions are caused by external constraints as exerted, e.g., by the protein, minor changes in the electronic excited state will reproduce the large intensities (Warshel and Barboy 1982). How to account for the uncoupling of the 11,12-HOOP modes and also for the high frequency of the uncoupled 11-HOOP at 921 cm^{-1}. Calculations using DFT theory showed that specific interaction of the 11,12 region of the retinal with a negative charge [possibly Glu181, Fig. 6.13 (b)] could not reproduce the behaviour. However, twists by approximately 40° of the C_{11}=C_{12} double bond and the C_{12}–C_{13} single bond in the same direction could cause such a decoupling (Yan et al. 2004). Such twists have been suggested to be the main cause for the energy storage in bathorhodopsin, amounting to approximately 120 kJ mol^{-1}. From such calculations it could also be estimated that indeed distortions can contribute considerably to the stored energy, which subsequently drives the conformational changes needed for receptor activation.

Can the RR spectrum of bathorhodopsin provide information on the retinal isomerisation? Figure 6.8 compares the fingerprint region of bathorhodopsin with that of the all-*trans* retinal protonated Schiff base, including the mode assigment based on specific ^{13}C isotopic labelling. The mode pattern is characteristic of the all-*trans* geometry, and very different from that of isorhodopsin (9-*cis*) and rhodopsin (11-*cis*). Thus, the spectrum clearly shows that in the low-temperature bathorhodopsin state, the chromophore has already been isomerised to all-*trans* (Mathies et al. 1987). We will show below that in bathorhodopsin generated at room temperature the isomerisation has also taken place.

However, before discussing more advanced techniques, another band has to be discussed, which has a bearing on the chromophore–protein interaction in particular. In Fig. 6.3 (A) there is a small band around 1660 cm^{-1}. A similar band can be also seen in all the spectra of the model compounds (Fig. 6.4). The band of rhodopsin, which was later located more precisely at 1656 cm^{-1}, shifts down to 1623 cm^{-1} for measurements in ^{2}H$_2$O (Palings et al. 1987). From this, and specifically from ^{13}C labelling of the chromophore at position 15, the band has been assigned to the C=N stretching vibration of the protonated Schiff base. The downshift upon ^{13}C-labelling has a straightforward explanation through the increase of the reduced mass, but the downshift upon deuteration deserves some additional comments. It is explained by a typical feature of molecular vibrations. The so-called C=N stretching mode not only contains the C=N stretching coordinate, but also the in-plane bending of the Schiff base NH group. The extent of the coupling depends on the proximity of the respective frequencies of the intrinsic vibrations. As for the NH bending this vibration is at lower frequency, the mode having predominantly C=N stretching character is shifted to a higher frequency

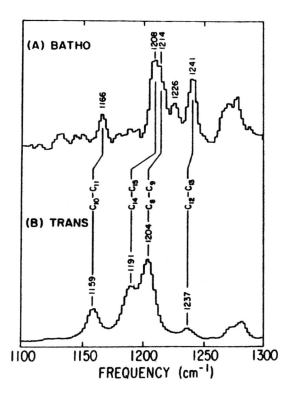

Fig. 6.8 Comparison of RR spectra of all-*trans* retinal protonated Schiff base and of bathorhodopsin in the fingerprint region (Mathies et al. 1987).

by this coupling, whereas the mode with predominantly NH bending character is shifted down.

Thus we are now able to explain qualitatively the deuteration-induced downshift of the C=N mode: for the N^2H bending the intrinsic vibration is at much lower frequency due to the doubling of the mass. Therefore, there is no coupling, and the C=N mode is shifted down and its frequency approximately corresponds to the "pure" C=N stretching mode (slightly coupled to the $C_{15}H$ bending mode). The NH bending frequency not only depends on the NH bending force constant but also on the interaction of the proton with the environment, i.e., on the strength of hydrogen bond that increases the NH bending frequency. Thus, strong hydrogen bonding will cause a larger upshift of the C=N mode, and correspondingly, a larger downshift upon deuteration. Therefore, the position and the isotopic shift of this mode provide important information on the molecular interactions (Deng and Callender 1987, Baasov et al. 1987).

As mentioned, the downshift amounts to approximately 35 cm^{-1} (Palings et al. 1987), which is among the largest found for protonated retinylidene Schiff bases, either in the protein or as model compounds. Therefore, the Schiff base proton must be involved in strong hydrogen bonding. This has been confirmed by the published structure of rhodopsin, although in an unexpected way: whereas it has been assumed that the hydrogen bond would involve the counterion to the protonated Schiff base Glu113, a fixed water molecule serves as the proton acceptor [Fig. 6.13 (b)]. On the other hand, this explains the considerably red-shifted absorption maximum, despite the strong hydrogen bond. Surprisingly, the band in bathorhodopsin is at a similar position (Fig. 6.6), and, in addition, the isotopic shift is similar. This clearly demonstrates that in bathorhodopsin the Schiff base is also protonated.

From the foregoing discussion, one has to conclude that the Schiff base proton is involved in a hydrogen bond similar to that in rhodopsin. Because it appears unlikely that the Schiff base would find another equally strong hydrogen bonding environment, one can conclude that the Schiff base has not moved in bathorhodopsin, despite the chromophore isomerisation. How can these points be reconciled? It is the strong HOOP modes that provide the explanation: as there is significant twisting of the chromophore in the neighbourhood of C_{10} to C_{13}, it can accommodate the isomerisation without any appreciable movement of the Schiff base. Thus, RR spectroscopy has provided important clues to the molecular events in rhodopsin, not only with respect to the chromophore, but also with respect to chromophore–protein interactions.

Owing to the secondary photoreactions mentioned above the intermediates lumirhodopsin and MI are difficult to study by low-temperature RR spectroscopy. The precursor of lumirhodopsin, the blue-shifted intermediate, BSI cannot be stabilised at low temperature. Therefore, time-resolved techniques are required to obtain RR spectra of these intermediates. In the rapid-flow experiments of rhodopsin described above, large amounts of rhodopsin were required even though the sample could be recycled due to the low-bleaching rate by the probing beam. For the intermediates, a large percentage of rhodopsin has to be excited by the pump beam, and therefore, much larger amounts are needed.

A major breakthrough has been achieved by miniaturising the flow-system, using a microscope for focussing the laser beams and a microfabricated flow-cell (Pan and Mathies 2001) (see section 4.4.2.1). In this way probe-only spectra of rhodopsin could be obtained with 12 nmol pigment, and pump-probe spectra with 60 nmol. High quality spectra of BSI, lumirhodopsin, and MI could be obtained with this technique. Needless to say the use of a CCD detector was essential in all these experiments. It is beyond the scope of this chapter to discuss these spectra in greater detail. A few striking results should suffice. Based on the HOOP intensities, it was concluded that in lumirhodopsin and MI the twist of the chromophore is largely relaxed, whereas in BSI, based on the 14-HOOP, a twist in the neighbourhood of C_{14} still persists. In lumirhodopsin, a critical rearrangement of the Schiff base occurs, as the deuteration-induced down-shift is very small (Pan and Mathies 2001, Pan et al. 2002) (see below for results obtained by FTIR experiments).

The isomerisation has been addressed with picosecond RR spectroscopy using a normal capillary rapid-flow system with overlapping pump and probe beams. It could be shown that within 200 fs, the isomerisation has taken place. Within 2–3 ps, vibrational cooling takes place, indicated by band narrowing and a blue-shift of the etylenic mode. The generation of "hot" molecules by fast photochemical processes due to the dissipation of excess energy is well known. The picosecond results indicate that photorhodopsin is essentially a hot bathorhodopsin (Kim et al. 2001).

6.1.3
Fourier Transform Infrared Studies of the Activation Mechanism of Rhodopsin

Here we will demonstrate the potential of IR difference spectroscopy for elucidating the functional mechanism of proteins. By forming difference spectra between functionally well-defined states of the protein, only those groups that are altered in the transition show up in the difference spectrum. As these have to be detected against the background of the unperturbed majority of molecular groups of the protein, the sensitivity has to be high. This is guaranteed by the advantages of FTIR spectroscopy.

Rhodopsin was among those biological systems that were first studied by IR difference spectroscopy (Siebert et al. 1983). The reason is obvious: it can be easily prepared in larger amounts and the reaction is triggered by light. Owing to the irreversible photoreaction, so far only static difference spectra have been measured, intermediates being stabilised by low temperature. Also, the problem of secondary photoreactions now arises. However, as the infrared probing beam does not interfere with the photoinduced reaction, the problem is less severe. By choosing the appropriate wavelength range and duration for the illumination, even using nanosecond-pulsed excitation, secondary photoproducts can be controlled. As the water content in the infrared sample has to be kept low, the sample form requires special attention. In the first experiments, hydrated film samples were used because of their higher IR transmission. However, in later experiments it was recognised that, in particular, the MI/MII transition, i.e., the transition to the active receptor conformation, critically depends on the ionic composition and concentration and on the pH of the aqueous solvent. These values cannot be precisely controlled in such samples. Therefore, the sandwich sample form is preferred.

Figure 6.9 (a) shows the absorption spectrum of a sandwich sample of rhodopsin in disc membranes. The amide I/II bands can be clearly identified. The band at 1735 cm^{-1} is caused by the C=O stretching mode of the phospholipid ester groups. The spectrum of the illuminated sample is superimposed, corresponding to the MII state. It is clear that, if there are spectral changes caused by the illumination, they must be very small as on this scale no differences can be seen. In order to elucidate possible spectral changes, the two spectra are sub-

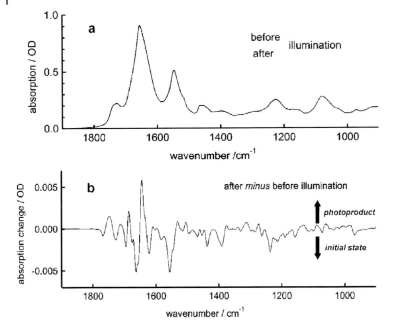

Fig. 6.9 FTIR spectra of rhodopsin in the sandwich cuvette: (a) absorption spectra before and after illumination, (b) difference spectrum, illuminated minus dark, ordinate scale enlarged by a factor of approximately 500, bands now clearly shown (R. Vogel and F. Siebert, unpublished work).

tracted. In practice, the absorption spectrum between the illuminated and non-illuminated sample is formed, i.e.,

$$\Delta A = -\lg[S_{ill}(\nu)/S_{dark}(\nu)] \tag{6.1}$$

where $S_{ill}(\nu)$ and $S_{dark}(\nu)$ are the single-beam spectra of the illuminated and non-illuminated sample, respectively. This is shown in Fig. 6.9 (b) with an extremely enlarged absorbance scale. Here clear bands become visible, having both positive and negative signs. The usual convention is that in the difference spectra negative bands belong to the initial (or dark) state, whereas positive bands are caused by the respective photoproduct. Thus, this difference spectrum describes the transition from the dark state to MII. The spectrum is highly reproducible and demonstrates the sensitivity of the method, which, as we will see, allows detection of the molecular changes of a single amino acid. It is important to mention that, in order to achieve this high sensitivity it was important to limit the spectral range to 1900 cm^{-1} (the lower frequency limit is given here by the BaF$_2$ window material from which the sandwich cuvette is fabricated). In addition, the use of highly sensitive liquid nitrogen-cooled semiconductor detectors (HgCdTe) is mandatory. Thus, with all these technical additions, the feasibility of FTIR difference spectroscopy is demonstrated. In contrast to the rather feature-less spectrum of a protein, many bands can be discerned here. As pointed out above, selectivity is realised as the only groups reflected are those which undergo molecular changes in this transition. From the many overlapping bands the ones selected are those which are caused by the molecular changes.

6.1.3.1 Low-temperature Photoproducts

In Fig. 6.10 we show, as an overview, the difference spectra of the intermediates bathorhodopsin, lumirhodopsin, MI, and MII. With the exception of the last, the intermediates had been stabilised at low temperature, the MII spectrum had been obtained at 10 °C, and pH 5, the low pH favouring MII. Before going into a detailed discussion, it should be mentioned that for these measurements (as for most of the data we present here) rhodopsin was in disc membranes, which

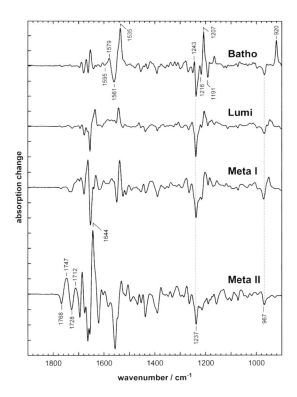

Fig. 6.10 Overview of FTIR difference spectra of bathorhodopsin, lumirhodopsin, metarhodopsin I, and metarhodopsin II. Bathorhodopsin and lumirhodopsin were stabilised at 80 K and 170 K, respectively and MI at 0 °C and pH 8 (R. Vogel and F. Siebert, unpublished).

were dried onto the special window of the sandwich cuvette, as described in the chapter on sampling techniques (Chapter 4). The thickness of the sandwich cuvette is approximately 4 µm. Typically, 512 scans are accumulated for each single-beam spectrum (dark and illuminated). Furthermore, the difference spectra of hydrated film samples published earlier (Siebert et al. 1983) are, up to the MI spectrum, essentially reproduced by the spectra shown here. In these static spectra the noise is very low, e.g., in the bathorhodopsin spectrum, the small wiggles between 1500 and 1300 cm^{-1} and also those around 1760 and 1730 cm^{-1} are reproducible and correspond, therefore, to molecular changes in the system. As always in difference spectroscopy, one has to take into account the possible (partial) overlap of positive and negative bands.

The great challenge in the interpretation of the FTIR difference spectra is the assignment of the bands to specific groups of the protein. Chromophore, amino acid side chains, and the peptide group of the protein backbone possibly contribute. Therefore, the interpretation of the spectra is considerably more complicated than that of the RR spectra. One important task will be the discrimination of chromophore bands from those of the protein. Here, the RR spectra of rhodopsin and of the intermediates may help. We will demonstrate this for the bathorhodopsin spectrum and use the RR spectra of rhodopsin and bathorhodopsin discussed above (Figs. 6.3 and 6.6). As obvious contributions of the chromophore we can identify the positive bands of bathorhodopsin at 1535 cm^{-1} (ethylenic stretching mode), at 1243 cm^{-1} (C_{12}–C_{13} stretching mode, this band overlaps partially with the negative band at 1238 cm^{-1}), 1207 cm^{-1} (C_{14}–C_{15} stretching mode), and the bands below 950 cm^{-1} (HOOP modes discussed above). Thus, the major positive bands are caused by the chromophore. This shows that the method is sensitive enough to detect the vibrations of the chromophore among all the contributions of the protein.

In addition, some of the major negative bands can be identified with the chromophore: the bands at 1238 cm^{-1} (C_{12}–C_{13} stretching mode), 1216 cm^{-1} (C_8–C_9 stretching mode), 1191 cm^{-1} (C_{14}–C_{15} stretching mode), and at 967 cm^{-1} (11,12-HOOP mode). It is tempting to assign the negative band at 1559 cm^{-1} to the ethylenic stretch of the 11-*cis* chromophore. However, its frequency is too high by far. According to the rule relating the frequency to the visible absorption maximum mentioned above the position would correspond to a λ_{max} of approximately 400 nm. Therefore, this broad band must have additional contributions.

The amide II modes of the protein backbone also absorb in this spectral range. Experience demonstrates that in functional difference spectra of proteins rather large bands are frequently observed in this spectral range, which we assign to amide II changes. Here, the negative band at 1559 cm^{-1} probably corresponds to the positive band around 1579 cm^{-1}. This would indicate that the peptide backbone of surrounding amino acids is distorted by the retinal isomerisation. Of course, no changes in secondary structure take place. Thus, it appears that the amide II mode is sensitive to such distortions. However, although this interpretation appears reasonable and is supported by the difference spectra of the later intermediates, for which similar bands are observed, an unequivocal assignment of these features to the amide II mode is still missing. It would require specific labelling of the peptide groups of rhodopsin with ^{13}C and/or ^{15}N. The assignment of the retinal modes has been confirmed by regenerating opsin with isotopically labelled chromophores (Bagley et al. 1985, Ganter et al. 1988).

If the protein backbone is distorted, one would expect that this also causes amide I changes. Indeed, in the bathorhodopsin difference spectrum several bands can be seen between 1600 and 1700 cm^{-1}. However, the RR spectra have shown that the C=N stretching mode of the protonated Schiff base of rhodopsin and bathorhodopsin are also located around 1657 cm^{-1}. Because of their nearly identical positions, one would expect that in the rhodopsin–bathorhodopsin difference spectrum these modes cancel each other out. This is supported by measurements with rhodopsin containing retinal deuterated or ^{13}C-labelled at C$_{15}$, which showed no effect of these labels on the rhodopsin–bathorhodopsin difference spectrum (Bagley et al. 1985). Only in the isorhodopsin–bathorhodopsin difference spectrum measured in ^{2}H$_2$O can the C=N stretching vibrations be seen, as the deuteration-induced downshifts are slightly different for those two states. Thus, the RR results are confirmed by the FTIR data. On the other hand, we can conclude that the bands in the rhodopsin–bathorhodopsin difference spectrum are caused by amide I changes, supporting the identification of amide II modes made above, because, concomitant with changes of the amide II bands changes of the amide I bands are also expected to occur.

Surprisingly, here strong bands also show up in the HOOP region. As they have the same position as in the RR spectra, they are assigned to the same HOOP modes. The high intensity in the RR spectra has been explained by excited state properties, which therefore, cannot explain the IR intensity. Using vibrational analysis based on quantum-chemical calculation it could be shown that HOOP modes gain high intensity if neighbouring C–C single bonds are distorted. Such a distortion causes coupling with in-plane skeleton modes, which, because of the considerable charge alternation along the conjugated C atoms in protonated retinal Schiff bases, introduces large IR intensities. This has been demonstrated specifically for the HOOP modes in the K intermediate of bacteriorhodopsin (Fahmy et al. 1989) (see also below). The negative band at 967 cm^{-1} could be assigned to the 11,12-HOOP mode. Because the hydrogens are *cis* to the double bond, the HOOP mode should be IR inactive or at least have low intensity. Thus, one has to conclude that in the dark state the chromophore is twisted in the neighbourhood of the 11,12-double bond. As the band is absent in rhodopsin containing the artificial chromophore 13-demethyl retinal, i.e., retinal lacking the methyl group at position 13 (Ganter et al. 1990), the twist must be caused by steric hindrance of the 13-methyl group, either with the hydrogen at position 10 or with the protein environment. This simple example demonstrates how system modification, here chromophore modification, and the resulting spectral alterations, can provide important structural and functional insights. Chromophore modification is a widely used technique in the study of retinal proteins (Ottolenghi and Sheves 1989), also in combination with FTIR difference spectroscopy (Siebert 1993, Vogel et al. 2005).

It is beyond the scope of this tutorial to discuss all the difference spectra in detail. In the lumirhodopsin difference spectrum, we will demonstrate how in a spectral region containing contributions from the chromophore and the protein, chromophore bands can be identified by isotopic labelling. As discussed, in the spectral region where the C=N stretching mode of the protonated Schiff base shows up, amide I modes are also present. Therefore, the chromophore band cannot be identified directly. The incorporation of a $^{13}C_{15}$ labelled chromophore should only shift the C=N stretching vibration, and the comparison with the spectrum of the unlabelled system should reveal the C=N modes of rhodopsin and lumirhodopsin. However, owing to the superposition by the amide I bands, a visual comparison is not easy and, as often in the analysis of label effects, difference spectra are subtracted, using bands shown not to be influenced by the label for normalisation. This is demonstrated in Fig. 6.11 (Ganter et al. 1988). It shows an enlarged view of the rhodopsin–lumirhodopsin difference spectra of unmodified rhodopsin, $^{13}C_{15}$ labelled rhodopsin, and the subtraction (labelled minus unlabelled) measured in H_2O (A) and ^2H_2O (B). It is obvious that the label causes spectral changes, however, the identification of the C=N modes is only possible in the subtraction (often called double difference spectrum). Here, the positive band at 1659 cm^{-1} is assigned to the C=N mode of rhodopsin, the larger negative band at 1636 cm^{-1} is assigned to the shifted C=N mode of rhodopsin and to the unshifted mode of lumirhodopsin. The positive band 1624 cm^{-1} is then attributed to the shifted mode of lumirhodopsin. The corresponding analysis of the data obtained for measurements in ^2H_2O assigns the unshifted bands of rhodopsin

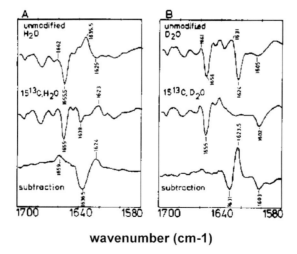

wavenumber (cm-1)

Fig. 6.11 FTIR difference spectra of lumirhodopsin in the C=N stretching region of the Schiff base: (A) (from the top) unmodified rhodopsin, rhodopsin containing 15-^{13}C-labelled retinal, weighted difference between the two spectra, (B) same spectra in ^2H_2O (Ganter et al. 1988).

and lumirhodopsin to 1624 and 1631 cm^{-1}, respectively, and the corresponding shifted modes to 1603 and 1624 cm^{-1}. From these results we can conclude that the deuteration-induced shift of the lumirhodopsin mode is only 5 cm^{-1} (from 1636 to 1631 cm^{-1}) indicating very weak coupling with the NH bending mode, and therefore, a considerable change to the environment of the Schiff base. Thus, as the photoreaction proceeds to the later intermediates larger changes in the system take place, which is also supported by the larger amide I bands in the rhodopsin–lumirhodopsin difference spectrum. The unusual properties of the C=N mode of lumirhodopsin were later confirmed by the rapid-flow RR spectra obtained with the microfabricated flow cell mentioned earlier. Thus, the analysis of the lumirhodopsin difference spectrum has provided important information on changes to the chromophore–protein interaction. It has also demonstrated the power of isotopic labelling if the induced spectral changes are carefully analysed. The formation of the so-called double difference spectra is a characteristic tool for the identification of bands shifted by isotopic labelling.

6.1.3.2 The Active State Metarhodopsin II (MII)

We will now turn to the analysis of the spectrum reflecting the transition to MII, i.e., the transition to the active state. Here, it is important to mention that the use of the sandwich cuvette is essential. Only with this sampling technique the same pK_a for the MI/MII equilibrium was obtained as measured for dilute samples (Vogel and Siebert 2003). In the MII spectrum (Figs. 6.9 and 6.10), the negative bands of the chromophore of rhodopsin can be recognised between 1300 and 1100 cm^{-1}. The lack of strong positive bands in this region is explained by Schiff base deprotonation, which reduces charge alternation in the conjugated system, and therefore, the IR intensity of the fingerprint modes. The ethylenic modes of rhodopsin and MII cannot be identified due to the superposition by amide II bands. Large, new bands show up above 1600 cm^{-1}. These bands are taken as a signature for the transition to the active state and are interpreted as amide I changes. The intense amide I bands reflect larger backbone changes of the protein as compared with the earlier intermediates. It is probable that they correspond to the relative movement of helices 3 and 6 detected by the spin-label technique, increasing the distance between the cytoplasmic ends (see the reviews on rhodopsin Okada et al. 2001; Sakmar et al. 2002), although a clear molecular interpretation of the amide I bands in the difference spectra is still missing.

The bands above 1700 cm^{-1} deserve special attention. They exhibit upon ^{1}H/^{2}H exchange an isotopic down-shift of 5–10 cm^{-1}, characteristic of the C=O stretching mode of protonated carboxyl groups. As these bands are also present at pH 8 if rhodopsin is solubilised in the detergent dodecylmaltoside (DM), which makes the transition to MII independent of pH, the bands cannot be caused by peripheral carboxyl groups as they would deprotonate at this pH. According to the amino acid sequence, and confirmed by the structure, the intra-membrane carboxyl groups are Asp83, Glu113, and Glu122. Glu113 has been identified by mutagenesis as the counterion of the protonated Schiff base in rhodopsin. It is therefore deprotonated and cannot contribute to the negative bands. In order to

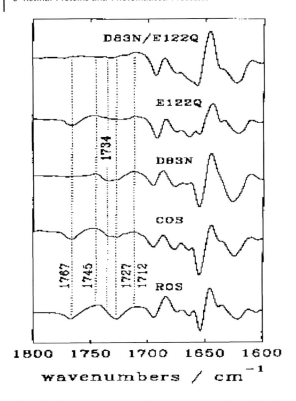

Fig. 6.12 Metarhodopsin II difference spectra of solubilised wild type (COS) and mutant rhodopsin, and of rhodopsin in disc membranes (ROS) (Fahmy et al. 1993).

test whether these bands can be attributed to single amino acid side chains, measurements of the mutants had to be carried out. This is shown in Fig. 6.12 (Fahmy et al. 1993).

Here, the measurements of the mutants were performed in the detergent DM to guarantee the formation of MII. The lower two spectra compare the results obtained for the wild type protein in membranes and in detergent. Although small deviations are present, overall the spectra are very similar. The small deviation around 1730 cm^{-1} could be later attributed to changes of the ester C=O stretching mode of the lipid phase, indicating an environmental change of one or a few lipid molecules upon receptor activation (Isele et al. 2000). The identification of this contribution demonstrates the great advantage of IR spectroscopy, which in principle is able to detect molecular changes of all constituents of the system. The upper three spectra show the influence of the single site mutations Asp83Asn, Glu122Gln, and of the double mutation. In the double mutant, only the positive band at 1712 cm^{-1} remains. Thus, the other difference bands can be attributed to Asp83 and Glu122, which undergo changes in hydrogen bonding,

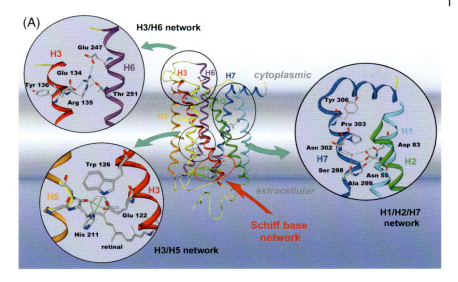

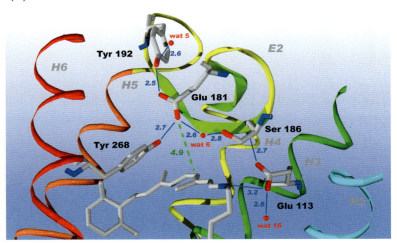

Fig. 6.13 (A) Structure of rhodopsin and enlarged views of certain regions demonstrating the location of Glu122 and of Asp83. (B) Retinal binding pocket of rhodopsin, demonstrating the location of the counterion Glu113.

but always remain protonated [see Fig. 6.13 (A) for location of these carboxyl groups within the protein]. This clearly demonstrates the high sensitivity of FTIR difference spectroscopy detecting changes of a single amino acid out of a protein with a molecular weight of 40 kD. Through mutation of Glu113 to alanine, the positive band at 1712 cm^{-1} could be attributed to protonation of the counterion Glu113 (Jäger et al. 1994) [see Fig. 6.13 (B) for the retinal binding pocket].

As a final example, we will show that, even though many details of spectra cannot be interpreted at a molecular level, the characteristic bands of MII can be used to monitor the presence of the active state independently of the absorption properties of the chromophore. It is known that the chromophore is hydrolysed in MII, and it is interesting to ask what the impact is of the chromophore release on the protein structure. Thus, the decay of MII is followed at 30 °C in the infrared. The half-life is about 2.8 min, so after 15 min, no MII remains. The corresponding time-dependent spectra are shown for pH 7.0 and 4.0 (Fig. 6.14, a) (Vogel and Siebert, 2001). It is obvious that at pH 7.0 the bands characteristic of the MII state decay, whereas at pH 4 the bands, after a minor decay persist over a long period. The final difference spectra (with respect to the dark state) are shown in Fig. 6.14 (B). The spectrum obtained at pH 4.0 has all the characteristics of the MII state. On the other hand, the spectrum obtained at pH 7.0 is very different as the MII bands have decayed. This shows that the state obtained at pH 7 represents an inactive state. Thus, pH can switch the opsin state from active (pH 4) to inactive (pH 7).

The spectra obtained at intermediate pH values therefore represent a pH-dependent mixture of the pH 4 spectrum (active state) and the pH 7 spectrum (inactive state). The corresponding fit yields a titration curve for the pH-dependent ratio of the active/inactive opsin states. For the determination of the correct ratio one had to take into account that the spectrum obtained at pH 4 still contains a small contribution of the pH 7 spectrum. This titration curve is shown in Fig. 6.14 (C). Essentially the same spectra and titration curve are obtained in the presence of hydroxylamine, which causes the cleavage of the retinal Schiff base because the oxime of retinal is formed ($-C_{15}OHH_2$). Therefore, the spectrum obtained at low pH cannot be explained by the presence of the, still bound, all-*trans* retinal. In this figure the usual MII/MI pH dependent ratio is also shown, which has the inactive–active transition at considerably higher pH. These results imply that the formation of the active state of rhodopsin does not require the presence of retinal. However, without the all-*trans* retinal being bound, the active state can only be obtained at non-physiological pH values. The bound all-*trans* retinal in MII is required to shift the active/inactive state equilibrium into a physiological pH range.

The idea that the protein itself can adopt multiple conformations is a common view in G-protein coupled receptor research. The binding of a ligand serving as an agonist (here all-*trans* retinal) shifts the equilibrium to the active state, whereas the binding of the inverse agonist (here 11-*cis* retinal) stabilises the inactive state. This important result demonstrates that infrared difference spectroscopy can provide novel insight into the functional aspects of rhodopsin states, for

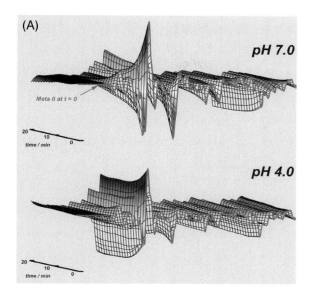

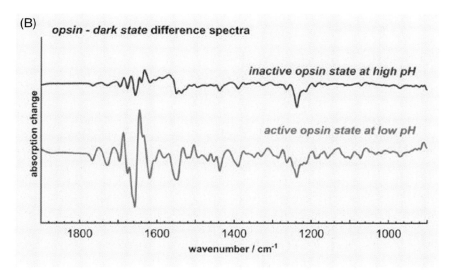

Fig. 6.14 (A) Time-resolved FTIR difference spectra monitoring the decay at pH 7 (top) and 4. At pH 4 the bands due to the active state do not seem to decay (Vogel and Siebert 2001). (B) FTIR difference spectra of the inactive, high pH (top) and active, low pH opsin states obtained after the decay of metarhodopsin II (Vogel and Siebert 2001).

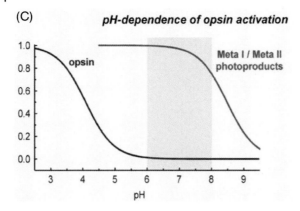

Fig. 6.14 (C) pH dependence of the metarhodopsin I/II equilibrium and of the opsin states (Vogel and Siebert, 2001).

which 3D structure data are not available and for which the chromophore cannot serve as a monitor. The findings are further relevant in a wider context as they point to mechanistic aspects that are most likely common to G-protein coupled receptors in general.

6.2
Infrared Studies of the Light-driven Proton Pump Bacteriorhodopsin

Bacteriorhodopsin from *Halobacterium salinarum* is a retinal protein. The structure is shown in the Introduction, Chapter 1 (Fig. 1.1) (see Lanyi and Luecke 2001; Haupts et al. 1999 for reviews). Superficially, there are several similarities to rhodopsin: seven transmembrane helices connected by cytosolic and extracellular loops make up the protein, and the chromophore, here all-*trans* retinal (Fig. 6.1), is connected via a protonated Schiff base to a lysine located on helix 7. As in rhodopsin, the counterion to the protonated Schiff base is located on helix 3, here Asp85. However, the arrangement of the seven helices is very different, and the protein functions as a light-driven proton pump, translocating protons from the cytosol to the extracellular space against a proton gradient. It is, therefore, a light energy converter creating a proton gradient, and therefore, it undergoes the cyclic photoreaction depicted in Fig. 6.15. The numbers subscript to the name of the intermediates (one letter) represent the respective absorption maxima. At room temperature, the photocycle (as the cyclic reaction scheme is called) lasts about 10 ms. Similarly as for rhodopsin, a large blue shift, here to 412 nm, is observed for two later intermediates, and it can be concluded that the Schiff base must be deprotonated in these M-states. In the subsequent N-state, the absorption maximum is again back in the 550 nm region, demonstrating that the Schiff base is reprotonated in this step. In the identification of these proton transfer

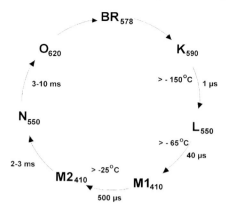

Fig. 6.15 Photocycle of bacteriorhodopsin. The absorption maxima of the intermediates, the transition temperatures, and the approximate reaction times are indicated.

steps it has been assumed that it is the Schiff base proton which is being pumped.

The molecular processes have been studied in a similar manner to those for rhodopsin, by RR and FTIR spectroscopy. In contrast to rhodopsin where only the structure of the inactive dark state is available, the structure of several intermediates of the photocycle could be determined by illumination of the crystals at low temperature, cryotrapping the intermediates. It is not the purpose of this chapter to repeat the description of similar experiments to those performed with rhodopsin. The present knowledge of the mechanism of this proton pump is reviewed in several excellent articles cited above, and also includes results obtained by vibrational spectroscopy (Mathies et al. 1987; Althaus et al. 1995; Gerwert 1993). RR spectroscopy has shown that the all-*trans* retinal chromophore is isomerised to the 13-*cis* geometry already in the J-state (precursor of the K-state), and that it is thermally converted back into the all-*trans* geometry in the O-state. Both static (K-state) and time-resolved techniques (J-, K-, and O-states) had been applied. In addition to the flow system described for rhodopsin, a sampling device using the rotating cuvette (see section 4.4.1.1) was especially successful in obtaining time-resolved RR spectra, as bacteriorhodopsin undergoes the cyclic photoreaction (Althaus et al. 1995). Here we will now focus on the application of time-resolved FTIR spectroscopy using the step–scan technique described in section 3.1.4.3.

It has been argued that, in order to accomplish vectorial proton transport from the cytosol to the extracellular space involving the Schiff base proton, two M-states must exist ("early" and "late"), one in which the proton is transferred from the Schiff base to the extracellular side of the protein, and the other in which the Schiff base will be reprotonated from the cytosolic side. As RR experiments provided no differences of the chromophore geometry in early and late M-states, it appears reasonable to assume that the protein conformations differ, which might be detected using time-resolved IR spectroscopy. Hence, because the cycling time is relatively short (around 10 ms), bacteriorhodopsin is particularly suitable for studies using time-resolved step–scan FTIR spectroscopy. The same sample can be excited as many times as is required for the complete step–scan run and the noise can be reduced using signal averaging. Thus, as a characteristic application of the step–scan technique we will demonstrate how this method is able to detect time-resolved chromophore and protein changes in bacteriorhodopsin. The highest time resolution obtained is 30 ns (Rödig et al. 1999). In addition, we will use the time-resolved spectra to introduce a methodology suitable for the analysis of the time-resolved spectra in terms of reaction pathway and reaction intermediates.

The time-resolved data have been obtained with a step–scan instrument using two transient recorders for acquisition of the time-resolved changes of the interferogram, and the sample has been excited by a frequency-doubled Nd:YAG laser, having a pulse width of 5 ns. The pulse energy has been reduced to approx. 1.5 mJ, and the sample area of the hydrated film sample had an diameter of approximately 6 mm. More details of the experimental procedures can be found in the original publication (Rödig et al. 1999). Here, we will only discuss the results obtained with a somewhat reduced time-resolution of 600 ns, as they are more relevant for the identification of the different M-states.

In Fig. 6.16 a 3D representation of the time-resolved spectra is shown (Rödig and Siebert, 1999) from which spectra at a given time-slice, or the time-resolved traces at a given wavenumber can be deduced. In Fig. 6.17 the corresponding extraction of the spectra is shown. A few important results can clearly be deduced. The band at 1190 cm^{-1} observed in the early spectra is indicative of the 13-*cis* geometry of the chromophore, confirming the light-induced all-*trans* $\rightarrow$ 13-*cis* isomerisation detected earlier by RR experiments (Mathies et al. 1987). The band at 983 cm^{-1} has been assigned to the 15-HOOP vibration using isotopic labelling. It shows that the chromophore in the early stages is twisted around the 14–15 single bond. Because it disappears at later times, it can be concluded that the twist relaxes. We will discuss some possible consequences of this twist below. The large negative band at 1526 cm^{-1} is assigned to the ethylenic mode of the initial state, and the positive band at 1512 cm^{-1} to that of the early photoproduct, indicating that the absorption maximum in the visible is red-shifted (see Fig. 6.15 for the photocycle). Using isotopically labelled retinals the negative band at 1640 cm^{-1} could be assigned to the C=N stretch of the protonated Schiff base of the initial state.

Of special importance is the band at 1761 cm^{-1} arising in the time range when M is formed, i.e., when the Schiff base becomes deprotonated. ^{2}H$_2$O causes a 5-cm^{-1} downshift. This, together with its position, strongly indicates that the band is caused by protonation of a carboxyl group (see the discussion of the MII spectrum of rhodopsin where a similar band could be identified). This has been confirmed by isotopic labelling (Engelhard et al. 1985): ^{13}C-labelled aspartic acid with the label position at the carboxyl group was introduced into the protein. This

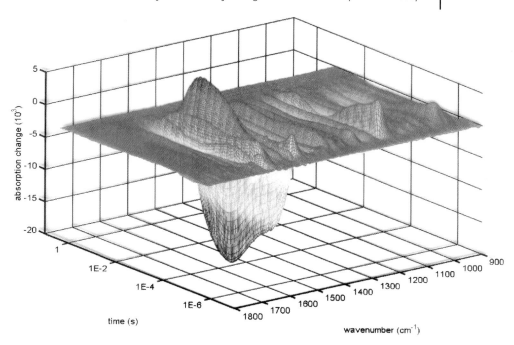

Fig. 6.16 3D representation of the time-resolved step–scan FTIR difference spectra of bacteriorhodopsin (Rödig and Siebert, 1999).

caused a 40-cm^{-1} downshift of the band, proving that the band must be caused by an aspartic acid. How can an amino acid in a protein be isotopically labelled? In this instance, the bacteria were grown in a minimum medium containing a mixture of amino acids. The expression of bacteriorhodopsin can be stimulated by switching off the oxygen supply. Simultaneously, the medium was exchanged for one containing the labelled aspartic acid. Aspartic acid is partially metabolised, and the bacteria can synthesise aspartic acid themselves. Therefore, the extent of labelling amounted to only 60%. As mentioned above, the counterion to the protonated Schiff base is Asp85, and it has been assumed that in the M state the counterion becomes protonated. This was later confirmed by studies of mutants (Braiman et al. 1988). The demonstration of the isotopic shift induced by introduction of labelled aspartic acids has been essential for proving the sensitivity of FTIR difference spectroscopy, allowing detection of changes of single amino acid side chains.

In order to address the question of the two M-states, the time-resolved spectra have to be analysed in a more quantitative manner. A general problem in physical chemistry is to derive a reaction scheme from time-resolved spectroscopic data. Depending on the actual situation, the problem can be fairly complex. The photocycle described above is only an approximation for the reaction path, e.g. back-

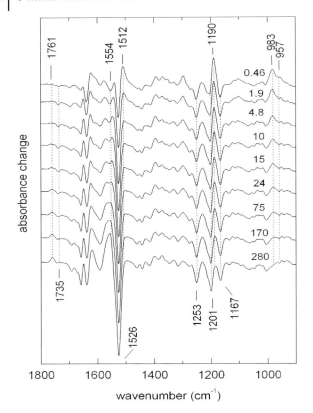

Fig. 6.17 Time-slice of the time-resolved difference spectra of bacteriorhodopsin shown in Fig. 6.16, times given in µs (Rödig et al. 1999).

reactions and branching have been proposed. Very often it is assumed that the reactions follow 1st order kinetics and that, therefore, the time course of the data (here of the spectra) can be approximated by a sum of exponentials. Thus, it is assumed that everywhere in the spectrum the same time-dependence is observed. In order to derive the time constants, the time-traces are described over the complete spectral range to a corresponding sum of exponentials, of which the prefactors $a_k(v)$ are termed amplitude spectra. This is therefore called a global fit.

$$S(v,t) = \sum_{k=1}^{n} a_k(v) \exp(-t/\tau_k) + a_0(v) \qquad (6.2)$$

$S(v,t)$ denote the time-resolved spectra.

There are several commercial software packages available to perform this nonlinear global fitting procedure. As a result, the time constants τ_k and the amplitude spectra $a_k(\nu)$ are obtained. Even if in this way a reasonably accurate description of the spectra can be often obtained, to derive a more complicated reaction scheme containing back-reactions and/or branching, additional assumptions have to be made. Here is not the place to discuss this complicated issue. We will just show that a simple reaction scheme is described by such a sum of exponentials with corresponding amplitude spectra.

We assume the following reaction:

$$P_1 \xrightarrow{k_1} P_2 \xrightarrow{k_2} P_3 \cdots \xrightarrow{k_{n-1}} P_n \tag{6.3}$$

which is termed the unidirectional reaction scheme, i.e., no back-reactions are included and all steps are first-order reactions. The last species may represent the first species, as is the situation for cyclic reactions. It is clear that this simple scheme is not adequate for more complicated reactions. However, any linear reaction scheme can be transformed into the scheme above. As a consequence, in general, the species P_i do not represent pure states with respect to their spectral properties but may be mixtures of pure states, that are in a fast equilibrium which is established as soon as P_i is formed. It is straightforward to show that the spectra of species P_i can be calculated from the amplitude spectra obtained by the global fit procedure (Rödig et al. 1999). This simple correlation is a great advantage of this method. It has been applied for the analysis of the time-resolved spectra described above (Figs. 6.16 and 6.17). The fitting procedure yielded the amplitude spectra shown in Fig. 6.18. Eight exponentials were necessary to describe the time-dependence.

In Fig. 6.19 the spectra of the P_i states are shown, labelled with the corresponding decay time constants. It is beyond the scope of this presentation to discuss all the different aspects of the spectra, and more details can be found in the original publication (Rödig et al. 1999). The first spectrum corresponds to the early red-shifted intermediate (K or KL), the next spectrum characterises the L-state, both states exhibiting the 15-HOOP band at 983 cm^{-1} mentioned above. This shows that the C_{14}–C_{15} single bond is twisted in these two states. The twist in the L-state could be responsible for the reduction of the pK_a of the Schiff base, which would, at least partially, initiate the proton transfer to Asp85. It is interesting that for the low-temperature L-state this HOOP band is not seen, and the consequences for the proton transfer step have bee discussed (Rödig et al. 1999). This example demonstrates that in many instances it is important to obtain time-resolved IR spectra. The proton transfer to Asp85 can be observed, although still with low amplitude, in the next spectrum, where a small positive band at 1761 cm^{-1} can clearly be discerned. The band increases in the next two spectra, which, in addition, show decreasing contributions from the HOOP band. Thus the three spectra with decay time constants of 40, 134, and 544 µs represent L–M mixtures with increasing contributions from M. By subtracting the L contributions, it could be shown that the size of the bands at 1616 and 1657 cm^{-1} is very small in the M-state with 40-µs decay time, considerably larger in the next M-state, and an additional small increase of these two bands is observed for M with a decay time of

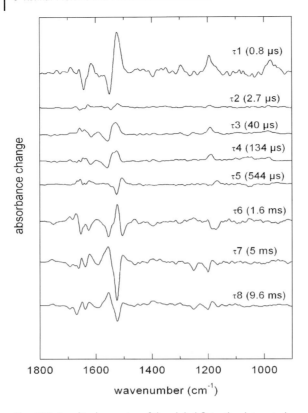

Fig. 6.18 Amplitude spectra of the global fit to the data set shown in Fig. 6.16 using eight exponentials (Rödig et al. 1999).

544 μs. These bands, which increase progressively, can be interpreted as amide I and amide II bands. Otherwise, the three spectra of the M-states are very similar. Thus, it appears that the M-states differ in the extent of conformational changes reflected by these two amide bands, confirming the hypothesis of M-states with differing protein conformation. These protein conformations may be required to accomplish a vectorial proton transport.

It is appropriate to discuss under which conditions this time-resolved step–scan technique can be applied to other systems. In the measurements described here, 8 or 16 signals have been averaged at each mirror position. Because of the short cycling time of bacteriorhodopsin, the repetition rate of the laser could be kept at 5 Hz, resulting in times of 1.6 and 3.2 s, respectively, for the acquisition of the time-resolved change of the interferogram at each mirror position. With a resolution of 8 cm^{-1} and a high-frequency cut-off of 1900 cm^{-1} the number of mirror positions (see section 3.1.4.3) amounts to approximately 540. Further, if one takes into account the time needed to change the mirror position, which is approximately 0.5 s, the total measuring time amounts to approximately 19 min

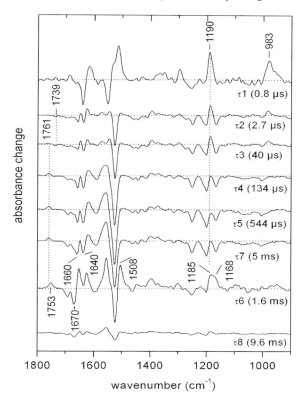

Fig. 6.19 Spectra of the intermediates calculated from the amplitude spectra of Fig. 6.18 (Rödig et al. 1999).

for 8 averages and 34 min for 16 averages. With such a run the signal to noise ratio is usually not sufficient, and the measurements have to be repeated until the number of flashes at each mirror position is around 64. Thus, the total measuring time is finally $8 \times 19 = 152$ and $4 \times 34 = 136$ min, respectively. This estimate holds for a time-resolution of approximately 600 ns. If the time-resolution is increased to 30 ns, the number of spectra to be averaged has to be increased by a factor of $(600/30)^{1/2} = 4.5$, increasing the total measuring time correspondingly. For bacteriorhodopsin, with its short cycling time, such measurements could still be performed. From this discussion it is clear that the complete time of the photoreaction is a critical parameter for the application of the time-resolved step–scan technique. If the reaction is not reversible, means would have to be developed to allow the quantitative exchange of the sample after each excitation.

As with the amide I changes in the FTIR difference spectra of rhodopsin, for the bands described here only a qualitative interpretation is possible: they indicate structural changes of the peptide backbone. For the MII spectra of rhodopsin we have argued that the strong amide I bands might reflect the tilting of helix

6 identified, for example, by the spin label technique (Farrens et al. 1996). Structural studies have indicated a tilt of helix 6 of bacteriorhodopsin in the late M-state (see reviews on bacteriorhodopsin Lanyi and Luecke 2001; Haupts et al. 1999). We have some indication that the amide I changes observed in bacteriorhodopsin could be caused by the distortion of the helix in the hinge region of the helical tilt (Hauser et al. 2002). Thus, it might be that the larger amide I changes observed in the two later M-states described here reflect the tilt of helix 6. However, direct proof is not available. Therefore, a better understanding of the amide I changes is highly desirable. In Chapter 5 we focussed on the molecular interpretation of the amide I band of proteins and peptides. Up to now the analysis has mainly been restricted to how the secondary structural elements and solvent molecules interacting with the peptide backbone influence the amide I band. However, if a more detailed description of the amide I band is available, which includes local geometries and distortions of the peptide backbone, it appears feasible that a molecular description of the amide I changes in the FTIR difference spectra becomes possible. In most instances, the protein conformations of the corresponding protein states do not differ in secondary structure but rather in the local rearrangements of the protein backbone.

6.3
Study of the Anion Uptake by the Retinal Protein Halorhodopsin Using ATR Infrared Spectroscopy

Halorhodopsin is a light-driven anion pump and shares many homologies with the light-driven proton pump bacteriorhodopsin. However, remarkably, the counterion to the Schiff base and proton acceptor in bacteriorhodopsin, Asp85, is replaced by a threonine, and Asp96, the proton donor for reprotonation of the Schiff base, is replaced by an alanine.

Halorhodopsin is found in two types of archaebacteria, in *Halobacterium salinarum* (termed HsHR) and in *Natronobacterium pharaonis* (termed NpHR). Although there are some differences in the amino acid sequence, most of the amino acids in the inner part of the protein, and especially those around the retinal chromophore, have homologue counterparts. Anions are pumped from the cell exterior to the cytosol, i.e., opposite to the direction of proton pumping in bacteriorhodopsin. The anion pumping is initiated by the light-induced all-*trans* → 13-*cis* isomerisation. The vectorial anion translocation is accomplished by first ejecting the anion into the cytosol and later by the uptake of the anion from the extracellular side. This latter step completes the photochemical reaction, which, as with bacteriorhodopsin, is a photocycle. Thus, there is a state in which the protein is anion-free. This state is called O, and it appears in about 1 ms, and it decays via a bimolecular reaction with the uptake of the anion. Apart from chloride, iodide, bromide, and nitrate are pumped by NpHR. In view of the very different shape of nitrate as compared with the mono-atomic anions, it appears that the path along which the anion moves through the protein must be unspecific.

6.3 Study of the Anion Uptake by the Retinal Protein Halorhodopsin

The crystal structure of HsHR has shown that in the dark state the anion is bound in the neighbourhood of the protonated Schiff base (Kolbe et al. 2000). Thus, the negative charge of the anion serves as the counterion for the protonated Schiff base, because Asp85, the counterion in bacteriorhodopsin, is no longer present. A review on halorhodopsin based on the structure of HsHR has recently been published (Essen, 2002).

An anion-free state of NpHR can also be produced by lowering the anion concentration below the binding constant. This is 1, 2.5, 3, and 16 mM, for bromide, chloride, iodide, and nitrate, respectively. By anion depletion, the colour of NpHR changes from purple to blue (blue NpHR), as the absorption maximum is red-shifted to 600 nm. This is in agreement with the anion being the counterion of the protonated Schiff base, as removal of the counterion usually causes a red-shift of the absorption maximum. Of particular interest, is whether the static, anion-free state produced by low anion concentration agrees with the anion-free O-state produced during the photocycle. For this, we compare the time-resolved difference spectrum of the O-intermediate with the difference spectrum obtained by following the binding of anions to blue NpHR. The ATR IR method is the most suitable for this (see section 4.2.1.2).

ATR samples of membrane proteins are usually prepared in the following way: the surface of the ATR crystal is overlaid with the suspension of membranes containing the membrane protein. The aqueous solvent is dried-off, causing the membranes to more-or-less stick firmly to the ATR surface. This process has been described well for bacteriorhodopsin and is shown in Fig. 6.20 (Heberle and Zscherb 1996). Initially, the spectrum is very similar to the spectrum of water. Upon drying-off the water, the bands for water at 3300 cm^{-1} (OH stretching), the broad band around 2130 cm^{-1} (combination band), and the broad fea-

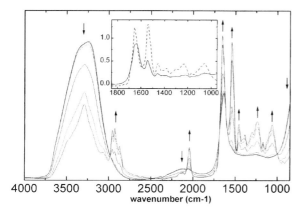

Fig. 6.20 ATR-FTIR spectra monitoring the adsorption of bacteriorhodopsin, containing purple membranes, to the surface of the ATR element while drying off the aqueous solvent. Insert shows the ATR-FTIR spectrum of the dried film (dashed line), and of the film after the addition of water (solid line) (Heberle and Zscherb 1996).

ture below 1000 cm^{-1} (intermolecular libration) disappear. However, the bands around 3290 cm^{-1} (amide A), 1650 cm^{-1} amide I, 1550 cm^{-1} amide II, and numerous bands below 1550 cm^{-1} caused by the protein and the membrane lipids increase, because the membrane comes into close contact with the ATR surface. Water also has an absorption band at 1650 cm^{-1} (OH bending mode). As the band intensity at this position increases upon drying, one can conclude that the water contributes less to the absorption as compared with the amide I band of bacteriorhodopsin. If the membrane stack on the ATR surface is now overlayed with aqueous buffer, the spectrum shown in the insert is obtained. Because the amide II band decreases considerably, swelling of the membrane stack takes place. As has been emphasised in section 4.2.1.2, the thickness of the aqueous layer above the membrane stack is not important because of the limited penetration depth of the IR beam. If the swelling has reached a stable state, binding studies can be performed by adding the respective compound to the buffer with no nonspecific distortion of the sample. This sample form guarantees the native conditions at defined pH and salt concentration. It has been used for time-resolved step–scan measurements of bacteriorhodopsin under well-defined pH conditions (Heberle and Zscherp 1996).

Films have been prepared from NpHR, similar to those for bacteriorhodopsin, with the anion binding to the anion-depleted form of halorhodopsin (i.e., blue NpHR) being followed. The corresponding spectra are shown in Fig. 6.21 (Guijarro et al. 2006). The uppermost spectrum shows the difference spectrum induced by chloride binding. Negative bands are due to blue NpHR, whereas positive bands reflect NpHR. The largest band is caused by the ethylenic mode (C=C stretching vibration) of the retinal chromophore, which has a lower position (1511 cm^{-1}) in blue NpHR as compared with NpHR (1525 cm^{-1}), in agreement with the different absorption maxima (600 versus 578 nm). It is remarkable that the chromophore in blue NpHR shows pronounced HOOP (hydrogen-out-of-plane) modes located around 960 cm^{-1}. As has been explained for rhodopsin and bacteriorhodopsin, the HOOP modes indicate that the retinal is twisted stronger in blue NpHR as compared with NpHR. Very unexpectedly, the binding of the anion is accompanied by large amide I bands, which indicate that the protein backbone experiences considerable rearrangements. It will be interesting to compare these backbone changes with those triggered by the chromophore isomerisation.

The next three spectra compare the molecular changes induced by chloride uptake with those induced by bromide and iodide. They are virtually identical. As could be shown (Guijarro et al. 2006), the nitrate binding spectrum is also very similar. This shows that the protein does not so much react on the size and form of the anion but on the negative charge. For a comparison of blue NpHR with the O-intermediate, the negative of this spectrum is shown in the lowest trace. Here the negative bands are due to the O-state, whereas the positive bands are caused by NpHR. This spectrum has been obtained with the time-resolved step–scan technique (see Guijarro et al. 2006). It is obvious that this spectrum is very similar to the anion-uptake spectra. The same amide I changes take place, and the O-intermediate is characterised by the HOOP modes. Thus, one can con-

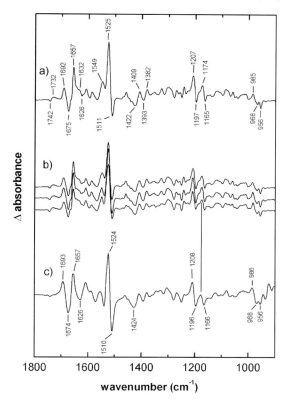

Fig. 6.21 Monitoring the binding of halide anions to halorhodopsin NpHR by ATR-FTIR difference spectroscopy: (a) binding of chloride; (b) binding of chloride, bromide, and iodide; (c) inverse of the difference spectrum of the O-intermediate, i.e., bands of O point downwards and bands of the dark state of halorhodopsin point upwards (Guijarro et al. 2006).

clude that the last step of the halorhodopsin photocycle, i.e., the decay of the O-intermediate and the uptake of the anion, is only a passive binding process. In agreement with this, the O-decay is slowed down by lowering the anion concentration, as is expected for a bimolecular reaction. Further conclusions on the anion translocation process can be found in the original publication (Guijarro et al. 2006).

6.4
Infrared Studies Using Caged Compounds as the Trigger Source

In section 4.4.1.3 Experimental Techniques, we described the basic idea of caged compounds and how they can be used to trigger reactions in an infrared cuvette, enabling the molecular changes evolving after the photolysis of the caged compound to be followed, i.e., after cleavage of the protecting group. Here we will describe some characteristic experiments. From the large number of caged compounds, those consisting of phosphorylated nucleotides in which the terminal phosphate group is protected, were the first to be used in reaction-induced IR spectroscopy. Caged ATP, ADP, and GTP belong to this group of caged compounds. They are particularly valuable, as many biological reactions are started by the corresponding uncaged compounds. Although triggering the Ca^{2+} ATPase from sarcoplasmatic reticulum had been among the first systems studied (Barth et al. 1990), the system is fairly complex and for tutorial purposes, a somewhat simpler protein appears to be more suitable.

The family of Ras proteins are GTP-binding proteins representing molecular switches. The GTP-bound form is the active state interacting with downstream effectors regulating various cellular responses. In the GDP-bound form, the protein is switched off. It exhibits an intrinsic GTPase activity, i.e., the conversion into the inactive state occurs spontaneously, although with the slow rate of 5.1×10^{-4} s.

In order to study the molecular changes triggered by the binding of GTP to the system, one has to take into account the molecular changes due to the photolysis process. The reaction sequence of the photolysis of caged GTP is shown in Scheme 6.1 (Cepus et al. 1998). The rate-limiting step for formation of active GTP is the decay of the intermediate 7, which is called the *aci*-nitro anion state of caged GTP. The photolysis has been monitored with time-resolved rapid-scan FTIR spectroscopy (Cepus et al. 1998), and some of the results are shown in Fig. 6.22. Spectrum A covers the time range from 10 to 26 ms, and spectrum B that

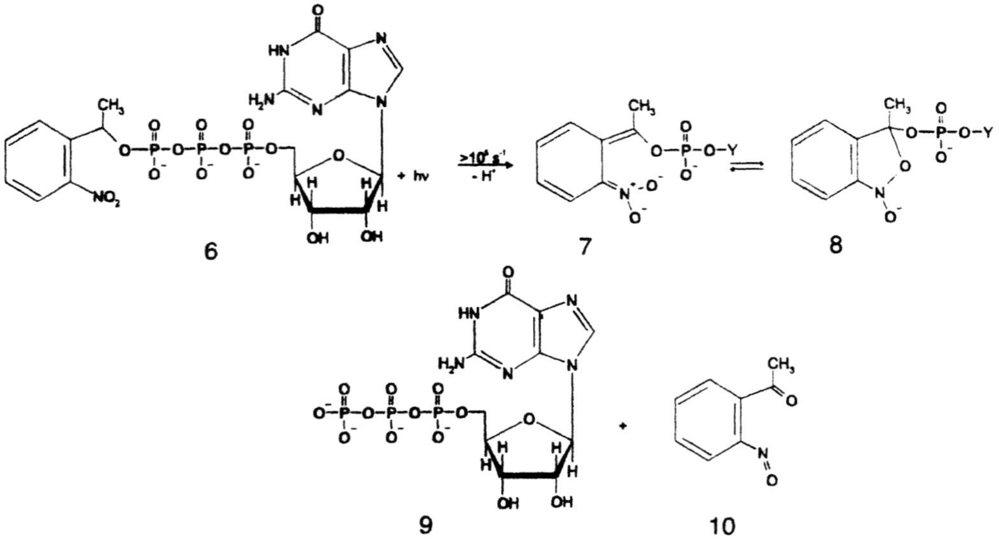

Scheme 6.1 Reaction sequence of caged GTP photolysis (Cepus et al. 1998).

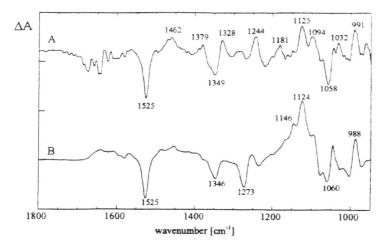

Fig. 6.22 Time-resolved rapid-scan FTIR spectra of the photolysis of caged GTP: A, time range from 10 to 26 ms; and B, time range from 86 to 105 ms (Cepus et al. 1998).

from 86 to 105 ms. It is important to mention that the measurements were performed in the presence of 250 mM DTT, that scavenges reaction product **10**, which reacts with SH groups and could in biological applications distort proteins. Spectrum A represents the difference spectrum between the *aci*-nitro anion state and the dark state of caged GTP, whereas spectrum B shows that between the final products and caged GTP. The final products are GTP and the reaction product of compound **10** with DTT. The three-band feature around 1124 cm^{-1} is characteristic of the three phosphate groups of GTP (and also of ATP). Specific ^{18}O-labelling at the α-, β-, and γ-phosphates has shown that the band composition is very complex. The bands at 1124 and 1093 cm^{-1} mainly represent α- and β-phosphates (PO$_2^-$) stretching vibrations coupled to the modes of γ-phosphate (PO$_3^{2-}$). Depending on the time range, the spectra in Fig. 6.22 have to be taken into account when measuring difference spectra of biological reactions triggered by GTP.

According to Allin and Gerwert (Allin and Gerwert 2001), the photolysis of caged GTP within the Ras protein can be described by the following, Eq. (6.3):

$$Ras \bullet cgGTP(A) \xrightarrow{h\nu} Ras \bullet aci - nitro - anion(B) + H^+ \rightarrow$$
$$Ras \bullet GTP(C) + oNAP \rightarrow Ras \bullet GDP(D) + P_i \quad (6.3)$$

where oNAP is compound **10** in Scheme 6.1 (*o*-nitrosoacetophenone), and P_i is inorganic phosphate. It is important to mention that caged GTP also binds to Ras. Therefore, there are no diffusion processes involved in the generation of Ras $\bullet$ GTP. The formation of the *aci*-nitro anion is too fast to be resolved by the time-resolved rapid-scan technique.

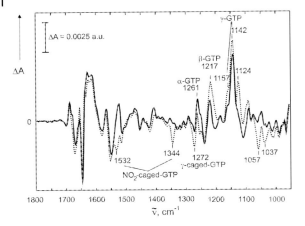

Fig. 6.23 Time-resolved rapid-scan spectra of the photolysis of Ras*caged-GTP. Difference spectrum between the initial state (A) and GTP bound to Ras (C) [Eq. (6.3)], dotted line, difference spectrum between the 1st photoproduct (B) and GTP bound to Ras (C), solid line [Eq. (6.3)] (Allin and Gerwert 2001).

In Fig. 6.23, the time-resolved spectrum between the initial state (A) and GTP bound to Ras (C) is shown (dotted line). In addition, the spectrum between the 1st photoproduct (B) and C is shown (solid line). The intermediate B could be captured with the rapid-scan technique, as the measurements had been performed at 260 K, the low temperature slowing-down the reaction to a rate constant of approximately 5 s^{-1}. Positive bands are caused by species C, whereas the negative bands are due to A (dotted line) and B (solid line). Therefore, the positive bands are in good agreement in the two spectra (taking into account some overlap with differing negative bands), whereas the caged GTP in particular shows some characteristic negative bands as indicated in the figure. As compared with GTP complexed to Mg^{2+} in aqueous solution (Fig. 6.22), there is a pronounced upshift of the main band at 1124 cm^{-1} to 1142 cm^{-1}. More insights into the assignment of these modes can be obtained by ^{18}O isotopic labelling, and a typical experiment is shown in Fig. 6.24.

Here, the spectra of the B → C transition are compared for unlabelled and labelled caged GTP, the oxygen atoms of the γ-phosphate being labelled, including the bridging oxygen between β- and γ-phosphate. Such experiments with additional labelling and extending the measurements to the C → D transition has allowed the following assignment to be derived:

It is clear that upon binding to the Ras protein considerable changes in the force constants take place. These changes are interpreted in terms of charge displacements induced by positive charges to the protein environment. It is concluded that these charge displacements result in weakening of the bond to the γ-phosphate, facilitating its hydrolysis. Thus, these studies have provided deeper

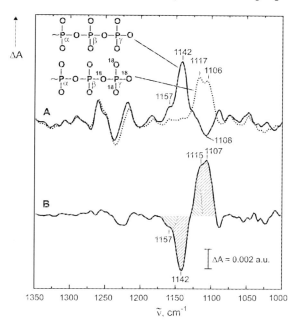

Fig. 6.24 Comparison of the spectrum of the B → C transition (Fig. 6.23) with ^{18}O-labelled caged GTP (Allin and Gerwert, 2001).

insights into the catalysis of GTP hydrolysis by Ras, although, as has been mentioned, the reaction is still very slow.

The hydrolysis is accelerated considerably by the physiologically interacting Ras-GAP protein, forming a complex with Ras. The mechanism of this acceleration has also been investigated by time-resolved rapid-scan FTIR spectroscopy. The basic results are shown in a 3D representation in Fig. 6.25 (Kötting and Gerwert 2005). The upper spectrum shows the intrinsic GTP hydrolysis reaction of Ras alone, whereas the lower spectrum shows the time-evolution starting with the photolysis of GTP within the GAP-Ras complex (note the different time scales). The much faster hydrolysis in the GAP-Ras system is evident (for these time-resolved difference spectra, the final products, i.e., Ras-GDP + P_i and GAP-Ras-GDP + P_i, respectively, have been taken as reference). In the lower representation, in addition to the formation of GAP-Ras-GTP, an intermediate, is observed at somewhat later times. This is evident from the faster decay of the band at 1143 cm^{-1} and the delayed rise of a band at 1186 and 1114 cm^{-1}. It is beyond the scope of this chapter to discuss the experiments in greater detail. Again using ^{18}O-labelling, the different phosphate modes of GTP and GDP have been assigned, and the observed frequencies in the GAP-Ras system seem to support the role of positive charges for catalysing the hydrolysis of GTP, provided here by amino acids of the bound GAP protein.

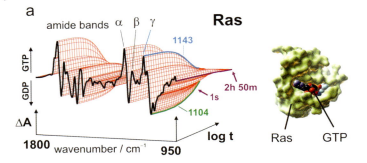

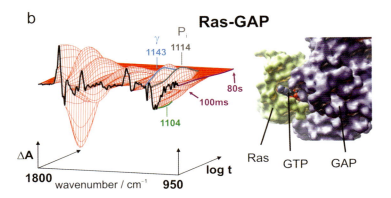

Fig. 6.25 Comparison of the photolysis reaction of Ras*caged-GTP (upper spectrum) and of GAP*Ras*caged-GTP (lower spectrum). The presence of the physiological GAP protein drastically accelerates the hydrolysis of GTP. Because of the higher time-resolution in the lower spectrum, intermediates of the reaction can be seen, i.e., Ras*GTP (at 1143 cm^{-1}, decaying with the hydrolysis of GTP) and the phosphate bound to the protein (at 1114 cm^{-1}), which is finally released to the bulk (Kötting and Gerwert, 2005).

References

Allin, C., Gerwert, K., **2001**, "Ras catalyzes GTP hydrolysis by shifting negative charges from γ- to β-Phosphate as revealed by time-resolved FTIR difference spectroscopy", *Biochemistry* **40**, 3037–3046.

Althaus, T., Eisfeld, W., Lohrmann, R., Stockburger, M., **1995**, "Application of Raman spectroscopy to retinal proteins", *Israel J. Chem.* **35**, 227–251.

Baasov, T., Friedman, N., Sheves, M., **1987**, "Factors affecting the C=N stretching in protonated retinal Schiff bases: a model study for bacteriorhodopsin and visual pigments", *Biochemistry* **26**, 3210–3217.

Bagley, K. A., Balogh-Nair, V., Croteau, A. A., Dollinger, G., Ebrey, T. G., Eisenstein, L., Hong, M. K., Nakanishi, K., Vittitow, J., **1985**, "Fourier-transform infrared difference spectroscopy of rhodopsin and its photoproducts at low temperature", *Biochemistry* **24**, 6055–6071.

Barth, A., Kreutz, W., Mäntele, W., **1990**, "Molecular changes in the sarcoplasmatic reticulum calcium ATPase during catalytic activity. A Fourier transform infrared (FTIR) study using photolysis of caged ATP to trigger the reaction cycle", *FEBS Lett.* **277**, 147–150.

Braiman, M. S., Mogi, T., Marti, T., Stern, L. J., Khorana, H. G., Rothschild, K. J., **1988**, "Vibrational spectroscopy of bacteriorhodopsin mutants: Light-driven proton transport involves protonation changes of aspartic acid residues 85, 96 and 212", *Biochemistry* **27**, 8516–8520.

Cepus, V., Ulbrich, C., Allin, C., Troullier, A., Gerwert, K., **1998**, "Fourier transform infrared photolysis studies of caged compounds", in *"Methods in Enzymology, Vol. 291. Caged Compounds"*, Marriot, G. (Ed.), Academic Press, San Diego, pp. 223–245.

Deng, H., Callender, R. H., **1987**, "A study of the Schiff Base modes in bovine rhodopsin and bathorhodopsin", *Biochemistry* **26**, 7481–7426.

Doyle, S. E., Castrucci, A. M., McCall, M., Provencio, I., Menaker, M., **2006**, "Nonvisual light responses in the Rpe65 knockout mouse: rod loss restores sensitivity to the melanopsin system", *Proc. Natl. Acad. Sci. USA* **102**, 10432–10437.

Engelhard, M., Gerwert, K., Hess, B., Kreutz, W., Siebert, F., **1985**, "Light-driven protonation changes of internal aspartic acids of bacteriorhodopsin: An investigation by static and time-resolved infrared difference spectroscopy using [4-^{13}C] aspartic acid labelled purple membrane", *Biochemistry* **24**, 400–4007.

Essen, L. O., **2002**, "Halorhodopsin: light-driven ion pumping made simple?" *Curr. Opin. Struct. Biol.* **12**, 516–522.

Eyring, G., Mathies, R. A., **1979**, "Resonance Raman studies of bathorhodopsin: evidence for a protonated Schiff base", *Proc. Natl. Acad. Sci. USA* **76**, 83–87.

Eyring, G., Curry, B., Broek, A., Lugtenburg, J., Mathies, R. A., **1982**, "Assignment and interpretation of hydrogen out-of-plane vibrations in the resonance Raman spectra of rhodopsin and bathorhodopsin", *Biochemistry* **21**, 384–393.

Fahmy, K., Großjean, M. F., Siebert, F., Tavan, P., **1989**, "The photoisomerization in bacteriorhodopsin studied by FTIR linear dichroism and photoselection experiments combined with quantumchemical theoretical analysis", *J. Mol. Struct.* **214**, 257–288.

Fahmy, K., Jäger, F., Beck, M., Zvyaga, T. A., Sakmar, T. P., Siebert, F., **1993**, "Protonation states of membrane-embedded carboxylic acid groups in rhodopsin and metarhodopsin II: a Fourier-transform infrared spectroscopy study of site-directed mutants", *Proc. Natl. Acad. Sci. USA* **90**, 10206–10210.

Farrens, D. L., Altenbach, C., Yang, K., Hubbell, W. L., Khorana, H. G., **1996**, "Requirement of rigid-body motion of transmembrane helices for light activation of rhodopsin", *Science* **274**, 768–770.

Ganter, U. M., Gärtner, W., Siebert, F., **1988**, "Rhodopsin-Lumirhodopsin phototransition of bovine rhodopsin investigated by FTIR difference spectroscopy", *Biochemistry* **27**, 7480–7488.

Ganter, U. M., Gärtner, W., Siebert, F., **1990**, "The Influence of the 13-methyl group of the retinal on the photoreaction of rhodopsin revealed by FTIR difference spectroscopy", *Eur. Biophys. J.* **18**, 295–299.

Gerwert, K., **1993**, "Molecular reaction mechanisms of proteins as monitored by time-resolved FTIR spectroscopy", *Curr. Opinin. Struct. Biol.* **3**, 769–773.

Guijarro, J., Engelhard, M., Siebert, F., **2006**, "Anion uptake in halorhodopsin from *Natronobacterium pharaonis* studied by FTIR spectroscopy: consequences for the anion transport mechanism", *Biochemistry* **45**, 11578–11588.

Haupts, U., Tittor, J., Oesterhelt, D., **1999**, "Closing in on bacteriorhodopsin: progress in understanding the molecule", *Annu. Rev. Biophys. Biomol. Struct.* **28**, 367–399.

Hauser, K., Engelhard, M., Friedman, N., Sheves, M., Siebert, F., **2002**, "Interpretation of amide I difference bands observed during protein reactions using site-directed isotopically labelled bacteriorhodopsin as a model system", *J. Phys. Chem. A* **106**, 3553–3559.

Heberle, J., Zscherp, C., **1996**, "ATR/FT-IR difference spectroscopy of biological matter with microsecond time resolution", *Appl. Spectrosc.* **50**, 588–596.

Holthues, H., Engel, L., Spessert, R., Vollrath, L., **2005**, "Circadian gene expression patterns of melanopsin and pinopsin in the chick pineal gland", *Biochem. Biophys. Res. Commun.* **326**, 160–165.

Isele, J., Sakmar, T. P., Siebert, F., **2000**, "Rhodopsin activation affects the environment of specific neighboring phospholipids: an FTIR spectroscopic study", *Biophys. J.* **79**, 3063–3071.

Jäger, F., Fahmy, K., Sakmar, T. P., Siebert, F., **1994**, "Identification of glutamic acid 113 as the Schiff base proton acceptor in the metarhodopsin II photointermediate of rhodopsin", *Biochemistry* **33**, 10878–10882.

Jung, K.-H., Spudich, J. L., **2004**, "Microbial rhodopsins: transport and sensory proteins throughout the three domains of life", in *CRC Handbook of Organic Photochemistry and Photobiology*, Horspool, W. H., Lenci, F. (Eds.), CRC Press, Boca Raton, pp. 1–12.

Kim, J. E., McCamant, D. W., Zhu, L., Mathies, R. A., **2001**, "Resonance Raman structural evidence that the cis-to-trans isomerization in rhodopsin occurs in femtoseconds", *J. Phys. Chem. B* **105**, 1240–1249.

Kolbe, M., Besir, H., Essen, L. O., Oesterhelt, D., **2000**, "Structure of the light-driven chloride pump halorhodopsin at 1.8 Å resolution", *Science* **288**, 1390–1396.

Kötting, C., Gerwert, K., **2005**, "Monitoring protein-protein interactions by time-resolved FTIR difference spectroscopy", in *"Protein-Protein Interactions, A Molecular Cloning Manual"*, 2nd edn, Golemis, E. A. (Ed.), Cold Spring Harbor, NY USA, pp. 279–299.

Lanyi, J. K., Luecke, H., **2001**, "Bacteriorhodopsin", *Curr. Opin. Struct. Biol.* **11**, 415–419.

Mathies, R. A., Oseroff, A. R., Stryer, L., **1976**, "Rapid-flow resonance Raman spectroscopy of photolabile molecules:rhodopsin and isorhodopsin", *Proc. Natl. Acad. Sci. USA* **73**, 1–5.

Mathies, R. A., Freedman, T. B., Stryer, L., **1977**, "Resonance Raman studies of the conformation of retinal in rhodopsin and isorhodopsin", *J. Mol. Biol.* **109**, 367–372.

Mathies, R. A., Smith, S. O., Palings, I., **1987**, "Determination of retinal chromophore structure in rhodopsins" in *"Biological Application of Raman Spectrometry: Vol. 2 – Resonance Raman Spectra of Polyenes and Aromatics"*, Spiro, T. G. (Ed.), Wiley & Sons, Chichester, pp. 59–108.

Nagel, G., Szellas, T., Huhn, W., Kateriya, S., Adeishvili, N., Berthold, P., Ollig, D., Hegemann, P., Bamberg, E., **2003**, "Channelrhodopsin-2, a directly light-gated cation-selective membrane channel", *Proc. Natl. Acad. Sci. USA* **100**, 13940–13945.

Okada, T., Ernst, O. P., Palczewski, K., Hofmann, K. P., **2001**, "Activation of rhodopsin: new insights from structural and biochemical studies", *Trends Biochem. Sci.* **26**, 318–324.

Ottolenghi, M., **1980**, "The photochemistry of rhodopsin", in *"Advances in Photochemistry, Vol. 12"*, Wiley-Interscience, New York, pp. 97–200.

Ottolenghi, M., Sheves, M., **1989**, "Synthetic retinals as probes for the binding site and photoreactions in rhodopsins", *J. Memb. Biol.* **112**, 193–212.

Palings, I., Pardoen, J. A., van den Berg, E. M. M., Winkel, C., Lugtenburg, J., Mathies, R. A., **1987**, "Assignment of fingerprint vibrations in the resonance Raman spectra of rhodopsin, isorhodopsin, and bathorhodopsin: implications for chromophore structure and environment", *Biochemistry* **26**, 2544–2556.

Palings, I., van den Berg, E. M. M., Lugtenburg, J., Mathies, R. A., **1989**, "Complete assignment of the hydrogen out-of-plane wagging vibrations of bathorhodopsin: Chromophore structure and energy storage in the primary photoproduct of vision", *Biochemistry* **28**, 1498–1507.

Pan, D., Mathies, R. A., **2001**, "Chromophore structure in lumirhodopsin and metarhodopsin I by time-resolved resonance Raman microchip spectroscopy", *Biochemistry* **40**, 7929–7936.

Pan, D., Ganim, Z., Kim, J. E., Verhoeven, M. A., Lugtenburg, J., Mathies, R. A., **2002**, "Time-resolved resonance Raman analysis of chromophore structural changes in the formation and decay of rhodopsin's BSI intermediate", *J. Am. Chem. Soc.* **124**, 4857–4864.

Rödig, C., Chizhov, I. V., Weidlich, O., Siebert, F., **1999**, "Time-resolved step-scan FTIR spectroscopy reveals differences between early and late M intermediates of bacteriorhodopsin", *Biophys. J.* **76**, 2687–2701.

Rödig, C., Siebert, F., **1999**, "Errors and artefacts in time-resolved step-scan FT-IR spectroscopy", *Appl. Spectrosc.* **53**, 893–901.

Sakmar, T. P., Menon, S. T., Marin, E. P., Awad, E. S., **2002**, "Rhodopsin: Insights from recent structural studies", *Annu. Rev. Biophys. Biomol. Struct.* **31**, 443–484.

Siebert, F., Mäntele, W., Gerwert, K., **1983**, "Fourier-transform infrared spectroscopy applied to rhodopsin, the problem of the protonation state of the retinylidene Schiff base re-investigated", *Eur. J. Biochem.* **136**, 119–127.

Siebert, F., **1993**, "Infrared spectroscopic investigations of retinal proteins", in *"Biomolecular Spectroscopy Part A"*, Clark, R. J. H., Hester, R. E. (Eds.), Wiley & Sons, Chichester, pp. 1–54.

Spudich, J. L., Yang, C.-S., Jung, K.-H., Spudich, E. N., **2000**, "Retinylidene proteins: structures and functions from archae to humans", *Annu. Rev. Cell Dev. Biol.* **16**, 365–392.

Vogel, R., Siebert, F., **2001**, "Conformations of the active and inactive states of opsin", *J. Biol. Chem.* **276**, 38487–38493.

Vogel, R., Siebert, F., **2003**, "New insights from FTIR spectroscopy into molecular properties and activation mechanisms of the visual pigment rhodopsin", *Biospectrosc.* **72**, 133–148.

Vogel, R., Siebert, F., Lüdeke, S., Hirshfeld, A., Sheves, M., **2005**, "Agonists and partial agonists of rhodopsin: retinals with ring modifications", *Biochemistry* **44**, 11684–11699.

Warshel, A., Barboy, N., **1982**, "Energy storage and reaction pathway in the first step of the vision process", *J. Am. Chem. Soc.* **104**, 1469–1476.

Yan, E. Y. C., Ganim, Z., Kazmi, M. A., Chang, B. S. W., Sakmar, T. P., Mathies, R. A., **2004**, "Resonance Raman analysis of the mechanism of energy storage and chromophore distortion in the primary visual photoproduct", *Biochemistry* **43**, 10867–10876.

7
Heme Proteins

Hemes, i.e., iron porphyrins, are ubiquitous protein cofactors and essential elements necessary for a variety of biological functions of proteins and enzymes (Messerschmidt et al. 2001). These functions include the transport of molecular oxygen (hemoglobin), the transfer of electrons (cytochromes), and the metabolism of substrates (heme enzymes). Recently, it has also been shown that heme proteins may act as sensors or signal transducers (Ignarro 2002; Jiang and Wang 2004).

Hemes are cyclic methine-bridged tetrapyrroles with the four pyrrole nitrogens coordinating the central iron (Fig. 7.1). Of the remaining two axial coordination

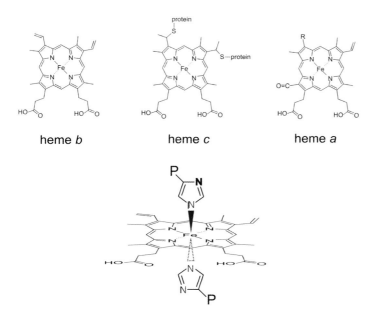

Fig. 7.1 Top: structural formulae of heme *b* (left), heme *c* (middle), and heme *a* (right). The substituent R in heme *a* is a farnesyl side chain. Bottom: axial coordination of heme *b* by two His ligands via the imidazole side chain (P denotes the linkage of the residues to the peptide chain).

Vibrational Spectroscopy in Life Science. Friedrich Siebert and Peter Hildebrandt
Copyright © 2008 WILEY-VCH Verlag GmbH & Co. KGaA, Weinheim
ISBN: 978-3-527-40506-0

sites, at least one is usually occupied by an amino acid side chain. The sixth coordination site either remains vacant (five-coordinated – 5c) or is occupied by an additional amino acid side chain or a small ligand such as water or oxygen (six-coordinated – 6c). The porphyrin macrocycles can exhibit different substitution patterns and accordingly most of the natural occurring hemes can be sorted into three groups, type-a, -b, and -c hemes. Type-b hemes, also known as protoporphyrin IX, possess two vinyl substituents whereas type-a hemes carry a vinyl and a formyl function. In addition, type-a hemes include a long-chain aliphatic substituent with the specific constitution depending on the organism. Type-c hemes are derived from protoporphyrin IX but the two vinyl substituents are used to form covalent linkages to the protein by addition of the thiol functions of cysteine residues, such that two thioether bridges are formed.

7.1
Vibrational Spectroscopy of Metalloporphyrins

The various substitution patterns affect the symmetry of the porphyrins and thus the vibrational properties. Metalloporphyrins would be of D_{4h} symmetry if they exhibited planar structures, had the same axial ligands of perfect axial symmetry, and if the macrocycle was substituted symmetrically. None of these conditions are fulfilled exactly but the properties of the D_{4h} point group represent a helpful starting point for analysing the vibrational spectra of hemes. In fact, hemes are one of the few cofactors in biology for which symmetry considerations may guide the vibrational analysis (Boxes 2B, 2C, Chapter 2) (Cotton 1990; Spiro 1983).

7.1.1
Metalloporphyrins Under D_{4h} Symmetry

We will first consider the electronic transitions of metalloporphyrins, which are derived from the π-molecular orbitals (MO) (Gouterman 1979). The highest occupied MOs have a_{1u} and a_{2u} symmetry and are very close in energy, whereas the lowest unoccupied MO is double degenerate (e_g). Thus, there are two promotions of the same symmetry (E_u) and similar energy, which are thus subject to strong configuration interactions. As a result, the transition dipoles combine additively and subtractively, leading to a strong electronic transition at ca. 400 nm (Soret- or B-band) and a substantially weaker one at ca. 550 nm (Q-band), respectively (Fig. 7.2). The ratio between the oscillator strengths of the B- and Q-bands is ca. 10.

Within the D_{4h} point group, we can distinguish between in-plane and out-of-plane normal modes. In-plane modes belong to the symmetry species A_{1g}, A_{2g}, B_{1g}, B_{2g}, and E_u. On the basis of the symmetry properties of the dipole moment and polarisability operator (Wilson et al. 1955), it can easily be seen that the A_{1g}, B_{1g}, and B_{2g} modes are Raman-active whereas the E_u modes are IR-active (Box 2C, Chapter 2). The A_{2g} modes are inactive in both the IR and the Raman spec-

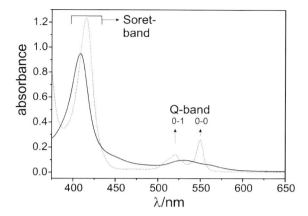

Fig. 7.2 Absorption spectrum of the Cyt-c in the reduced (dotted line) and oxidised form (solid line) in the regions of the Soret- and Q-transitions.

trum. We identify the A_{1u}, A_{2u}, B_{2u}, B_{1u}, and E_g modes as out-of-plane normal modes. Among them only the A_{2u} and E_g modes are IR- and Raman-active, respectively. The remaining modes display no IR or Raman activity.

To analyse the RR activity, we have to take into account that the electronic transitions that give rise to the B- and Q-bands are localised in the heme plane. Within the D_{4h} symmetry, therefore, only in-plane modes are RR active. In the RR experiment we have to distinguish between A- and B-term scattering. A-term scattering dominates for excitation in resonance with strong, symmetry-allowed electronic transitions (see Section 2.2.3., Box 2C), as is the situation for the Soret transition. In fact, RR spectra obtained with excitation lines within the Soret absorption band predominantly display vibrational bands mainly originating from the totally symmetric A_{1g} modes, among which the most prominent ones are found between 1300 and 1700 cm^{-1} (Fig. 7.3) (Spiro 1983). These modes mainly include the stretching coordinates of the tetrapyrroles macrocycle. The high intensities of these bands are associated with the large oscillator strength of the Soret transition, whereas the excited state displacements, the second determinant for the A-term RR intensity, are relatively small for the C–N and C–C stretching coordinates owing to the large size of the porphyrins.

The B-term scattering mechanism holds for modes that belong to the irreducible representations obtained by the direct product of the electronic transitions involved (vide supra), i.e.,

$$E_u \times E_u = A_{1g} + A_{2g} + B_{1g} + B_{2g} \tag{7.1}$$

Upon excitation in resonance with the Q-band, the B_{1g} modes are associated with the strongest RR intensity as their vibronic coupling strength (H_{RS}, see Box 2C) is much stronger than for B_{2g} or A_{1g} modes. Additionally, bands originating from

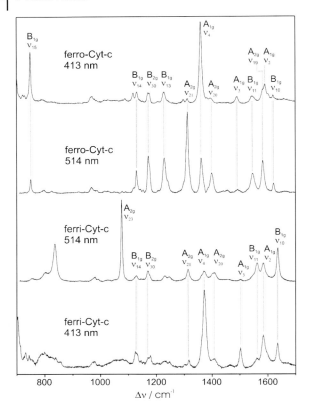

Fig. 7.3 RR spectra of ferro-Cyt-c and ferri-Cyt-c in aqueous solution at pH 7.0, excited at 413 nm (Soret-band) and 514 nm (Q-band). Weidinger and P. Hildebrandt, unpublished. The band assignment follows the study of Hu et al. (1995) (cf. Table 7.2).

the A_{2g} modes are observed that are Raman-inactive under off-resonance conditions. Excitation profiles in the Q-band region display maximum intensity for these modes usually for the $0 \to 1$ vibronic transition although in some instances a second maximum of the excitation profile is also found for the $0 \to 0$ transition (Spiro 1983).

The RR bands that result from the modes of the various symmetry species can be distinguished on the basis of the depolarisation ratios [Eqs. (2.59–2.61)]. For the B_{1g} modes, the off-diagonal elements of the polarisability tensor in addition to α_{zz} are zero and $\alpha_{xx} = -\alpha_{yy}$. Thus, $\bar{\alpha}$ and γ_{as} are zero, and Eq. (2.60) yields

$$\rho_{dp} = \frac{I_{perp}}{I_{para}} = \frac{3\gamma_s^2}{4\gamma_s^2} = \frac{3}{4} \tag{7.2}$$

whereas for B_{2g} modes the only non-zero component is α_{xy} such that $\bar{\alpha}$ and γ_{as} are also zero and the depolarisation ratio adopts the same value, i.e., 0.75. In contrast to these depolarised modes, the A_{1g} modes are polarised. Because the diagonal elements of the tensor and thus $\bar{\alpha}$ is non-zero and only γ_{as} is zero we obtain

$$\rho_p = \frac{I_{\text{perp}}}{I_{\text{para}}} = \frac{3\gamma_s^2}{45\bar{\alpha}^2 + 4\gamma_s^2} < \frac{3}{4} \tag{7.3}$$

The depolarisation ratio for A_{1g} modes is therefore distinctly smaller than for B_{1g} and B_{2g} modes as $\bar{\alpha}^2$ cannot be negative. A_{2g} modes, however, display a completely different picture because the depolarisation ratio is much larger than one. This anomalous polarisation results from the asymmetric character of the polarisability tensor, which leads to $\alpha_{xy} = -\alpha_{yx}$.

The different depolarisation ratios and the preferential enhancement of A_{1g} versus A_{2g}, B_{1g}, and B_{2g} modes with Soret- versus Q-band excitation are very helpful for the vibrational assignment. However, it should be noted that A- and B-term enhancement mechanisms are not exclusive to excitation in resonance with the Soret- and Q-band, respectively. Also, in the latter, A_{1g} modes are detectable but they exhibit ca. 100-times weaker intensity as compared with Soret-band excitation as the magnitude of the A-term enhancement approximately scales with the square of the oscillator strength. Thus, the Q-band excited RR spectrum is dominated by modes that gain intensity via the B-term mechanism. Also for these modes, the absolute RR intensity is distinctly lower than for the A_{1g} modes with Soret-band excitation. As a practical consequence, high-quality RR spectra are obtained with much lower porphyrin or heme protein concentrations when using excitation lines in resonance with the Soret transition as compared with the Q-transition. Conversely, vibronic coupling is also operative for Soret-band excitation but the resulting intensities are much weaker than those for the A_{1g} modes and, hence, the B-term enhanced modes lead to relatively weak bands.

7.1.2
Symmetry Lowering

The considerations discussed above hold for a D_{4h} symmetry of an idealised metalloporphyrin, which, however, is only an approximation for the actual symmetry of true metalloporphyrins or heme cofactors in proteins. The symmetry of natural porphyrins is lower than D_{4h} for the following reasons. Firstly, different side chains at the individual pyrrole groups cause an asymmetric substitution pattern, as is the situation for nearly all natural occurring hemes, i.e., hemes *a*, *b*, and *c* (Fig. 7.1). In *b*-type hemes, the substitution by the two vinyl groups induces RR activity (A-term) into the E_u modes, which are Raman-forbidden and IR-allowed under D_{4h} symmetry. This can readily be understood as the asymmetric disposition of the vinyl groups destroys the centre of inversion. In type-*a* hemes, the symmetry is further lowered as one of the vinyls is replaced by a formyl group, which removes the *x,y* equivalence. Thus, the degenerate E_u modes split into two components, which, in fact, are detectable in the RR spectra under Soret-band excitation (Babcock 1988).

Both the vinyl and the formyl substituents are coupled to the conjugated π-electron systems of the porphyrin and thus the internal modes of these substituents also gain resonance enhancement. These are particularly the C=C (vinyl) and C=O (formyl) stretching modes that are found between 1600 and 1700 cm^{-1}.

In c-type hemes, the thioether bridges lower the symmetry of the porphyrin as is reflected by the RR activity of the E_u modes. Moreover, the internal coordinates of the thioether bridges couple with the internal coordinates of the porphyrin leading to additional RR-active modes. The most prominent one is a band at ca. 700 cm^{-1}, which includes large contributions from the C–S stretching.

Deviations from the ideal planar porphyrin structure may also induce RR activity into out-of-plane modes, which are RR-inactive under D_{4h} symmetry (Spiro 1983; Kitagawa and Ozaki 1987). Doming of the tetrapyrroles macrocycle destroys the x,y plane of reflections resulting in the C_{4v} point group. Then the (in-plane) A_{1g} and (out-of-plane) A_{2u} species are combined to yield the totally symmetric A_1 species (polarised) of the C_{4v} point group. These modes are expected to gain RR intensity via the A-term mechanism, i.e., upon Soret band excitation. Correspondingly, B_{1u} and B_{2u} modes transform into B_1 and B_2 modes (depolarised), respectively, and may become RR-active via vibronic coupling (B-term scattering). Ruffling of the porphyrin that causes a change from the D_{4h} to the D_{2d} symmetry transforms the B_{2g} and A_{2u} into B_2 species (depolarised), which would also be B-term enhanced. Although both ruffling and doming make A_{2u} RR-active, these structural distortions result in various symmetry species and thus different polarisation behaviour, which allows the different types of structural distortions to be distinguished.

7.1.3
Axial Ligation

Whereas symmetry lowering due to doming or ruffling of the heme can be brought about by interactions with the protein environment, loss of the heme mirror plane also results from an asymmetric axial ligation pattern of metalloporphyrins. This also has the effect that metal–ligand stretching vibrations can become visible in the RR spectra (Spiro 1983; Kitagawa and Ozaki 1987). Under D_{4h} symmetry, the symmetric metal–ligand stretching mode of a symmetrically ligated metalloporphyrin (L–Me–L; cf. Fig. 7.1) is not symmetry-forbidden, but it is rarely observed although it belongs to the A_{1g} symmetry species. The asymmetric stretching is only IR-active. If one ligand is missing and the symmetry is lowered to C_{4v}, there is only one Me–L stretching (A_1), which upon coupling to the in-plane electronic transition is RR-active, given that the excited state displacement of the Me–L stretching coordinate is not zero. This is particularly so for ferrous heme proteins in which the heme iron is only ligated by a histidine. This Fe–His stretching appears with considerable intensity upon Soret-band excitation (*vide infra*).

In six-coordinated iron porphyrins carrying a π-electron acceptor as an axial ligand (L_π), the Fe–L_π stretching may couple to the in-plane electronic transition of the porphyrin. The coupling sensitively depends on the electron density distribution in the porphyrin and thus on the charge densities on the central ion and its axial ligands. For ferrous hemes, i.e., Fe(II), and particularly in the presence of an electron donating axial ligand, electron density is transferred into the low lying π^*-orbitals of the porphyrin (which also has a consequence on the force constants of the C–N stretchings of the porphyrin, *vide infra*). Ligands that possess empty π-orbitals such as O_2, CO, or NO have the opposite effect as they can withdraw electron density from the iron d_π-orbitals and eventually from the π-orbitals of the porphyrin. Thus, there is an intimate electronic coupling between the π-orbitals of the ligand and the porphyrin, which can qualitatively explain the RR activity of the Fe–L_π stretchings for these ligands with Soret-band excitation. This mechanism also accounts for the resonance enhancement of the Fe–L_π bending vibrations and for the intra-ligand stretching.

An alternative enhancement mechanism for Me–L stretching modes is operative when the excitation line is in resonance with a charge-transfer (CT) transition (Spiro 1983). These transitions have a much lower oscillator strength than the $\pi \rightarrow \pi^*$ transitions of the porphyrine such that they may be obscured by the dominating B- or Q-bands. For iron porphyrins there are a variety of possible CT transitions but those that are considered to be most relevant for RR enhancement refer to the promotion of an electron from the a_{1u} or a_{2u} orbitals of the porpyhrin to the partially filled d-orbitals of the Fe. In contrast to the in-plane $\pi \rightarrow \pi^*$ transitions of porphyrins, the CT transitions lie in the z-direction. Owing to the different orientations, such CT transitions can be identified by polarised absorption spectroscopy even when they are close to the much stronger porphyrin in-plane transitions. This has in fact been shown for cytochrome P-450 crystals for which the CT could be identified at the short-wavelength side of the Soret-absorption band (Hanson et al. 1977). Excitation in resonance with this CT transition then causes the enhancement of the Fe–ligand (thiolate) stretching (Champion et al. 1982). Nevertheless, the RR intensities of the Fe–ligand stretching modes are relatively low compared with those of the porphyrin modes, such that an unambiguous identification is aggravated.

7.1.4
Normal Mode Analyses

The high symmetry of metalloporphyrins greatly facilitates the vibrational analysis. Thus, this class of compounds represents one of the few examples for which fairly reliable *empirical* normal modes of analysis have been achieved. Such studies were first carried out for Ni octaethylporphyrin (NiOEP), which was originally assumed to adopt an almost perfect D_{4h} symmetry (Abe et al. 1978). In fact, this approximation provided reasonable results even though the actually symmetry was found to be lower in later studies (Li et al. 1989).

In a first approximation, which was already able to describe the main features of the experimental spectra in a satisfactory manner, the pyrrole substituents were treated as single point masses such that the number of normal modes was substantially reduced. Within the D_{4h} symmetry, this approximation results in 71 in-plane normal modes including 9 A_{1g}, 8 A_{2g}, 9 B_{1g}, 9 B_{2g}, and 18 (double degenerate) E_u modes. The polarisation properties and the specific RR (*A*-term, *B*-term) and IR activity of the bands allowed the modes with different symmetry to be distinguished, thereby reducing ambiguities in the vibrational assignments. In addition, the experimental data set was significantly enlarged by including the IR and RR spectra of various isotopomers (Abe et al. 1978). In later studies, which included the ethyl substituents of NiOEP and also took into account the actual lower symmetry, the normal mode analysis of the in-plane modes was refined and the out-of-plane modes were also treated (Li et al. 1989; Li et al. 1990).

Nowadays, quantum chemical methods have also become applicable for molecules of the size of metalloporphyrins, even when open shell metal ions (e.g., Fe) are included. Specifically, scaled quantum mechanical force fields obtained by density functional theory (DFT) represent a promising alternative for calculating vibrational spectra (see Section 2.1.4). For Ni-porphine, a porphyrin that lacks all alkyl, vinyl, or propionate substituents at the pyrrole rings, a detailed comparison of empirical and DFT vibrational analyses demonstrated that DFT can provide a highly accurate prediction of the vibrational frequencies (± 5 cm^{-1}) and a semiquantitative description of Raman and IR intensities (Kozlowski et al. 1999). These calculations do not require a large set of isotope shifts for appropriate adjustments of the force constants, and nevertheless afford somewhat more accurate results than empirical normal modes analyses. Moreover, DFT calculations provide further information about the structural distortions of the porphyrin.

7.1.5
Empirical Structure–Spectra Relationships

For many metalloporphyrins, high-resolution crystal structures are available that constitute a sound basis for correlating structural data with RR spectral parameters. Thus, it was found that the frequencies of many modes between 1300 and 1700 cm^{-1} ("marker bands") display a distinct dependency on the size of the porphyrin core, expressed by the distance from the centre of the porphyrin core (C_t) to the pyrrole nitrogens (Fig. 7.4) (Parthasarathi et al. 1987). Specifically, these are the modes ν_3, ν_2 (A_{1g}), ν_{38}, ν_{37} (E_u), ν_{11}, ν_{10} (B_{1g}), and ν_{19} (A_{2g}), which include in particular the C–C stretching coordinates of the tetrapyrrole. When the porphyrin macrocycle has to expand, for instance due to the substitution of a small-sized metal ion by a larger one, the force constants of these coordinates decrease leading to a downshift of the corresponding frequencies. These correlations are linear according to

$$\nu = K(A - d_{\text{Ct-N}}) \tag{7.4}$$

where the frequency ν and the core size ($d_{\text{Ct-N}}$) is expressed in cm^{-1} and Å, respectively. *A* and *K* are mode-specific empirical constants that have been derived from plotting the RR frequencies versus $d_{\text{Ct-N}}$, as determined from the crystal

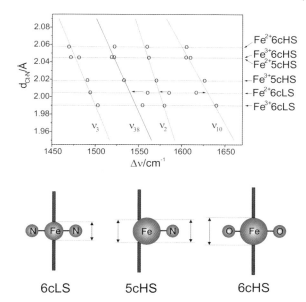

Fig. 7.4 Top: correlation between the porphyrin core size d_{Ct-N} and the frequencies of selected heme modes for iron porphyrin complexes. The solid lines represent best fits to a larger set of experimental data from a variety of metalloporphyrins. The data were taken from Parthasarathi et al. (1987). Bottom: illustration of the variation of the core-size with the spin- and ligation-state. "N" and "O" denote a strong (e.g., His) or a weak (e.g., water) ligand, respectively.

structures. These correlations are very useful for studying heme proteins as they provide indirect but very reliable information about the oxidation, spin- and coordination-state of the heme iron.

Let us first consider the ferric heme in which the iron has a d^5-electronic configuration. In an octahedral ligand field, the five d-orbitals split into three degenerate e_g- and two degenerate t_{2g}-orbitals. The splitting results in two main possibilities for distributing the five d-electrons: on the one hand all electrons are put into the three e_g-orbitals, such that only one electron remains unpaired ($S = 1/2$; low spin, LS); on the other hand each electron can be placed into an individual orbital corresponding to the maximum possible S of 5/2 (high spin, HS). If two strong axial ligands are coordinated to the heme iron, the energy gap between the e_g- and t_{2g}-orbitals may become too large, such that doubly occupied orbitals represent the energetically favoured configuration (six-coordinated LS, 6cLS; cf. Figs. 7.1 and 7.4). Conversely, weak ligands are associated with a small energy gap and thus the six-coordinated HS (6cHS) configuration is preferred. In the HS state, the five orbitals require a larger space than in the LS form and thus the effective ion radius increases such that the porphyrin macrocycle has to expand to accommodate the iron (Fig. 7.4). This corresponds to an increase of the C_t–N distance from 1.989 to 2.045 Å. The concomitant weakening of the force constants leads to frequency shifts of most of the marker modes by up to 20 cm^{-1}. The HS configuration is also obtained when the heme iron is coordinated by only one axial ligand (five-coordinated HS, 5cHS). In this instance, however, the only axial ligand pulls the heme somewhat out of the porphyrin plane, such the effective space requirements are lower than for a 6cHS configuration and the macrocycle expands only to a C_t–N distance of 2.019 Å. Correspondingly, the downshift of the marker bands compared with the 6cLS configuration is smaller (ca. 10 cm^{-1}).

The same considerations also hold, to a first approximation, for ferrous hemes in which, due to the additional electron in the d-orbitals, the iron occupies a slightly larger space when comparing the same spin- and coordination-states of the ferric heme. However, the higher electron density may cause additional effects on the force constants, which may lead to deviations from the linear relationships in Fig. 7.4. For instance, strong electron donating ligands may lead to the transfer of electron density from the iron to the porphyrin orbitals and thus alter the force constants. This effect is particularly pronounced, for instance, for the B_{1g} mode v_{11} in ferrous cytochrome P-450 in which the heme iron is axially coordinated by an electron-rich thiolate ligand. The increased electron density in the porphyrin then leads to a v_{11} frequency that is lower than the v_{38} frequency, in contrast to the predictions of Fig. 7.4 (Anzenbacher et al. 1989). Also mode v_2 (A_{1g}) is sensitive to electronic effects. The most pronounced dependence on the electron density in the porphyrin, however, is observed for the mode v_4 (A_{1g}). This mode displays only a small dependence on the core-size changes but responds specifically to a change in the oxidation state of the heme iron. In ferric hemes it is typically found between 1370 and 1375 cm^{-1} but it is lowered to 1358–1363 cm^{-1} in the ferrous state. A further electron density transfer to the porphyrin by electron rich axial ligands may lower the frequency down to 1340–1345 cm^{-1}, as has been found in ferrous heme carrying a thiolate axial ligand (e.g., cytochrome P450) (Spiro 1983; Anzenbacher et al. 1989; Hildebrandt 1992).

In contrast, frequency upshifts of v_4 are observed in the presence of axial ligands that can accept electron density. This is so for ferrous heme complexes with O_2 or CO and the resultant v_4 frequency can be even higher than in "normal" ferric hemes (Spiro 1983; Hildebrandt 1992).

Further deviations from the linear C_t–v relationships may occur for strongly distorted heme structures and unusual spin states (Howes et al. 1999; Alden et al. 1989). However, for the majority of the heme proteins the inspection of the marker bands between 1300 and 1700 cm^{-1} allows a reliable determination of the oxidation, spin, and coordination state of the heme iron.

7.2
Hemoglobin and Myoglobin

Myoglobin and hemoglobin are oxygen transport and storage proteins in aerobic organisms (Messerschmidt et al. 2001). In both proteins, the oxygen-binding unit is a *b*-type heme (iron protoporphyrin IX) in which only one axial coordination site is occupied by a histidine (proximal site). The second (distal) site can be used for reversible binding of molecular oxygen. Oxygen binding and release is not associated with a change in the redox state of the heme, which always remains in the reduced form. Myoglobin (Mb) is a single peptide chain of ca. 18 kDa, and acts as oxygen storage in muscle tissue. Hemoglobin is composed of two pairs of two subunits ($\alpha_2\beta_2$), thereby forming a complex quaternary structure. The function of hemoglobin is intimately related to the interplay between the four subunits, which modulates the oxygen binding affinity of the fours hemes. This allosteric process corresponds to a conformational transition between a relaxed state *R* (low oxygen affinity) and a tense state *T* (high oxygen affinity), such that Hb can efficiently bind oxygen at high oxygen partial pressure in lung tissue and release oxygen at low oxygen partial pressure in the cells where oxygen is utilised for energy conversion (Cantor and Schimmel 1980).

Mb and Hb were two of the first proteins for which high-resolution crystal structures were obtained. However, many structural details relevant for the functioning of these proteins have subsequently been elucidated by spectroscopic techniques, among which RR and IR spectroscopy played a prominent role. The vibrational spectroscopy of this class of proteins that will be discussed in the following sections is, in many respects, also relevant for the interpretation of the RR and IR spectra of other *b*-type heme proteins, such as peroxidases, catalases, or cytochromes P-450.

7.2.1
Vibrational Analysis of the Heme Cofactor

The cofactors of Mb and Hb are not covalently attached to the protein matrix. Hence, they can readily be extracted and replaced by synthetic protoporphyrin isotopomers. In this way, the vibrational analyses of the RR spectra of Mb and Hb have been supported and reliable assignments have been achieved on the basis of empirical force fields (Hu et al. 1996) (Table 7.1).

Mb and Hb exist in three stable states, which can be distinguished by the characteristic vibrational band pattern in the marker band region (Spiro and Strekas 1974; Rousseau and Ondrias 1983; Spiro 1985). In met-myoglobin (met-Mb), the heme iron is in the (inactive) ferric form and coordinated by a His and a water molecule leading to a 6cHS configuration. Thus, the corresponding marker bands are at 1373 (v_4), 1483 (v_3), 1544 (v_{11}), 1563 (v_2), and at 1608 cm^{-1} (v_{10}) (Hu et al. 1996). Reduction of the heme is associated with the loss of the aquo-ligand resulting in a 5cHS heme. In this deoxy-myoglobin (deoxy-Mb), the heme iron moves slightly out of the porphyrin plane and correspondingly the oxidation marker band v_4 is down-shifted by 15 cm^{-1} reflecting the increased electron density in the porphyrin. The frequencies of the remaining marker bands that respond to the porphyrin core size changes are not so different from those of met-Mb, as the increased space requirements of Fe in the ferrous state are somewhat compensated by the out-of-plane displacement.

Table 7.1 Vibrational assignments for met-Mb, deoxy-Mb, and oxy-Mb in the marker band region.[a]

Mode	Met-Mb(Fe^{3+})[b]	Deoxy-Mb(Fe^{2+})[c]	Oxy-Mb(Fe^{2+})[c]
$\nu(C=C)$[d]	1621	1622	1620
ν_{10} (B_{1g})	1608	1607	1640
ν_{37} (E_u)	1583	1586	1606
ν_{19} (A_{2g})		1552	1586
ν_2 (A_{1g})	1563	1565	1583
ν_{11} (B_{1g})	1544	1546	1564
ν_{38} (E_u)[e]	1521, 1511		
ν_3 (A_{1g})	1483	1473	1508
$\delta(=CH_2)_s$[d]	1421	1427	1432
ν_{28} (B_{2g})	1426		
ν_{29} (B_{2g})	1402	1397	1400
ν_{12} (B_{1g})	1389		
ν_4 (A_{1g})	1373	1358	1374
ν_{41} (E_u)	1341	1338	1345
ν_{21} (A_{2g})		1305	1308
$\delta(CH=)$[d]	1316		
$\delta(CH=)$[d]	1301		

[a] Band frequencies are given in wavenumbers.
[b] Assignments and data taken from Hu et al. (1996).
[c] Assignments made according to data taken from Spiro and Strekas (1974).
[d] $\nu(C=C)$, $\delta(=CH_2)_s$, and $\delta(CH=)$ denote the vinyl C=C stretching, CH_2 deformation, and the CH deformation, respectively.
[e] The vinyl substituents remove the degeneracy of the E_u modes such that two components may be observed. This effect is more pronounced for type-a hemes (see Section 7.1.2).

Deoxy-Mb is capable of binding molecular oxygen at the vacant coordination site (oxy-myoglobin, oxy-Mb). Oxygen binding brings the iron back into the heme plane and the porphyrin core contracts corresponding to an upshift of the marker bands by ca. 20 cm^{-1} compared with deoxy-Mb (Table 7.1). Also, the oxidation marker band ν_4 shifts up by ca. 15 cm^{-1} even though the oxidation state of the heme iron remains formally unchanged. However, the electron accepting capacity of molecular oxygen causes a transfer of electron density from the heme to the ligand (*vide supra*) (Table 7.1).

The met-, deoxy-, and oxy-states of hemoglobin display fairly similar spectra to the corresponding counterparts of Mb. Furthermore, the individual heme groups in Hb cannot be distinguished. There are small but detectable differences in the RR spectra of Mb and Hb isolated from different species (Desbois et al. 1984). These differences specifically refer to the modes originating from the vinyl substituents. The most prominent substituent mode is the C=C stretching, which is observed between 1620 and 1635 cm^{-1}. The frequency mainly depends on the orientation of the vinyl group with respect to the porphyrin. Whereas for a coplanar orientation, the vinyl stretching is found at ca. 1620 cm^{-1}, it shifts to higher frequencies when the substituent is tilted out of the porphyrin plane (Smulevich et al. 1996). Various vinyl orientations may be brought about by interactions with adjacent amino acid side chains. The vinyl stretching bands, therefore, may serve as a local probe for the heme pocket structure, as has also been demonstrated for other b-type heme proteins.

Further modes involving the vinyl groups are rather weak and difficult to identify except for the C-vinyl bending. These modes give rise to well-separated but low-intensity bands at ca. 410 and 440 cm^{-1} with slight frequency variations between the met-, deoxy-, and oxy-forms (Hu et al. 1996). Reliable correlations between the frequencies of these modes and structural parameters of the substituents have not yet been established. This is also true for the corresponding bending mode involving the propionate side chains that is observed at ca. 375 cm^{-1}.

7.2.2
Iron–Ligand and Internal Ligand Modes

In Mb and Hb, a His coordinates to the heme iron on the proximal side of the porphyrin. In the absence of an axial ligand on the distal side, the heme adopts a 5cHS configuration, which in the ferrous form (deoxy-Mb, deoxy-Hb) allows detection of the Fe–His stretching mode in the RR spectrum when using Soret-band excitation. This mode is typically found between 215 and 255 cm^{-1} depending on the intermolecular interactions of the His ligand (Stein et al. 1980). For a non-hydrogen bonded His, the mode is observed at ca. 218 cm^{-1}. Upon formation of a hydrogen bond between the N(γ) proton of the imidazole ring and a proton acceptor, the frequency is increased by 10–20 cm^{-1} as has been noted, for instance, in ferrous 5cHS cytochrome c peroxidase (Smulevich et al. 1988). In the limiting situation, i.e., upon deprotonation of the imidazole ring, the band shifts up to even greater than 250 cm^{-1}. Thus, this mode can be used as a local probe for the protein environment in the heme pocket.

The Fe–His stretching mode has been specifically used for analysing the allosteric $R \rightarrow T$ transition in Hb in order to elucidate the structural basis for the increased oxygen binding affinity. In the R-state of deoxy-Hb, the Fe–His stretching is observed at ca. 222 cm^{-1} (Matsukawa et al. 1985). Inducing the T-state by effectors such as inositol hexaphosphate leads to a downshift by ca. 7 cm^{-1} in deoxy-Hb and thus the frequency is very similar to that of monomeric deoxy-Mb.

In the oxy-forms, the Fe–His stretching is not resonance enhanced and thus can no longer be detected. Instead, the Fe–O$_2$ stretching at ca. 570 cm^{-1} gains resonance enhancement through coupling with the in-plane electronic transition as the excitation profile follows the Soret-band envelope (Spiro 1983). Unlike the Fe–His stretching, the $R \rightarrow T$ transition does not affect the frequency of the Fe–O$_2$ stretching (Kitagawa and Ozaki 1987). The O–O stretching mode appears to be more difficult to detect in oxygen-binding heme proteins.

A spectroscopically more "interesting" ligand is CO, which readily binds to the distal side of deoxy-Hb and deoxy-Mb. The binding behaviour of CO and its effect on the electron density distribution in the heme is similar to O_2. However, CO offers a variety of advantages over O_2, as both the Fe–CO stretching and the Fe–CO bending can be observed in the RR spectra. In addition, the C=O stretching gives rise to a strong IR-active mode, which, due to its frequency at ca. 1950 cm^{-1}, can even be detected in the absolute IR spectra without interference with IR bands of the protein. Thus, it is possible to use the CO ligand as a local probe for the structure of the heme pocket and to gain more insight into the mode of binding of diatomic ligands to hemes, which in turn is important for elucidating the molecular functioning of oxygen binding proteins (Hb, Mb) and of NO sensing proteins.

The frequencies of the Fe–CO and CO stretching modes determined in various heme proteins and model compounds display an inverse linear relationship, which results from the opposite effect of back bonding on the Fe–C and C=O bonds. These effects, in turn, depend on the electron donating capacity of the proximal ligand (Spiro et al. 2001). Thus, the analysis of the Fe–CO and CO stretching modes may contribute to the elucidation of structural details on the proximal side of the heme, such as hydrogen bond interactions with the proximal His ligand. On the other hand, there may be steric interactions of the CO ligand on the distal side of the heme, which may cause deviations from the upright geometry and thus also affect the vibrational frequencies. In fact, the perturbation of the Fe–CO geometry is not completely clarified for the CO complex of Mb as spectroscopic, theoretical, and crystallographic data are not fully reconciled.

7.2.3
Probing Quaternary Structure Changes

The CO complexes of deoxy-Mb and deoxy-Hb (MbCO, HbCO) are also of interest for analysing functionally important conformational transitions in these heme proteins and, in a wider context, to understand the conformational manifolds and dynamics of proteins in general. The concept of these studies is based on the flash photolysis of MbCO or HbCO, which leads to the dissociation of the CO ligand from the heme with a high quantum yield. The subsequent relaxation processes are then probed either by IR (CO stretching) or by RR spectroscopy using violet (heme modes) or UV excitation (protein modes, *vide infra*). Pioneering experiments in this field have been carried out by Frauenfelder and coworkers, who studied MbCO in frozen matrices, monitoring the CO stretching by IR spectroscopy (Frauenfelder et al. 1986). The most striking result that they found was that after photodissociation the "free" CO displays various stretching bands, depending on the temperature and the delay time between the pump and the probe event. Moreover, complex relaxation phenomena involving non-exponential kinetics with non-Arrhenius behaviour have been observed. These findings indicate that the "free" CO senses various interactions in the heme pocket corresponding to different conformational states that are arranged within a conformational hierarchy. Different CO stretchings have also been found for the bound CO in temperature-dependent studies. These substates, which are very similar in energy, differ only in subtle structural details but may exhibit different kinetics for the ligand binding. On the basis of systematic IR spectroscopic studies on MbCO, it was possible to map the energy landscape of the protein conformations.

Similar experiments were designed to elucidate protein structural changes of ligand binding and release in Hb albeit at ambient temperature (Friedman 1994). Again, photodissociation of the HbCO complex was the starting point, causing the sudden change of the electronic configuration of the heme from the 6c CO-bound to the 5c deoxy form. This transition occurs in the femtosecond time regime, but does not include a relaxation of the heme geometry to that of the equilibrated deoxy-Hb (Dasgupta and Spiro 1986). This conclusion was derived from RR spectroscopic experiments with pulsed laser excitation in resonance with the Soret-transition. The spectra indicate a heme geometry with the Fe ion still located in the porphyrin plane. Evidently, the out-of-plane displacement of the metal ion, as is characteristic of the deoxy form, occurs on time scales longer than 10 ns, concomitant with protein structural changes that remove steric constraints on the heme. The non-relaxed deoxy heme is ready to rebind the CO ligand within the pico- and nanosecond time scales, leading to ca. 50% of the hemes being ligated by CO.

Of particular interest, however, are the relaxation processes of the protein that start with the empty heme pocket. The subsequent structural rearrangements correspond to the transition from the *R*- to a *T*-type state and take place within tenths of microseconds, followed by rebinding of the CO that enters the empty heme pocket from the solvent (Jayaraman et al. 1995). In this study, pump–probe time-resolved RR spectroscopy was employed with probe wavelengths adjusted to the electronic transitions of the heme (e.g., 436 nm) and the aromatic residues of the protein (e.g., 230 nm). In the latter, the Raman bands of Tyr and Trp are selectively resonance enhanced whereas the Raman bands of the remainder of the protein provide only a negligible contribution to the spectrum. Thus, it was possible to monitor the spectral changes of RR bands that are indicators of structural changes of the heme and the protein. These bands include the Fe–His stretching (see Section 7.2.2.), the oxidation, and core-size marker bands (see Section 7.1.5), and the specific modes of the Tyr and Trp residues (Fig. 7.5). The time-dependent spectral changes of the Trp and Tyr modes were determined by subtracting the UV-excited RR spectrum of the unphotolysed HbCO complex from those measured at variable delay times following the photolysis pulse. With increasing delay time, these time-resolved difference spectra approach that constructed from the RR spectrum of deoxy-Hb, which, in principle, correspond to a time-resolved difference spectrum at infinite delay time. The temporal evolution of the most intense difference bands of Tyr and Trp allowed for a global fit of four exponential functions corresponding to a sequential kinetic reaction scheme with decay times from ca. 20 ns to 500 µs at pH 7.4.

242 | *7 Heme Proteins*

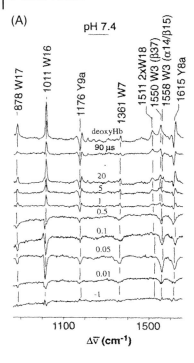

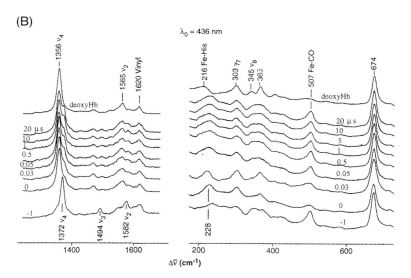

Fig. 7.5 (legend see p. 243)

In the same way and in the same time range, the time-dependent changes of the oxidation marker band v_4 and the Fe–His stretching mode are determined. In the latter, it was possible to determine three transitions between species that are characterised by Fe–His stretching frequencies of 228, 224, 222, and 216 cm^{-1}. The corresponding decay constants are consistent with those derived from the UV RR spectra, implying that the protein structural changes are paralleled by changes in the hydrogen-bond interactions of the His ligands. In contrast, the relaxation of the heme geometry appears to be completed with the formation of the first thermal intermediate as has already been shown in early nanosecond pump–probe experiments (Dasgupta and Spiro 1986).

On the basis of the time-resolved RR spectroscopic results obtained with different excitation lines, a fairly detailed picture has been shaped of the protein and cofactor dynamics of the $R \rightarrow T$ transition over a wide time scale. The approach allows identification of intermediate species that are not accessible by other spectroscopic techniques. In this way, RR spectroscopy can contribute substantially to the understanding of the molecular functioning of Hb.

Some of the Tyr and Trp modes are known to be sensitive towards hydrogen bond interactions and environmental effects and thus may be used to monitor the structural changes in the immediate vicinity of these residues (Rodgers et al. 1992). As the tetrameric Hb includes 12 Tyr and 6 Trp residues, difference spectra are required to document the spectral changes following the photodissociation of the CO ligand. The assignment of the difference signals in the UV RR spectra to individual Tyr and Trp residues was guided by the well-known three-dimensional structures of Hb in the R- and T-state (Messerschmidt et al. 2001). Inspecting the structural differences between the equilibrium R- and T-states, it is possible to reduce the number of aromatic residues that are likely candidates for the observed spectral changes. In addition, the experiments have been extended to site-specific mutant proteins (Nagai et al. 1999) and hybrid Hb variants that have been constructed on the basis of subunits containing selectively labelled $^{13}C_6$-Tyr (Wang et al. 2000). As a consequence, the spectral changes of the Tyr and Trp modes could be attributed to individual residues with high confidence. In this sense, the progress in knowledge on the structure–dynamics relationships of Hb represents an instructive example of the enormous potential of modern vibrational

Fig. 7.5 Time-resolved RR spectra of HbCO at various delay times following photolysis with a 419-nm pulse (10 ns). (A) Difference spectra obtained by subtracting the time-resolved RR spectra of HbCO from those of the photolysis product using excitation with 230-nm pulses. In this way, the spectra display the selectively enhanced bands of Tyr and Trp residues that undergo structural changes after HbCO photolysis. The top spectrum represents the static difference spectrum between deoxy-Hb and HbCO. (B) Time-resolved RR spectra obtained with 436-nm excitation in the region of the marker bands (left) and the Fe–His and Fe–CO stretching modes (right). The spectra are reproduced from Jayaraman et al. (1995) with permission.

spectroscopic techniques, specifically when they are combined with molecular and structural biological methods.

7.3
Cytochrome *c* – a Soluble Electron-transferring Protein

Cytochrome *c* (Cyt-c) is a small (ca. 12 kDa) soluble protein that is ubiquitous in plants, fungi, bacteria, and higher organisms (Scott and Mauk 1995). It carries a covalently bound heme (type-*c*) anchored to two Cys side chains by addition of the thiol functions to the vinyl substituents. A His and a Met side chain serve as axial ligands affording a 6cLS configuration in both the reduced and the oxidised form of the heme (Fig. 7.6). By switching between both oxidation states, Cyt-*c* functions as an electron carrier in energy transduction. In aerobic organisms, it transfers electrons from the cytochrome *c* reductase to cytochrome *c* oxidase where, finally, four electrons are utilised to reduce oxygen to water (see Section 7.4). These two enzyme complexes are embedded in the mitochondrial membrane whereas Cyt-*c* is assumed to diffuse along the membrane surface from the electron donor to the electron acceptor site. For a long time, it has been thought that electron transfer is the only function of Cyt-*c* but recently compelling evidence has been provided for its role in apoptosis (Jiang and Wang 2004).

There are an enormous number of vibrational spectroscopic studies on Cyt-*c*, which are not just motivated by its central role in bioenergetics. In addition, this well-characterised protein is particularly suited as a model system to analyse fundamental issues in biological electron transfer, protein–membrane interactions, or protein folding, as well as to testing and to developing novel experimental approaches.

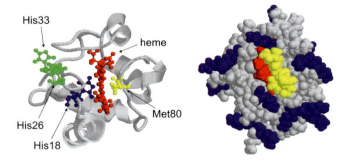

Fig. 7.6 Crystal structure of ferric horse heart Cyt-*c* (pdb 1akk). The picture on the left side indicates the heme (red) and the native axial Met80 (yellow) and His18 (blue) ligands. The His residues 33 and 26 that replace the Met80 ligand in the B2 state are displayed in green. The space filling representation on the right side shows the hydrophobic patch (yellow) close to the exposed heme edge (red) and the surface lysine residues (blue).

7.3.1
Vibrational Assignments

The RR spectra of Cyt-c proteins (cf. Fig. 7.3) from the various sources are fairly similar and careful comparative studies have revealed only subtle differences (<2 cm^{-1}) for some of the bands (Shelnutt et al. 1981). In the frequency region between 1000 and 1700 cm^{-1} (including the marker band region), there are also no substantial differences compared with other heme proteins with 6cLS heme configurations in both the reduced (ferro-Cyt-c) and the oxidised form (ferri-Cyt-c). However, below 1000 cm^{-1} the RR spectrum of Cyt-c displays a unique band pattern, which partly results from the covalent attachment of the heme to the protein matrix. Spiro and coworkers presented an empirical vibrational analysis on the basis of RR spectra of Cyt-c enzymatically reconstituted with isotopically labelled hemes (Hu et al. 1993) (Table 7.2). It was shown that the high complexity of the low-frequency region results from the modes that include the internal coordinates of the porphyrin substituents and out-of-plane modes that become RR-active due to the lowering of the planar D_{4h} symmetry.

Table 7.2 Vibrational assignments for ferric and ferrous Cyt-c in the marker band region.[a]

Mode	Ferro-Cyt-c[b]	Ferri-Cyt-c[b]
ν_{10} (B$_{1g}$)	1626	1635
ν_{37} (E$_u$)	1610	1598[c]
ν_2 (A$_{1g}$)	1596	1585
ν_{19} (A$_{2g}$)	1587	1582[d]
ν_{11} (B$_{1g}$)	1551	1561
$2\nu_{15}$	1501	
ν_3 (A$_{1g}$)	1496	1501
ν_{29} (B$_{2g}$)	1403	1407
ν_{20} (A$_{2g}$)	1400	1401[c]
ν_4 (A$_{1g}$)	1364	1371
$\nu_{15} + \nu_{24}$	1354	1338
δ(CH)[e]	1317	1316
ν_{21} (A$_{2g}$)	1314	
δ(CH)[e]	1302	1303

[a] Band frequencies are given in wavenumbers. Unless indicated otherwise, the data refer to yeast iso-1 Cyt-c. Note that the band frequencies between yeast iso-1 and horse heart Cyt-c may differ by ca. 1 cm^{-1}.
[b] Assignments and data taken from Hu et al. (1993).
[c] Assignments and data taken from Döpner et al. (1998).
[d] Assignments and data (horse heart) taken from and Spiro and Strekas (1974).
[e] δ(CH) denotes the CH deformation of the thioether substituent.

Symmetry lowering due to the out-of-plane distortions of the saddle-like shaped heme also gives rise to a softening of the selection rules for A_{2g} modes. Although they are expected to be only detectable with Q-band excitation, some also exhibit fairly striking RR activity in the Soret-excited spectra in Cyt-c (cf. Fig. 7.3).

One of the strongest peaks at Soret-band excitation is observed at ca. 690 cm^{-1}. A careful inspection reveals that it is actually composed of three components. Two of these (682 and 692 cm^{-1}) originate from the C–S stretchings of the thioether bridges, whereas the third one is assigned to the A_{1g} mode ν_7 (700 cm^{-1}). In addition, specific protein–heme interactions in Cyt-c lead to an unusual vibrational signature between 300 and 425 cm^{-1} that is unparalleled in other c-type heme proteins. In this region, the Soret-excited spectrum displays a set of 8 relatively intense bands. For ferro–Cyt-c, these unusually sharp bands are well resolved as they partly exhibit band widths as narrow as 5 cm^{-1}. Among them, only the bands at 347 and 360 cm^{-1} are assigned to modes of the tetrapyrroles macrocycle (ν_8, ν_{50}). The six remaining bands originate from modes with large contributions from the bending coordinates of the thioether bridges, giving rise to pairs of conjugate bands, i.e., 372 and 382 cm^{-1}, 394 and 401 cm^{-1}, and 413 and 421 cm^{-1}. The "splitting" of these substituent modes can be understood in terms of slightly different orientations of the two thioether bridges leading to different couplings with internal coordinates of the porphyrin. The narrow band widths of the individual band components may thus reflect the rigid embedment of the cofactor in the protein matrix. This interpretation is in fact supported by a comparison with the RR spectrum of ferri–Cyt-c, which shows the same band pairs albeit less clearly resolved due to the larger band widths.

The similar frequencies and relative intensities of the individual components suggest that the protein undergoes only small structural changes upon oxidation, which is in line with the crystal structure data. The broader band widths, however, may be related to a slightly more flexible heme pocket structure, which in fact is reflected by the significant lower stability of the ferri- versus the ferro-form. This view is supported by the fact that structural changes of ferri–Cyt-c upon increasing the temperature, changing the pH or binding to charged or hydrophobic surfaces (*vide infra*) "destroy" the characteristic vibrational signature in this region (*vide infra*).

7.3.2
Redox Equilibria in Solution

Upon reduction of the heme, the expected frequency downshifts of the oxidation- and core-size marker bands are observed, but also the low frequency region varies substantially. In solution at pH 7.0, reduction is readily achieved by ascorbate due to the relatively positive redox potential of Cyt-c (+0.25 V). Thus, dithionite, which is widely used for chemical reduction of cofactors in proteins, can be avoided. This is certainly advantageous in view of the unwanted side reactions of dithionite (formation of aggressive radicals) specifically under the action of laser irradiation in the violet region. Oxidation is more difficult to accomplish. Potassium ferricyanide, the typical oxidant, has the tendency to release cyanide, which can replace the native axial Met80 ligand. Again, this process is strongly promoted under laser light irradiation. However, even more severe, keeping Cyt-c in the oxidised state during the RR experiment is not trivial, by far, as, like many other heme proteins, Cyt-c has a considerable tendency to undergo photoreduction. Continuous movement of the sample through the laser beam (e.g., rotating cuvette) and low laser powers are essential to avoid this unwanted process. The mechanism of photoreduction in Cyt-c and in other heme proteins (e.g., cytochrome c oxidase) has not yet been clarified.

A gentler way to control the redox state of Cyt-c in solution is based on electrochemical reduction and oxidation via mediators. This approach, originally developed for spectroelectrochemical studies with UV–vis absorption detection, has been adapted and optimised for IR spectroscopy (see Section 4.2.1.3). The difference spectra display the bands of the heme and the protein that change with the redox state (Schlereth and Mäntele 1993). It has been shown that the contribution of the heme bands is relatively weak in the frequency range above 1200 cm^{-1} and most of the signals can be attributed to protein bands (Fig. 7.7). Among them, the amide I bands between 1620 and 1700 cm^{-1} are the most interesting ones as their positions are directly related to the different types of secondary structure elements. Assignments of these bands to specific sections of the polypeptide chain, however, are not trivial and or unambiguous, despite the availability of three-dimensional structures for both redox states. Using appropriate optics and detectors (see Section 3.1.1.2), the IR-spectroelectrochemical studies can also be extended to the far-infrared region, corresponding to a region between 1000 and 150 cm^{-1} (Fig. 7.8) (Berthomieu et al. 2006). In this region, the contributions

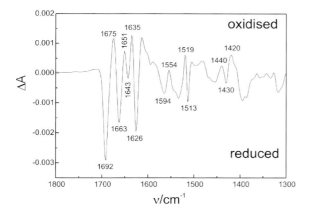

Fig. 7.7 Redox-induced IR difference spectrum of Cyt-c in solution (pH 7.0) measured from a spectroelectrochemical cell. The difference spectrum was obtained by subtracting the IR spectrum measured at +0.1 V from that measured at +0.35 V. Positive and negative signals refer to the oxidised and reduced form, respectively. (N. Wisitruangsakul and P. Hildebrandt, unpublished)

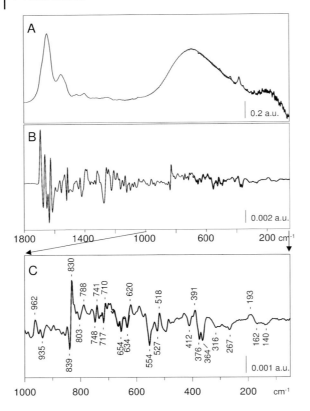

Fig. 7.8 IR spectra of Cyt-c in aqueous solution at pH 7.0. (A) Absolute IR absorption spectrum. (B) Redox-induced difference spectrum of Cyt-c ("reduced" minus "oxidised"). (C) Expanded view of the difference spectrum (B) in the region between 1000 and 50 cm^{-1}. The spectra are reproduced from Berthomieu et al. (2006) with permission.

from the protein matrix are weaker and most of the difference signals can be attributed to heme modes. IR spectroelectrochemistry not only provides structural information complementary to RR spectroscopy, but also allows precise control of the redox equilibrium and coupled processes (Schlereth and Mäntele 1993).

7.3.3
Conformational Equilibria and Dynamics

A large number of vibrational spectroscopic studies on Cyt-c have focussed on the different conformational states that can be stabilised by varying external parameters such as pH or temperature, or by binding to denaturants, amphiphiles, anionic surfaces (Scott and Mauk 1995). Most of the underlying conformational transitions are reversible and include structural changes in the heme pocket and on the level of the protein tertiary and secondary structure. These structural changes can, therefore, be probed by a variety of techniques, including vibrational spectroscopy.

A particularly large manifold of conformational states exists for ferri–Cyt-c, evidently due to the relatively weak Fe–S(Met80) bond in the oxidised state of the heme. Consequently, all non-native states of ferri–Cyt-c lack the Met80 ligand whereas the His18 ligand can only be removed from the heme under very drastic conditions. The Met ligation of ferri–Cyt-c is only stable in a relatively narrow pH range. Above a pH of ca. 8.0, Lys residues may replace this ligand and at pH values higher than 10, a hydroxide occupies this coordination site. With increasing acidification of the solution, first a His residue replaces Met80 (pH < 6.0) and then water can serve as the axial ligand. Furthermore, these non-native 6c species are typically in equilibrium with the 5c heme in which the axial coordination site remains vacant. All these species can be readily distinguished on the basis of the RR spectra as the various spin- and ligation states display characteristic vibrational signatures in the "marker band" region (Myer et al. 1983; Döpner et al. 1998; Boffi et al. 1999; Oellerich et al. 2002). Furthermore, the substantial changes in the heme–protein interactions accompanying the ligand exchange are reflected in the low-frequency region of the spectra.

The bis-His-coordinated (6cLS) state is of particular interest as it has been shown to play a crucial role in the folding mechanism of ferri–Cyt-c. In the presence of guanidinium hydrochloride (GuHCl, 6 M), even at pH 7.0, this species is formed. The ligand replacing Met80 may either be His33 or His26 (Yeh et al. 1998). This transition is remarkable as in the native structure both residues are located relatively remotely from the heme pocket, such that the underlying structural changes are fairly substantial. To probe the dynamics of the folding mechanism, time-resolved RR spectroscopy has been employed using the heme as a reporter group for large-scale and local protein folding events. The starting point for these experiments was the unfolded state of ferri–Cyt-c obtained in mildly acidic solutions (pH $\approx$ 4) in the presence of GuHCl (ca. 4 M). Folding was initiated by rapid mixing (see Section 4.4.2) with a buffer solution of pH 7.0 such that the resultant pH was ca. 5 and the GuHCl concentration was lowered below 1 M (Yeh et al. 1998). The folding process was monitored on the basis of the transient RR spectra from ca. 100 μs to the long millisecond time scale (Fig. 7.9). The analysis of the marker band region allowed identification of three intermediates that correspond to a 6cLS, 5cHS, and 6cHS configuration, characterised by the ν_3 mode at 1502, 1492, and 1482 cm^{-1}, respectively. These three species have been attributed to heme states in which the position of Met ligand is occupied by His33 or His26 (6cLS) or a water molecule (6cHS), or it remains vacant (5cHS). The time evolution of the marker bands indicates that transitions between the three species and the subsequent decay to the native (6cLS) form proceeds within milliseconds. On this time scale, however, the most pronounced spectral changes in the low-frequency region have already been completed, as indicated by comparing the RR spectra of the unfolded form and the time-resolved spectrum obtained after 0.1 ms.

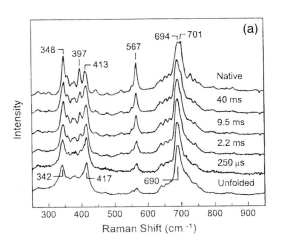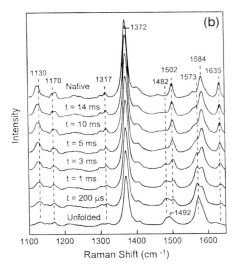

Fig. 7.9 Time-resolved RR spectra of ferri–Cyt-c (413 nm) in the low frequency (a) and marker band region (b) following the initiation of folding. The spectra of the unfolded species refer to 4.4 M GuHCl solutions at pH 4.7 (a) and pH 3.6 (b), whereas folding was induced upon rapid mixing with aqueous (GuHCl-free) solutions to yield a solution of 0.7 M GuHCl at a pH of ca. 5. The spectra are reproduced from Yeh et al. (1998) with permission.

Taking the low-frequency region of the spectra as an indicator of the fold of the polypeptide chain at least in the vicinity of the heme, one can distinguish between two main phases of the folding process. There is a rapid "nascent phase" (<100 µs) in which a largely folded but yet misligated protein is formed, and a slow (long ms) "ligand exchange phase" in which a rapid equilibrium between a 5cHS (-/His18), a 6cHS (water/His18), and a 6cLS (His-3/His18) is eventually converted into the native Met/His-ligated form. The kinetic and the structural data that are obtained from these time-resolved RR studies are consistent with and complementary to those derived from other spectroscopic techniques (e.g., circular dichroism and fluorescence spectroscopy) that are capable of probing the conformational changes on different levels of the structural hierarchy (i.e., the tertiary and secondary structure of proteins). The results accumulated by the various techniques, including RR spectroscopy, essentially contribute to a better understanding of the folding dynamics of Cyt-c and of soluble proteins in general (see also Section 5.6).

The results demonstrate that a far-reaching unfolding of the protein is not the prerequisite to formation of the three non-native ligation states detected in GuHCl-containing solutions. Also, "milder" conditions such as interactions with liposomes, micelles, and amphiphiles readily promote the ligand exchange even at pH 7.0. The RR spectra of these heme species measured in complexes with phospholipid vesicles, sodium dodecyl (SDS) micelles and monomers are essentially identical to those recorded from Cyt-c in the presence of GuHCl (Oellerich et al. 2002, 2003, 2004). On the other hand, for ferri–Cyt-c bound to phospholipid vesicles, no changes to the secondary structure but only a loosening of the tertiary structure have been inferred from the IR spectroscopic analysis of the amide I band region and the exchange of the amide bond protons (Heimburg and Marsh 1993). For binding to SDS micelles and monomers, the secondary structure changes are distinctly smaller than in the presence of GuHCl (Oellerich et al. 2002). This finding is somewhat surprising as specifically the formation of the bis-His ligated 6cLS form requires a major (tertiary) structural change to bring the peptide segment 30(20)–49 into close proximity to the heme pocket to allow for binding of His33 (His26) to the heme. The non-native 6cLS, 5cHS, and 6cHS species that exhibit no or only minor secondary structure changes are, therefore, denoted as "B2" states as distinct from their counterparts in the largely unfolded protein ("U" states) (Fig. 7.10). In complexes with phospholipids and SDS, the

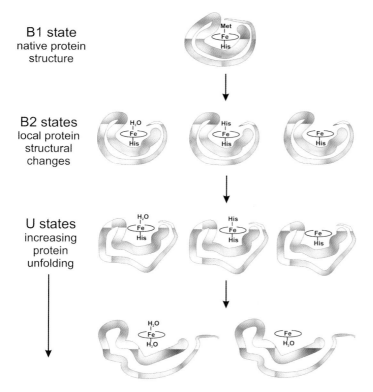

Fig. 7.10 Illustration of the various stages of the structural changes of Cyt-c from the native state to the unfolded protein. Increasing polypeptide unfolding in the U-states first causes the removal of the Met80 ligand from the heme and finally also the rupture of the Fe–His18 bond (Oellerich et al. 2002).

B2 states are in equilibrium with the native form of ferri–Cyt-*c*, denoted as state "B1", depending on the lipid composition, the ionic strength, the temperature, or the lipid/protein ratio. All native and non-native species can be readily distinguished in the marker band region, and thus these equilibria may be analysed in a quantitative manner by applying the component analysis (see Section 4.5) to the experimental RR spectra (Oellerich et al. 2004; Oellerich et al. 2003).

Cyt-*c* carries a large number of lysines in a ring-shaped arrangement around a hydrophobic patch including the exposed heme edge (Fig. 7.6). At pH 7.0, these residues are positively charged such that Cyt-*c* can readily bind to anionic surfaces such as the head groups of phospholipids or SDS micelles (Oellerich et al. 2002). Alternatively, binding may occur via the hydrophobic patch interacting with the non-polar chains of the amphiphiles. Both modes of binding are found upon complex formation with phospholipid vesicles and SDS and, in each instance, cause the formation of the B2 states, albeit with different B2/B1 ratios and different distributions among the B2 substates (Oellerich et al. 2003, 2004).

For binding of ferri–Cyt-*c* to SDS micelles, the dynamics of the conformational transitions have been monitored by freeze-quench RR spectroscopy (see Section 4.4.2.2) (Oellerich et al. 2003). The first step that is observed in the short millisecond range is the dissociation of the Met80 ligand from the heme iron, as indicated by the growing-in of the marker bands of a 5cHS species (Fig. 7.11). Subsequently, the His33 (His26) coordinates to the heme iron leading to a coordination equilibrium between a 6cLS and a 5cHS form. Thus, these processes occur with rate constants that are comparable to those observed in the ligand exchange phase of ferri–Cyt-*c* folding.

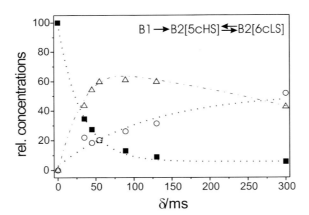

Fig. 7.11 Changes of the relative concentrations of the various ferric Cyt-*c* species following binding of ferri–Cyt-*c* to SDS micelles (data taken from Oellerich et al. 2003). The data were evaluated from the spectral amplitudes of the individual species as determined by component analysis. The solid squares, open triangles, and open circles refer to B1, B2[5cHS], and B2[6cLS], respectively. The dotted and dashed–dotted lines represent fits according to the kinetic model in the inset of the figure.

In contrast to ferri–Cyt-c, the reduced form of Cyt-c is significantly more stable and unfolding requires much harsher conditions, i.e., higher temperatures, more extreme pH values, and higher denaturant concentrations. However, the counterparts of the mono-His-ligated (5cHS) and bis-His-ligated (6cLS) species are also found under denaturing conditions (Yeh et al. 1998) and upon binding to phospholipid vesicles (Oellerich et al. 2002). In addition, there are two mono-His-ligated 5cHS species differing with respect to hydrogen bond interactions of the proximal His ligand. This is revealed by the frequencies of the Fe–His stretching (see Section 7.2.2) and the mode v_4, which are reporter bands for the electron density changes in the porphyrin. An increased hydrogen bond strength of the His ligand with an adjacent amino acid, as reflected by the frequency upshift of the Fe–His stretching from 201 to 227 cm^{-1}, causes an increase in the electron density in the porphyrin macrocycle as indicated by a decrease of the v_4 frequency from 1359.9 to 1353.7 cm^{-1} (Droghetti et al. 2006).

Studying the binding of Cyt-c with membrane models, i.e., phospholipid vesicles, is not only important for elucidating the conformational manifold and dynamics of proteins. The biological functions of Cyt-c are intimately linked to membrane interactions. For an efficient transport of electrons, Cyt-c has to remain in contact with the inner mitochondrial membrane and to diffuse along the membrane surface between the two membrane bound enzyme complexes cytochrome c reductase and cytochrome c oxidase (*vide supra*). Also the role of Cyt-c in apoptosis is linked to membrane interactions as its involvement in the apoptotic signal transduction cascade requires the transfer of Cyt-c across the mitochondrial membrane into the cytosol (Jiang and Wang 2004). Thus, the conformational changes of Cyt-c detected upon binding to phospholipid vesicles may also be relevant under physiological conditions. For the reaction between Cyt-c and cytochrome c oxidase, this view is supported by an RR study on the fully oxidised complex between both proteins that is stabilised at low ionic strength (Döpner et al. 1999). The RR spectra of these complexes differ from the sum of the spectra of the individual proteins, indicating structural changes in the heme sites induced upon complex formation. The component analysis of the spectra reveals in fact that the bound Cyt-c exists in an equilibrium between the native state B1 and the 6cLS form of state B2. The study included a variety of Cyt-c mutants with modifications in the binding domain. It was found that the relative concentration of B2 formed in the protein–protein complex correlates with the enzymatic activity of cytochrome c oxidase towards Cyt-c oxidation. These results point to the functional role of the conformational state B2 for the biological redox process.

7.3.4
Redox and Conformational Equilibria in the Immobilised State

Soon after the discovery of the surface-enhanced Raman (SER) effect (Sections 2.3 and 4.3), applications to biological systems were being explored. Among the first biomolecules to be tested was Cyt-c (Cotton et al. 1980), which offers two specific advantages. Firstly, an efficient adsorption on Ag surfaces is ensured due to the ring-like arrangement of positive charges on the front surface of the protein (*vide supra*) and the high molecular dipole moment pointing through the centre of the ring of lysines. Secondly, both the Q- and the Soret-band lie in the broad range of the plasmon resonances of SER-active nanostructured Ag materials, resulting in a high quantum efficiency of the surface-enhanced resonance Raman (SERR) effect. These two factors make Cyt-*c* a particularly useful test protein for developing SERR spectroscopic techniques and for studying fundamental interfacial processes of redox proteins. This latter aspect specifically refers to the character of the metal/electrolyte interface as a model system for biological interfaces (Cotton et al. 1980; Murgida and Hildebrandt 2004, 2005).

In the first studies, the SERR spectra of Cyt-*c* adsorbed on "bare" Ag surfaces (electrodes, colloids) were shown to differ from the RR spectra of the protein in solution, furnishing the discussion about possible denaturation of the adsorbed protein and thus questioning the applicability of SERS to biological systems. Since then, the conditions, under which reversible conformational changes and irreversible denaturation take place, have become largely understood at least for Cyt-*c* (Fig. 7.12).

In contact with aqueous solutions, the potential of zero charge is ca. -0.7 V such that Ag colloids and Ag electrodes (at potentials > -0.7 V) carry a positive charge, which is overcompensated by specifically adsorbed anions (e.g., chloride, citrate) in the inner Helmholtz layer (see Section 4.3.4). This alternating charge distribution corresponds to very high electric fields ($>10^9$ V m^{-1}, *vide infra*). For the Ag electrodes, the strength of the electric field can be controlled by vary-

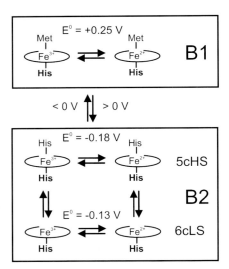

Fig. 7.12 Reaction scheme for the potential-dependent redox and conformational equilibria of Cyt-*c* electrostatically immobilised on electrodes (Wackerbarth and Hildebrandt 2003).

ing the electrode potential, i.e., it is lowered when the electrode potential approaches the potential of zero charge and it increases in the opposite direction. As the electric field strength is one of the parameters that govern the structure of proteins adsorbed in the electrical double layer, structural changes of the immobilised Cyt-*c* depend on the electrode potential. The reduced B1 state, i.e. the native Cyt-*c*, has a redox potential of +0.25 V (see Section 7.3.3) (Wackerbarth and Hildebrandt 2003). Therefore, at relatively negative electrode potentials (<0.0 V) corresponding to low electric fields, the native structure of Cyt-*c* is preserved and in the SERR spectrum, it exclusively displays the characteristic vibrational signature of the reduced B1 state.

This SERR spectrum is nearly identical to the RR spectrum ferro–Cyt-*c* in solution except for subtle differences in frequencies (± 1 cm^{-1}) and relative intensities. Switching the electrode potential to values above the redox potential causes the oxidation of B1 but the increased electric field strength destabilises the ferric form of B1, which is then converted into the B2 states. Similarly, the SERR spectra of these states are nearly identical to the RR spectra of the corresponding species observed in complexes with phospholipid vesicles or SDS micelles (see section 7.3.3). Decreasing the electrode potential again eventually leads to the reduction of the two B2 species, which exhibit distinctly more negative redox potentials than the state B1, i.e., ca. -0.13 V and -0.18 V for the 5cHS and 6cLS species, respectively. The lowering of the redox potentials by ca. 400 mV as compared with the B1 state can easily be understood in terms of the different ligation pattern and the structural changes in the heme pocket. At negative electrode potentials (<0.0 V), the reduced forms of B2 are not stable and are thus converted back into the reduced B1 state.

These conformational transitions that have been analysed in great detail by SERR spectroscopy (Wackerbarth and Hildebrandt 2003) are fully reversible and include only those species that are also observed upon binding to biological interfaces (e.g., phospholipid vesicles). However, the immobilised Cyt-*c* may also suffer from irreversible structural changes that are observed with increasing dwell times at positive electrode potentials (>+0.3 V) and high photon fluxes in the SERR experiments. Evidently, these conditions promote an unfolding of the polypeptide chain such that the back conversion to the native (B1) state is inhibited. Also, at very negative electrode potentials, i.e., close to the potential of zero charge, denaturation may occur. Presumably, the low charge density in the electrical double layer may favour interactions of the metal with non-polar amino acid residues, such that even originally buried hydrophobic peptide segments are brought into contact with the surface.

A more promising approach for SERR spectroscopic studies is based on coating the SER-active surfaces with biocompatible monolayers that are formed by self-assembly of functionalised alkyl thiols (SAM, see Section 4.3.6). For immobilisation of Cyt-*c*, these thiols are terminated by carboxyl functions, which, at pH 7, are partially deprotonated and thus are capable of binding Cyt-*c* via electrostatic interactions (Murgida and Hildebrandt 2001a). Under these conditions, the elec-

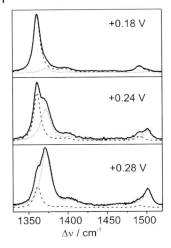 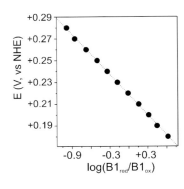

Fig. 7.13 SERR spectra of Cyt-c (left) immobilised on an Ag electrode coated with an SAM of carboxyl-terminated thiols involving 11 carbon atoms (C11-SAM). The spectra were measured as a function of the electrode potential using 413-nm excitation. The dashed and dotted lines are the component spectra of the reduced and oxidised B1 state. The amplitudes of the component spectra were converted into relative concentrations [Eq. (7.5)] to determine the redox potential in a Nernstian plot (right). The potentials have been corrected for the interfacial potential drop across the SAM/electrode interface. (Murgida and Hildebrandt 2001a)

tric field experienced by the immobilised Cyt-c is much lower than at Ag surfaces coated with specifically adsorbed anions. This is particularly true for a high surface coverage by Cyt-c that reduces the overall negative charge density and thus lowers the local field at the binding site. Under these conditions, a thorough potential-dependent SERR spectroscopic analysis of the conformational and redox equilibria has been carried out for Cyt-c immobilised on carboxyl-terminated SAMs of varying thickness (Fig. 7.13). The SERR spectra measured as a function of the electrode potential can be analysed solely in terms of two components, the reduced and the oxidised B1 state. Even without a component analysis, the potential-dependent changes of the distribution between both states are evident: at an electrode potential of +0.18 V, the reduced form prevails as indicated by the dominant band at ca. 1360 cm^{-1} (v_4). With increasing potential, the corresponding mode of the oxidised form at ca. 1370 cm^{-1} grows in and eventually becomes the strongest band (+0.28 V). The quantitative analysis of the spectra affords the amplitude ratio of the two-component spectra at each potential. The amplitude ratio I_{B1ox}/I_{B1red} is not identical to the concentration ratio c_{B1ox}/c_{B1red} but is related to this quantity via the SERR cross sections $\sigma_{SERR,B1red}$ and $\sigma_{SERR,B1ox}$ according to

$$\frac{c_{B1ox}}{c_{B1red}} = \frac{\sigma_{SERR,B1red}}{\sigma_{SERR,B1ox}} \cdot \frac{I_{B1ox}}{I_{B1red}} \quad (7.5)$$

As a good approximation one may assume that the ratio of the SERR cross sections is the same as that of the RR cross sections. The latter quantity can be determined by comparing the RR intensities of ferri– and ferro–Cyt-c in solutions of equal concentrations, measured under identical conditions. Then the concentration ratio of the reduced and oxidised B1 in SERR experiments is readily obtained from Eq. (7.5) and can be plotted as a function of the electrode potential. The data reveal an ideal Nernstian behaviour but the redox potential determined in this way shows a negative offset compared with the redox potential of native Cyt-c in solution. This offset, which steadily varies with the chain length of the thiols, originates from the potential drop across the Ag/SAM interface. Within the framework of a simple electrostatic model (see Section 4.3.4) one can correct the experimental determined redox potentials for this potential drop. Then, the corrected redox potentials are chain-length independent and identical to the redox potential in solution.

When the protein coverage on the SAM-coated electrodes is lowered, the effective electric field strength increases and the native structure of Cyt-c is destabilised. SERR spectra measured under these conditions reveal increasing contributions of the conformational state B2 upon decreasing the length of the alkyl chains. The electrostatic model allows rationalising of these observations in terms of an electric-field (E_F) dependent conformational equilibrium between the states B1 and B2 (Fig. 7.14) (Murgida and Hildebrandt 2001, 2004, 2005). The data in Fig. 7.14, which further include the results obtained for Cyt-c on electrodes coated by SAMs with phosphonate head groups and for Cyt-c on anion-coated electrodes, follow a van't Hoff type relationship

$$\left[\frac{\partial \ln K(E_F)}{\partial E_F} \right]_{p,T} = \frac{N_A \Delta \mu}{RT} \tag{7.6}$$

where N_A is the Avogadro number, R the gas constant, T the temperature, and $\Delta \mu$ refers to the difference between the dipole moments of B2 and B1. The electric-field-dependent equilibrium constant $K(E_F)$ is defined by

$$K(E_F) = \frac{[B2]}{[B1]} \tag{7.7}$$

The data reveal a ca. 3% larger dipole moment for B2 as compared with B1.

The electric fields at the protein binding sites that are estimated from the electrostatic model are in the order of from 10^8 to 10^9 V m^{-1}. Similar values are also obtained by analysing the vibrational Stark effect on the C=N stretching of nitrile-terminated SAM using SER or surface-enhanced infrared (SEIRA) spectroscopy (Oklejas et al. 2003; Murgida and Hildebrandt 2006). The field strengths at the electrochemical interface are thus comparable to those estimated for biological membranes in the vicinity of the charged and polar head groups (Clarke 2001), thereby further supporting the view that electrode/SAM interfaces may represent models for biological interfaces.

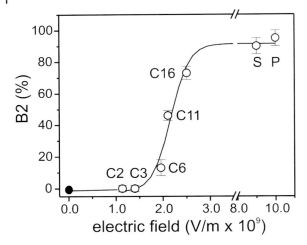

Fig. 7.14 Electric-field dependence of the B2/B1 conformational equilibrium of ferri–Cyt-c expressed in terms of the percentage of B2. The data (hollow circles) were obtained from SERR measurements of Cyt-c immobilised on Ag electrodes covered with various coatings. Ci refers to SAMs of carboxyl-terminated monolayers, where "i" indicates the number of the carbon atom of the thiol; P denotes the C11 SAM with a phosphonate instead of a carboxyl head group; and S stands for an Ag electrode covered with sulfate ions. All spectra were measured at an electrode potential of +0.22 V in contact with a solution containing no Cyt-c (pH 7.0), corresponding to a low surface coverage of Cyt-c. The black circle indicates the reference point in solution, i.e., 0% B2. The solid line is obtained by a fit of Eq. (7.6) to the experimental data (Murgida and Hildebrandt 2001a).

The redox equilibrium of immobilised Cyt-c has also been studied on SAM-coated Au electrodes. For this metal, SERR spectroscopy is not applicable as plasmon resonances for gold are above 600 nm, which are out of resonance with the electronic transitions of the heme. Instead, Au surfaces are appropriate for employing ATR-SEIRA difference spectroscopy (Ataka and Heberle 2003). Difference spectra are obtained by subtracting a spectrum measured at a reference potential E_{ref} (+0.1 V) from the spectra measured at variable potentials E_m (>+0.1 V) (Fig. 7.15). With increasing potential difference ($E_m - E_{ref}$), the difference signals increase reflecting the conversion from the reduced Cyt-c prevailing at +0.1 V to the oxidised form. The amplitudes of the signals, plotted as a function of the E_m, may be analysed in terms of the Nernstian equation leading to a redox potential of +0.25 V, which is consistent with that determined for the B1 state on Ag electrodes and for Cyt-c in solution (*vide supra*). Also the positions of the SEIRA difference bands agree with those observed for redox-induced difference spectra of Cyt-c in solution, which further confirms the conclusion that the native structure of the protein is also preserved on Au electrodes (cf. Fig. 7.7). However, there are fairly substantial differences in the amplitudes of the signals due to the specific selection rules in the SEIRA experiments. The surface enhancement of the

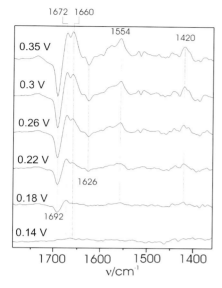
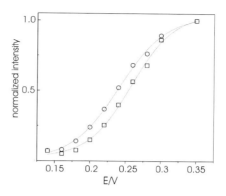

Fig. 7.15 Potential-dependent SEIRA difference spectra ("oxidised" minus "reduced") of Cyt-c immobilised on an Au electrode coated with a C11-SAM (SAM of carboxyl-terminated thiols including 11 carbon atoms; left side). The reference spectra was measured at +0.1 V. On the right side, the normalised intensities of the bands at 1672 cm^{-1} (circles) and 1660 cm^{-1} (squares) are plotted as a function of the electrode potential. The dotted lines represent the fits of the Nernstian equation to the experimental data. (N. Wisitruangsakul and P. Hildebrandt, unpublished)

IR bands is sensitive to the angle of the dipole moment of the molecular oscillator with respect to the surface. The enhancement, therefore, has the maximum value for a perpendicular orientation of the dipole whereas for a parallel orientation the enhancement is zero.

Let us consider the consequence for the amide modes that give rise to the most prominent signals in the SEIRA spectra of Cyt-c. The main contribution to the amide I originates from the C=O stretching. In an α-helix segment, for example, the individual C=O groups are all aligned in the propagation direction of the helix such that particularly strong enhancement for the amide I band is expected when the α-helix is standing upright on the surface. Conversely, no contribution is expected from helices oriented parallel to the surface. For the amide II (N–H in-plane bending), we have exactly the opposite situation as the dipole moment is oriented parallel to the helix. Consequently, the intensity pattern of the amide bands is controlled by the orientation of the protein with respect to the surface. This constraint, on the one hand, causes a loss of information because the amide I band analysis does not allow the determination of the secondary structure. On the other hand, novel information about the (changes of the) protein orientation can be extracted from SEIRA spectra (Ataka and Heberle 2003, 2004).

Electrostatic immobilisation is not the only mode of binding Cyt-*c* to coated electrodes. In addition, Cyt-*c* may also be immobilised to Ag and Au electrodes via covalent cross linking, hydrophobic interactions, or coordinative bonds, such that the bound protein can be studied by SERR and SEIRA spectroscopy (Murgida and Hildebrandt 2005; Ataka and Heberle 2004). Hydrophobic binding occurs via the non-polar patch comprising the amino acids 81–85 in the centre of the rings of surface lysines (Fig. 7.6). In contact with a hydrophobic surface (e.g., methyl-terminated SAMs), this patch is likely to be pulled away from the heme and the formation of the conformational state B2 is favoured. Coordinative binding is achieved through pyridyl-terminated SAMs, which directly anchor to the heme iron by substituting the Met80 ligand by a pyridyl group. For covalent attachment of Cyt-*c*, we start with carboxylate-terminated SAMs. Upon addition of EDC, covalent bonds between the SAM head groups and lysine residues of Cyt-*c* are formed (see Section 4.3.6). Here the orientation is not expected to be uniform as there are a relatively large number of potential attachment sites. The non-Nernstian behaviour in fact confirms this view.

7.3.5
Electron Transfer Dynamics and Mechanism

Cyt-*c* has been widely used for studying the fundamental aspects of biological electron transfer processes. Most of the experiments have been carried with Cyt-*c* in aqueous solution using chemically modified protein variants that are able to undergo photoinduced electron transfer (ET) reactions (Winkler and Gray 1992). The underlying concept is based on the covalent binding of photo-oxidisable groups [e.g., Ru(II)trisbipyridine] on amino acid residues on the surface of the protein. Attachment sites may either be surface Lys or Cys residues of the wild-type protein or appropriate amino acids (Cys, His) that are introduced at selected positions by site-directed mutagenesis. Upon photoexcitation, the Ru(II) complex is converted into a long lived excited triplet state, which can donate an electron to the ferric heme. Subsequently, the heme is re-oxidised by re-transferring an electron to the Ru(III) in the electronic ground state. These processes are usually studied by pump–probe transient absorption experiments and the results have contributed substantially to the better understanding of long range ET in biological systems.

In principle, these processes can also be monitored by pump–probe time-resolved RR spectroscopy as shown for Cyt-*c* variants with Ru(II) labels at Lys72 and Cys102, which are relatively close to the heme group (Simpson et al. 1996). It was originally expected that RR spectroscopy might provide additional information about the structural relaxation of the heme concomitant to the ET step. However, the transient RR spectra did not include any indications for intermediate states, e.g., a reduced heme with a non-relaxed porphyrin structure. This negative result may imply that the structural relaxation of the heme is faster than the time resolution of the experiment (<10 ns), or that the spectral region that was probed (i.e., the marker band region) is too insensitive to reflect the reorganisation process. A serious drawback of this approach, however, is the relatively harsh conditions that are required to achieve sufficiently high changes in the relative concentrations of the oxidised and reduced species as a prerequisite for a reliable spectra analysis. Accordingly, substantially higher pump pulse energies had to be chosen as compared with transient absorption experiments. Consequently, photoactive species are formed that affect the overall ET process. As a result, the kinetics derived from both techniques is very different. It appears to be that the use of the experimentally demanding pump–probe RR spectroscopy is not justified for probing this type of ET reactions.

However, regardless of using transient absorption or RR spectroscopy, studies on photoinduced ET reactions in solution are associated with a principle limitation. Most of the natural biological redox processes occur at charged interfaces and involve proteins that are either embedded in the membrane or immobilised on the surface of the membrane. In any case, the reaction conditions are very different compared with the proteins in solution. Specifically, local electrostatic fields at membranes are likely to have a pronounced effect not only on protein structures (*vide supra*) but also on the dynamics of ET reactions and thus on the overall mechanism of the redox process.

In this respect, coated electrodes represent a promising alternative as these devices may be considered as simple membrane models, specifically as far as the electrostatic field effects are concerned. The ET kinetics of electrostatically immobilised Cyt-*c* on coated Au electrodes has been studied by electrochemical methods (Avila et al. 2000) which, however, are not capable of identifying the nature of the electron transferring species. Hence, surface enhanced vibrational spectroscopic methods offer the advantage of monitoring the ET dynamics while simultaneously probing the structures of the species involved, a prerequisite to understanding the reaction mechanism of interfacial processes.

Time-resolved SERR spectroscopy has been employed for Cyt-*c* immobilised on electrodes coated with SAMs of carboxyl-terminated thiols (Murgida and Hildebrandt 2001b). The experiments have been carried out at high protein coverage such that the only process that is probed is the heterogeneous ET of the state B1 (*vide supra*). Upon employing a rapid potential jump from E_i to E_f, the redox equilibrium at the initial potential E_i is perturbed and the new equilibrium now referring to the final potential E_f is established via ET reactions between the adsorbed Cyt-*c* and the electrode. If E_i is equal to the redox potential E^0, the driving force for the ET is 0.0 eV, such that the analysis of the relaxation process affords the formal heterogeneous ET rate constant k_{ET}^0. A series of such time-resolved spectra are shown in Fig. 7.16. At the initial potential of +0.11 V, the adsorbed Cyt-*c* (B1 state) is completely reduced, as indicated by the lack of the 1373-cm^{-1} band that is characteristic of the oxidised form. Following a potential jump to the redox potential of +0.25 V, the time-resolved SERR spectra of Cyt-*c*, obtained with increasing delay time, show a growing-in of the 1373-cm^{-1} band at the expense of the band at ca. 1360 cm^{-1}. All spectra can be simulated on the basis of the component spectra of the ferric and the ferrous form of B1. At infinite delay time corresponding to the stationary spectrum at E_f, the SERR spectrum reflects both the

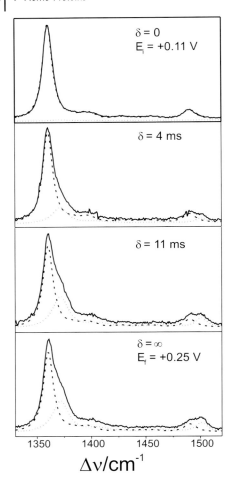

Fig. 7.16 SERR spectra of Cyt-c immobilised on an Ag electrode coated with a C2-SAM (SAM of carboxyl-terminated thiols including 2 carbon atoms) obtained with 413-nm excitation. The top and the bottom spectra are stationary spectra measured at the initial (E_i) and final electrode potentials (E_f). The second and third spectra are time-resolved SERR measured after different delay times δ relative to the potential jump from E_i to E_f. The dashed and dotted lines are the component spectra of the reduced and oxidised state B1, respectively. The potentials indicated in the figure have been corrected for the interfacial potential drop (Murgida and Hildebrandt 2001b).

reduced and the oxidised B1 at equal concentrations. Note that the amplitudes of the component spectra, however, are different, which is due to the different RR cross sections of both species. The time-dependent changes of the relative concentration of ferrous B1 can be analysed in terms of a one-step relaxation process to determine k_{ET}^0 (see Section 4.4.3) (Fig. 7.17).

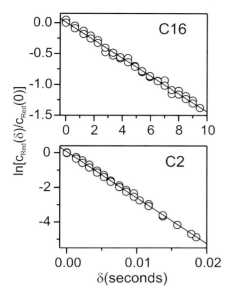

Fig. 7.17 Changes in the relative concentrations of the reduced B1 following a potential jump from the initial (negative) potential to the redox potential (final potential). The data were determined from time-resolved SERR measurements (see Fig. 7.16) for Cyt-c immobilised on Ag electrodes coated with C16- and C2-SAMs (SAMs of carboxyl-terminated thiols including 16 and 2 carbon atoms, respectively) (Murgida and Hildebrandt 2001b).

When the SAM length is reduced from 15 to 10 methylene groups, corresponding to a decrease of the ET distance d from 24 to 19 Å (Fig. 7.18), k_{ET}^0 increases from ca. 0.07 s^{-1} to 41 s^{-1}. These data can be rationalised in terms of the exponential distance-dependence of the electron tunnelling mechanism

$$k_{ET}^0 = A \cdot \exp(-\beta d) \qquad (7.8)$$

using values for the pre-exponential factor A and the tunnelling decay factor β as determined for (intramolecular) long-range ET processes in solution. For the longest chain length, the reorganisation energy of the heterogeneous ET has also been determined by time-resolved SERR spectroscopy, using overpotential- and temperature-dependent measurements (Murgida and Hildebrandt 2002). In the more accurate overpotential-dependent measurements the final potential E_f is stepwise varied such that the potential difference

$$\Delta E = |E^0 - E_f| \qquad (7.9)$$

and thus the driving force for the reduction ($E^0 > E_f$) or the oxidation ($E^0 < E_f$) increases.

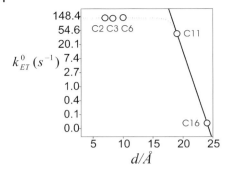

Fig. 7.18 Distance-dependence of the formal heterogeneous ET rate constant for the B1 state of Cyt-c immobilised on Ag electrodes coated with carboxyl-terminated SAMs of different chain lengths (C*i* refer to carboxy-terminated thiols including *i* carbon atoms) (Murgida and Hildebrandt 2001b). The data were determined by time-resolved SERR spectroscopy with potential jumps from negative potentials to the redox potential (see Figs. 7.16 and 7.17). The solid line connects the data points in the electron tunnelling regime where an exponential distance-dependence of the ET rate constant is expected (Eq. 7.8). The dotted line connects the data points for which electron tunnelling is not rate-limiting.

$$k_{ET}(\Delta E) = k_{ET}^0 \frac{1 - erf\frac{\lambda - F\Delta E}{2\sqrt{\lambda RT}}}{1 - erf\frac{\lambda}{2\sqrt{\lambda RT}}} \tag{7.10}$$

The reorganisation energy of 0.22 eV is similar to the value obtained from temperature-dependent measurements but is much smaller than that determined for the ET process of Cyt-c in solution. This discrepancy, however, can readily be understood taking into account that a large contribution to the reorganisation energy originates from the solvent environment. For the immobilised protein, this contribution is likely to be much smaller than for the fully solvated Cyt-c. Energy costs for solvent reorganisation as the result of the ET are particularly high for solvent molecules in the vicinity of the heme crevice. This part of the protein surface constitutes the binding domain for the SAM surface and thus exhibits only a limited accessibility to solvent molecules in the immobilised state.

Time-resolved SERR spectroscopy reveals a peculiar behaviour of Cyt-c immobilised at carboxylate-terminated SAMs of alkyl chains with less than 10 methylene groups. Hence, the rate constants that are derived from the spectroscopic experiments deviate from the expected behaviour and only moderately increase with decreasing ET distance until they level off at ca. 130 s^{-1} (Fig. 7.18). These results indicate that electron tunnelling is no longer rate-limiting and, in fact, the rate constant was found to be overpotential-independent in this regime. Thus, another process controls the heterogeneous ET process at short distances. This process does not include a conformational transition of the redox site as the time-resolved

SERR spectra do not provide any indication of an intermediate state. In fact, electrochemical studies of Cyt-c immobilised on SAM-coated Au electrodes reveal a similar tendency, although the limiting value for the experimentally determined rate constant is higher than that derived from the SERR experiments on Ag electrodes (Avila et al. 2000).

To explore this puzzling behaviour, the time-resolved SERR experiments have been carried out in ^2H_2O, revealing an increasing apparent kinetic isotope effect from 1.0 to 4.0 on decreasing the chain length from 19 to 6.3 Å. Concomitant temperature-dependent time-resolved SERR studies afford a significantly larger activation enthalpy of 34.3 kJ mol^{-1} in ^2H_2O versus 24.2 kJ mol^{-1} in H_2O at an ET distance of 7.6 Å. These effects may be (partly) due to the higher viscosity of ^2H_2O versus H_2O (Murgida and Hildebrandt 2001b, 2002, 2005). A more detailed molecular description of this rate-limiting event is not yet available but it probably involves a rearrangement of the protein–SAM complex to achieve an optimum ET geometry.

One would expect that time-resolved SEIRA spectroscopy will further contribute to the elucidation this process as this technique probes structural and reorientational changes of the immobilised protein following the potential jump. So far, step-scan and rapid-scan SEIRA spectroscopy have been employed in the electron tunnelling regime to probe the dynamics of the ET and the concomitant protein structural changes (Fig. 7.19). Hence, the time-evolution of amide I

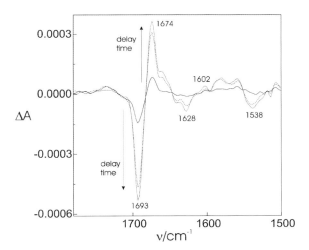

Fig. 7.19 Rapid-scan SEIRA difference spectra of Cyt-c immobilised on Au electrodes coated with C16-SAMs (SAM of carboxyl-terminated thiols including 16 carbon atoms). The difference spectra ("oxidised" minus "reduced"), which are obtained by subtraction of the reference spectrum measured at +0.1 V, refer to the spectra recorded after 1 s (solid line), 8 s (dashed line), and 20 s (dotted line) following a potential jump from +0.1 to +0.246 V. (N. Wisitruangsakul and P. Hildebrandt, unpublished)

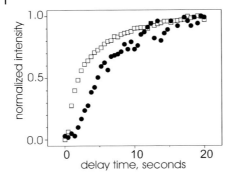

Fig. 7.20 Temporal evolution of the SEIRA intensities of the 1693- and 1672-cm^{-1} bands (open squares) and of the 1660-cm^{-1} band (solid circles) of Cyt-c adsorbed on C16-SAM (SAM of carboxyl-terminated thiols including 16 carbon atoms) coated Au electrode following a potential jump from +0.1 to +0.246 V (see Fig. 7.19).

bands, which give rise to strong signals in the stationary SEIRA difference spectra (*vide supra*), is monitored. After fitting exponential functions to the data, rate constants are obtained that are consistent with the data obtained from time-resolved SERR spectroscopy (Fig. 7.20) (Murgida and Hildebrandt 2006).

The most simple description of the kinetic behaviour is based on two-step sequential reaction scheme including the structural rearrangement or reorientation of the adsorbed Cyt-c (k_S) followed by electron tunnelling (k_{ET}). Whereas k_{ET} increases exponentially with decreasing distance [see Eq. (7.8)], k_S shows the opposite tendency. This latter behaviour may be attributed to the increasing electric field strength upon approaching the electrode. Thus, one expects that the overall ET rate constant runs through a maximum and then decreases with further increase in the electric field strength. This expectation is in fact confirmed by extending the time-resolved SERR measurements of Cyt-c to coated electrodes for which the local electric field strength was further increased through increasing the charge density in the interaction domain (Murgida and Hildebrandt 2004).

Further support for the electric-field dependent two-step redox process comes from the time-resolved RR experiments under very low electric field strengths. These conditions can be established by using methyl-terminated SAMs to which Cyt-c binds via the hydrophobic patch at the heme crevice. The lack of a charged binding domain drastically reduces the electric field strength experienced by the bound protein, and the ET process proceeds with a rate that approaches the value expected if electron tunnelling was also the rate-limiting step at distances shorter than 15 Å (Rivas et al. 2002).

Both for hydrophobic binding (methyl-terminated SAMs) or for electrostatic binding under very high electric field strengths (phosphonate-terminated SAMs, sulfate-coated electrodes) (Murgida and Hildebrandt 2004; Wackerbarth and Hildebrandt 2003), a significant portion of the immobilised Cyt-c is converted into the conformational state B2, such that the ET reactions are intimately coupled to conformational transitions. The dynamics of these complex processes can be disentangled by time-resolved SERR spectroscopy revealing very different timescales for the two limiting situations. Whereas at low electric field strengths (hydrophobic binding) the rate constants of conformational transitions between the states B1 and B2 may approach values higher than 10^4 s^{-1}, they are slowed down by ca. three orders of magnitude at very high electric field strengths, e.g., for Cyt-*c* bound to sulfate-coated Ag electrodes.

7.3.6
The Relevance of Surface-enhanced Vibrational Spectroscopic Studies for Elucidating Biological Functions

The similarities in the potential distribution at biomembranes and SAM-coated electrodes offers new possibilities for the study of biological processes under conditions that are more closely related to the natural environment. The importance of SAM-coated electrodes as functional membrane models has substantially increased with the development and adaptation of (time-resolved) SERR/SER and SEIRA spectroscopy as these techniques provide structural and dynamic information about immobilised biomolecules with enormous sensitivity and selectivity. The detailed studies that have been carried out specifically with Cyt-*c* have shed light on the behaviour of the protein under the influence of electric fields, and thus allow important conclusions to be made concerning the molecular functioning of Cyt-*c* at the mitochondrial membrane (Murgida and Hildebrandt 2004, 2005) and biological ET processes in general.

On the other hand, one should not overestimate the analogies between electrochemical and biological interfaces, as some properties of biomembranes that are known to be functionally important for steering biological processes cannot adequately be mimicked by SAMs on electrodes. These are, for example, the dynamic behaviour of the lipid bilayer and the fairly heterogeneous compositions of membranes. Furthermore, one should keep in mind the potential and limitations of the spectroscopic approaches. Firstly, SERR spectroscopy is not a fully noninvasive method as the exciting laser beam may initiate or accelerate unwanted processes, brought about by local heating in the laser beam or by photochemical reactions (see Section 4.1.2.2). Thus, great care has to be taken (i.e., low photon flux, moving electrodes) to suppress these processes as far as possible. Secondly, interpretation of the spectroscopic data is not necessarily simplified due to the inherent selectivity of the methods. Difference signals in the SEIRA spectra, for example, may be due to either structural or orientational changes, and the assignment of the signals to specific modes of the protein also has to take into account the distance-dependence of the surface enhancement. In general, an unambiguous interpretation is aggravated by the lack of other comparably sensitive spectroscopic methods that may provide complementary structural information.

Thus, SERR and SEIRA spectroscopies are not yet standard techniques that can be routinely applied to any protein. Comprehensive studies are required in each example, such that one may eventually obtain reliable structural and kinetic results from the spectra. However, the potential of surface enhanced vibrational spectroscopies is a long way from being fully exhausted and is not restricted to fundamental life sciences. Protein immobilisation represents an essential element in emerging biotechnological fields that requires the development of micro- and nanobioelectronic devices. Thus in this instance, powerful analytical tools are needed and in this respect the importance of surface-enhanced vibrational spectroscopies may strongly increase in the future.

7.4
Cytochrome c Oxidase

Cytochrome c oxidases (CcO) are the terminal enzymes in the respiratory chains of mitochondria and many bacteria (Michel et al. 1998). These enzymes are integral membrane proteins, with the number of subunits depending on the organism. The enzymes include various cofactors that serve for the catalytic reduction of oxygen to water and the coupling of ET with the proton translocation across the membrane. As a result, a proton gradient is built up that is finally used to drive ATP synthesis. In the mammalian and in many bacterial enzymes these cofactors include three redox sites. The binuclear Cu_A centre serves as the electron entry site for electrons delivered by Cyt-c. The electrons are further transferred to a heme a and subsequently to the heme a_3–Cu_B complex, which is capable of binding molecular oxygen. Here the catalytic process takes place with the stepwise reduction of oxygen to water. The number of subunits differs substantially between bacterial enzymes (3–4 subunits) and the mammalian enzyme (13 subunits) but only the subunits I–III represent the minimal set (core subunits) that are preserved in most copper–heme oxidases. For two representatives of these aa_3–oxidases, the three-dimensional structures have been determined revealing a very similar arrangement of the cofactors in subunits I and II. However, despite these well-resolved structural data, the molecular functioning of the enzyme is not yet fully understood. Vibrational spectroscopies are indispensable tools to elucidate the structure–dynamics–function relationships of oxidases. In the following sections we will describe various approaches that have been applied to aa_3–heme–copper oxidases, and point out the contributions that have been made to the understanding of the molecular function of these enzymes.

7.4.1
Resonance Raman Spectroscopy

Excitation throughout the visible spectral region leads to an exclusive enhancement of the RR bands of the two hemes, whereas contributions from the Cu sites remain largely obscured under these conditions (see Section 8.1). As with Cyt-c, the ferric hemes can be easily photoreduced by the exciting laser beam, such that moving samples (e.g., rotating cuvettes, flow systems) and low photon fluxes

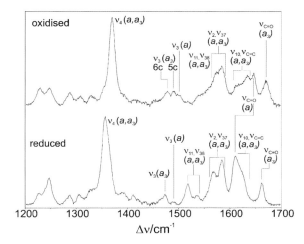

Fig. 7.21 RR spectra of CcO (*Rhodobacter spheroides*) in the fully oxidised and fully reduced state, measured with 413-nm excitation. A detailed assignment is listed in Table 7.3. (See Hrabakova et al. 2006)

are essential (see Section 4.1.2.2). The most intense spectra are obtained by excitation in the Soret-band region (Babcock 1988) (Fig. 7.21). The two heme groups in aa_3–oxidases possess the same chemical constitution (type-*a* heme) but a different coordination pattern, which is reflected in the spin marker band region of the RR spectra. The analysis of this region (1450 and 1700 cm^{-1}), however, is rather difficult due to the large number of contributing and partly overlapping bands of two heme groups. This high band density is particularly severe for type-*a* hemes for which, because of the relatively low symmetry, the two E_u modes ν_{37} and ν_{38} split into two RR-active components. In addition, the stretching modes of the two conjugated vinyl and formyl substituents gain resonance enhancement upon Soret-band excitation. Together with the remaining A_{1g}, B_{1g}, and A_{2g} modes, each heme may contribute up to 12 modes to this spectral region. To disentangle this complex vibrational band pattern, different excitation lines may be employed as the Soret transitions of heme *a* and heme a_3 are slightly displaced. In the fully oxidised enzyme, these transitions are at ca. 414 and 426 nm for heme a_3 and heme *a*, respectively. Thus, excitation at the low and high energy side leads to a preferential enhancement of the RR bands of heme *a* and heme a_3, respectively.

This approach does not allow for a complete spectral discrimination of the RR bands of one or the other heme. It is also not appropriate for the fully reduced state for which the differences in the electronic transitions are smaller, as both hemes show $0 \rightarrow 0$ transitions at ca. 440 nm. Hence an alternative strategy greatly simplifies the assignment. The HS heme a_3 is capable of binding diatomic ligands such as CN$^-$ or CO, which leads to the conversion from the HS to the 6cLS form. Moreover, these inhibitor complexes allow the creation of mixed-

Table 7.3 Vibrational assignments for the ferrous and ferric forms of hemes a and a_3 of beef heart CcO in the marker band region.[a]

Mode	Ferric CcO[b]		Ferrous CcO[b]	
	Heme a	Heme a_3	Heme a	Heme a_3
ν(C=O)[c]	1649	1674	1609	1665
ν(C=C)[c]	1627	1620	1626	1619
ν_{10} (B$_{1g}$)	1639	1609	1614	1606
ν_{37} (E$_u$)	1604	1585		
ν_2 (A$_{1g}$)	1590	1575	1586	1580
ν_{11} (B$_{1g}$)	1546	1520	1519	
ν_{38} (E$_u$)[d]	1567, 1528	1558, 1505	1568, 1544	1562, 1532
ν_3 (A$_{1g}$)	1499	1480	1491	1474
ν_4 (A$_{1g}$)	1373	1373	1354	1356

[a] Band frequencies are given in wavenumbers.
[b] Assignments and data taken from Heibel et al. (1993) and Babcock (1988).
[c] ν(C=C) and ν(C=O) are the stretching modes of the vinyl and formyl substituents, respectively.
[d] Splitting of the E$_u$ mode due to the removal of the x,y degeneracy as a result of the asymmetric porphyrin substitution.

valence states. Thus, the spectral discrimination is strongly enhanced such that is possible to construct approximate component spectra of the individual hemes by subtraction procedures (Argade et al. 1986). The approximate spectra may be further refined by applying a band fitting analysis to the spectra measured with different excitation lines (Heibel et al. 1993). Because for each band, the frequency and half-width is independent of the excitation line, the degrees of freedom is reduced in the fitting procedure and a reliable global simulation of all spectra can be achieved. The vibrational assignments for the ferric and ferrous hemes a and a_3 in the marker band region are listed in Table 7.3 (cf. Fig. 7.21).

The porphyrin modes in the marker band region largely follow the linear core-size frequency relationships (see Section 7.1.5) except for those modes which have also been found to deviate from the expected behaviour in b- and a-type hemes. These are specifically the modes ν_{11} and ν_2 in the ferrous states which appear at higher frequencies than predicted solely from the core-size-dependence.

The marker bands of heme a_3 are of particular interest as the distal coordination site serves for the binding of oxygen at the starting point of the catalytic cycle. In the oxidised form, the RR bands of heme a_3 are at positions characteristic of a 6cHS configuration (i.e., ν_3 at ca. 1480 cm^{-1}). The distal ligand is most likely a hydroxide bridging the heme a_3 and the Cu$_B$ centre (Michel et al. 1998).

In the reduced state, a RR spectroscopic distinction between 5cHS and 6cHS hemes is not possible as the core size and thus the marker band frequencies (e.g., v_3 at ca. 1470 cm^{-1}) are fairly similar in both instances (Parthasarathi et al. 1987). However, hydroxide coordination to a ferrous heme is highly unlikely and, hence, it is more plausible to assume that the distal coordination site is vacant and ready for binding oxygen. This interpretation is indirectly confirmed by the detection of the Fe–His stretching of heme a_3, which should only gain RR intensity in a mono-His ligated ferrous heme (*vide supra*). The frequency of 214 cm^{-1} (Babcock 1988; Argade et al. 1986) indicates the absence of hydrogen bond interactions with the imidazole ring.

The formyl stretching modes are sensitive probes for the local environment of type-a hemes as they reflect hydrogen-bond interactions of the C=O group. With increasing hydrogen bond strength, the C=O stretching shifts to lower frequencies, as shown in studies on model compounds (Babcock 1988). Thus, the high frequency of ca. 1670 cm^{-1} for the formyl stretching in heme a_3 points to a largely hydrophobic environment. This frequency is only reduced by ca. 10 cm^{-1} upon reduction, which can be understood in terms of the increased electron density in the porphyrin rather than by a change in the hydrogen bond interactions. For ferric heme a, however, the C=O stretching is as low as ca. 1648 cm^{-1} and shifts down even more to ca. 1610 cm^{-1} in the reduced form. These findings indicate a redox-linked transition from a hydrogen bond of medium strength to a rather strong hydrogen bond interaction. On the basis of these results, it had been proposed that the formyl group of heme a is involved in the redox-coupled proton transfer, although this hypothesis could not be confirmed in later studies (Babcock 1988).

7.4.2
Redox Transitions

The redox transitions of CcO include the sequential reduction of four metal sites. These transitions are coupled with the translocations of protons that are either used for the catalytic transformation of oxygen to water ("chemical protons") or are actively pumped across the membrane ("vectorial protons"). The transfer of protons occurs via two channels (D, K) that have been identified on the basis of the crystal structure and site-directed mutagenesis studies (Michel et al. 1998). The individual ET and proton transfer steps are linked via (anti-) cooperative effects between the redox centres and various protonable groups (Xavier 2004). Thus, the redox potentials for most of the redox sites are strongly pH-dependent and are also influenced by the overall level of reduction of the enzyme.

Spectroelectrochemistry with UV–vis and IR detection (see section 4.2.1.3) represents a sensitive approach to analysing these processes (Gorbikova et al. 2006; Richter et al. 2005; Hellwig et al. 1998, 1999; Rich and Breton, 2002; Nyquist et al. 2001). Compared with soluble proteins such as Cyt-c, considerable experimental difficulty is associated with the much larger size of the CcO and the presence of detergents required for solubilising the membrane-bound protein. To overcome unacceptably long equilibration times, the size of the working electrode has to be enlarged and specific redox mediator cocktails have to be chosen. In

7 Heme Proteins

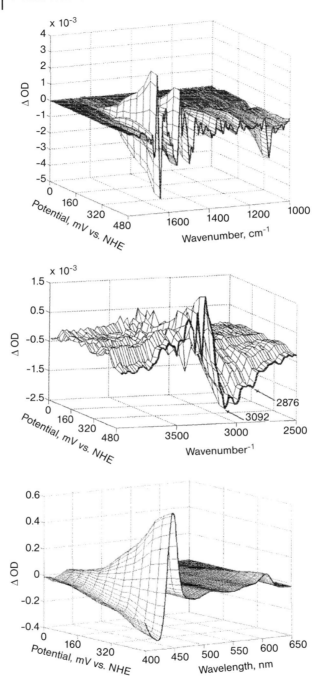

Fig. 7.22 (legend see p. 273)

addition, the experiments have to be carried out under strictly anaerobic conditions to avoid interference with the catalytic reactions. The combination of IR spectroelectrochemistry and the ATR technique appears to be particularly powerful.

On the basis of this approach, a comprehensive analysis of CcO from *Paracoccus denitrificans* has been presented by Gorbikova et al. (Gorbikova et al. 2006), who measured the UV–vis and IR spectra over a wide potential range (+0.48– 0.0 V), which covers all relevant redox transitions. The difference spectra generated with respect to the reference potential measured at −0.3 V cover the UV–vis region from 400 to 650 nm and the IR region from 1000 to 4000 cm^{-1} (Fig. 7.22). The UV–vis region reflects the changes in the electronic transitions of heme a and heme a_3 due to the redox process. Information about the redox state changes are also included in the IR spectral region but are not restricted to the heme centres. In addition, it covers the copper sites even though the assignment of the underlying vibrational mode is not definite.

The potential-dependence of the intensities at the various wavelength and wavenumber positions was globally analysed, such that it was possible to group a large number of bands according to their potential-dependence. In this way, four groups of characteristic spectral markers were obtained. One group, which displays a pH-independent Nernstian behaviour of a one-electron redox couple, is attributed to the Cu_A site with a redox potential of +0.25 V. All other groups describe pH-dependent redox transitions with redox potential variations of up to 100 mV between pH 6.5 and pH 9.0. Of particular interest are the bands that describe the heme a and heme a_3 redox transition, which are both strongly coupled via anti-cooperative interactions.

The strength of the IR spectroelectrochemical analysis is that redox *and* protonation sites are probed simultaneously. The protonable groups are, in general, carboxylate side chains of Glu and Asp, and the C=O stretchings of the COOH function give rise to strong absorption bands between ca. 1720 and 1740 cm^{-1} that are readily detectable in the difference spectra. Moreover, some of these bands have already been assigned on the basis of protein variants in which individual Glu and Asp residues have been substituted (Hellwig et al. 1998). The concomitant detection of protonation and redox marker band changes then provides a hint as to the involvement of specific Asp and Glu residues in the redox-linked translocation of protons. Whether or not this translocation refers to a vectorial proton transport has to be checked by separate measurements of the proton pump activity (Richter et al. 2005).

Fig. 7.22 Spectroelectrochemical titrations of CcO monitored by IR and UV–vis absorption spectroscopy. The figure displays the intensity contour as a function of the wavenumber (wavelength) in the IR (top and middle) and UV–vis region (bottom). The data are reproduced from Gorbikova et al. (2006) with permission.

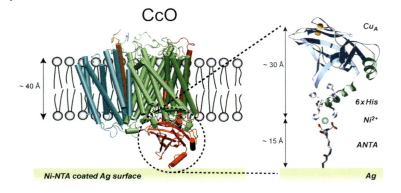

Fig. 7.23 Illustration of reconstituted C*c*O immobilised on a metal electrode. The His-tag of the protein is bound to the NTA-functionlised surface via complexation by Ni or Zn (Ataka et al. 2004; Hrabakova et al. 2006).

The application of surface enhanced vibrational spectroscopies for analysing the redox process of C*c*O is more demanding as the well-defined immobilisation of membrane proteins still represents a challenge. In one approach, C*c*O has been immobilised to Au or Ag electrodes while maintaining its natural membrane environment (Fig. 7.23). The stepwise assembly of the reconstituted enzyme on the Au electrode has been monitored through SEIRA spectroscopy, but no redox-difference spectra could be measured (Ataka et al. 2004). It was, therefore, concluded that the His-tagged C*c*O is electrically decoupled from the electrode. Analogous SERR experiments with the same device assembled on an Ag electrode revealed, in fact, only a very slow ET process that proceeds on the minute time scale and does not lead to a complete reduction of all redox sites (Hrabakova et al. 2006). To analyse the redox processes of membrane-bound C*c*O by surface enhanced vibrational spectroscopy, alternative immobilisation strategies are required that ensure a better electronic communication between the protein and the electrode.

7.4.3
Catalytic Cycle

The enzymatic process of C*c*O starts with binding of molecular oxygen to the reduced heme a_3. This reaction is very fast and has been monitored by pump–probe time-resolved experiments with UV–vis, IR, and RR spectroscopic detection. In these experiments, the ferrous CO complex is photolysed by the pump pulse. Under anaerobic conditions, re-binding of CO can be analysed to elucidate details of the protein dynamics (Heitbrink et al. 2002; Dyer et al. 1994). The studies are guided by the same concepts that have been widely used for hemoglobin and myoglobin (see Section 7.2.3).

To probe the enzymatic reduction of oxygen, these experiments are carried out in oxygen-saturated buffer solution such that binding of oxygen is more likely than the recombination with CO (Varotsis et al. 1993; Ogura et al. 1996; Han et al. 2000; Kitagawa, 2000). For these experiments, RR spectroscopic detection is particularly powerful as it allows probing of the vibrational modes of the various iron-oxo complexes of heme a_3, which are formed during the catalytic process. The first species (A) that is probed displays an RR band at ca. 570 cm^{-1} attributable to the Fe(II)–O$_2$ stretching, showing the expected ca. 30-cm^{-1} downshift for the ^{18}O$_2$ isotope (Fig. 7.24). Within ca. 0.5 ms the A form is converted into the P form (607-nm intermediate), originally thought to include a peroxo ligand. This species is characterised by an RR band at 800 cm^{-1}. The subsequent F form ("580-nm intermediate") displays a band at a somewhat lower frequency (785 cm^{-1}). In view of the similar frequencies and ^{18}O/^{16}O shifts, both bands are attributed to a ferryl-oxo bond stretching. This finding is surprising as the

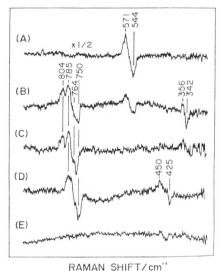

Fig. 7.24 Time-resolved RR spectra of the CcO–CO complex following photolysis in the presence of molecular oxygen. The figure displays the difference spectra obtained at the different delay times by subtracting the spectra measured in the presence of ^{18}O$_2$ from those measured in the presence of ^{16}O$_2$. The positive and negative peaks refer to the samples including ^{16}O$_2$ and ^{18}O$_2$, respectively. The difference spectra A, B, C, D, and E refer to delay times of 0.1, 0.27, 0.54, 2.7, and 5.4 ms, respectively. All spectra were measured with 428-nm excitation. The data are reproduced from Ogura et al. (1996) with permission.

overall charge distribution in the catalytic centre must be different in the P state as compared with the F state that is formed after accepting the third electron. Thus, the differences between the two Fe(IV)=O species must be related to the immediate protein environment.

It had been proposed that in the P state an additional reduction equivalent is "borrowed" from a nearby Tyr residue. The resultant Tyr radical is then re-reduced via the electron delivered to the catalytic site in the F state. The different electrostatic interactions in the catalytic site may then account for the different Fe(IV)=O stretching frequencies and electronic transitions.

The final intermediate that has been detected following the reduction by the fourth electron is an Fe(III)–OH complex characterised by an Fe–OH stretching frequency of ca. 450 cm^{-1}. The low frequency and the relatively small ^{2}H/H isotopic effect suggests a ferric HS heme a_3 and a strongly hydrogen bonded hydroxide ligand.

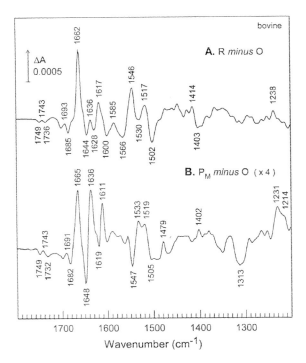

Fig. 7.25 Perfusion-induced ATR IR difference spectra of the fully reduced (top) and the P state (bottom), both referring to the spectrum of the oxidised CcO. The spectra are reproduced from Iwaki et al. (2003) with permission.

Studying the catalytic process of CcO by IR spectroscopy is more difficult. One approach exploits the fact that the P and F intermediates can be stabilised at alkaline pH. Under these conditions the P intermediate is formed by CO in the presence of O_2. Hence, CO acts as the reducing agent that brings the oxidised enzyme to the mixed-valence state. The IR spectroscopic characterisation of this state is carried out by means of the ATR technique on the basis of difference spectra obtained prior and after perfusion of the protein film with CO/O_2-containing alkaline buffer solutions (Iwaki et al. 2003). In the case of the F intermediate, an H_2O_2-containing solution is employed. In both experiments, the formation of the intermediates is not stoichiometric. Thus, the actual intermediate content must be quantified by monitoring the UV–vis absorption bands at 607 nm (P form) and 580 nm (F form) as spectral references. The "intermediate" minus the "oxidised" IR difference spectra obtained under the optimised conditions were then compared with the "reduced" minus the "oxidised" IR difference spectra to identify the bands characteristic of the intermediate states (Fig. 7.25). Some of the signals in the difference spectrum "P state" minus "ferric state" have been attributed to the heme a_3 group, reflecting the structural and electronic perturbation from the ferrous to the ferryl state. Complementary to the RR results, these studies suggest a change in the heme a_3–protein interactions in the P state, specifically including a nearby carboxyl side chain. These distortions are relaxed upon conversion into the F state, which is consistent with the view of Tyr radical recombination.

7.4.4
Oxidases from Extremophiles and Archaea

Owing to the specific biological significance and the availability of three-dimensional structure data, most of the spectroscopic studies were carried out with aa_3–heme–copper oxidases. However, nature provides a large manifold of terminal oxidases that differ with respect to the electron donor (e.g., quinol instead of cytochrome *c*) and the type and number of cofactors. Structure–function relationships are, therefore, different in these enzymes, which have been isolated specifically from extremophilic bacteria or archaea (Messerschmidt et al. 2001).

To study these enzymes, one may employ similar strategies as optimised for the investigation of aa_3-type oxidases, although the considerable complexity of some of these enzymes represents a challenge for spectroscopists. Various "simple" members of this enzyme family may just have different types of heme cofactors (e.g., ba_3, bb_3), which can even be advantageous for discriminating the contributions of the various redox sites in the vibrational spectra. However, other enzymes actually represent supercomplexes with fused subunits involving additional redox sites (e.g., caa_3 oxidase from *Thermus thermophilus*), thereby substantially complicating the interpretation of the spectra. Hence, close cooperation with molecular biology is required such that the individual building blocks can be studied separately.

References

Abe, M., Kitagawa, T., Kyogoku, Y., **1978**, "Resonance Raman spectra of octaethylporphyrinato-Ni(II) and *meso*-deuterated and ^{15}N-substituted derivatives. II. A normal coordinate analysis", *J. Chem. Phys.* **69**, 4526–4534.

Alden, R. G., Crawford, B. A., Doolen, R. Ondrias, M. R., Shelnutt, J. A., **1989**, "Ruffling of Ni(II) octaethylporphyrin in solution", *J. Am. Chem. Soc.* **111**, 2070–2072.

Anzenbacher, P., Evangelista-Kirkup, R., Schenkman, J., Spiro, T. G., **1989**, "Influence of thiolate ligation on the heme electronic structure in microsomal cytochrome P-450 and model compounds – resonance Raman spectroscopic evidence", *Inorg. Chem.* **28**, 4491–4495.

Argade, P. V., Ching, Y. C., Rousseau, D. L., **1986**, "Resonance Raman spectral isolation of the *a* and *a3* chromophores in cytochrome oxidase", *Biophys. J.* **50**, 613–620.

Ataka, K., Heberle, J., **2003**, "Electrochemically induced surface-enhanced infrared difference absorption (SEIDA) spectroscopy of a protein monolayer", *J. Am. Chem. Soc.* **125**, 4986–4987.

Ataka, K., Giess, F., Knoll, W., Naumann, R., Haber-Pohlmeier, S., Richter, B., Heberle, J., **2004**, "Oriented attachment and membrane reconstitution of His-tagged cytochrome *c* oxidase to a gold electrode: *In situ* monitoring by surface-enhanced infrared absorption spectroscopy", *J. Am. Chem. Soc.* **126**, 16199–16206.

Ataka, K., Heberle, J., **2004**, "Functional vibrational spectroscopy of a cytochrome *c* monolayer: SEIDAS probes the interaction with different surface-modified electrodes", *J. Am. Chem. Soc.* **126**, 9445–9457.

Avila, A., Gregory, B. W., Niki, K., Cotton, T. M., **2000**, "An electrochemical approach to investigate gated electron transfer using a physiological model system: Cytochrome *c* immobilized on carboxylic acid-terminated alkanethiol self-assembled monolayers on gold electrodes", *J. Phys. Chem. B*, **104**, 2759–2766.

Babcock, G. T., **1988**, "Raman scattering by cytochrome oxidase and by heme a model compound", in *"Biological Applications of Raman Spectroscopy"*, Vol. III, Spiro, T. G. (Ed.), Wiley, New York, chap. 7.

Berthomieu, C., Marboutin, L., Dupeyrat, F., Bouyer, P., **2006**, "Electrochemically induced FTIR difference spectroscopy in the mid- to far infrared (200 µm) domain: A new setup for the analysis of metal-ligand interactions in redox proteins", *Biopolymers* **82**, 363–367.

Boffi, A., Das, T. K., Della Longa, S., Spagnuolo, C., Rousseau, D. L., **1999**, "Pentacoordinate hemin derivatives in sodium dodecyl sulfate micelles: Model system for the assignment of the fifth ligand in ferric heme proteins", *Biophys. J.* **77**, 1143–1149.

Cantor, C. R., Schimmel, P. R., **1980**, *"Biophysical Chemistry"*, Vol. III, Freeman, New York.

Champion, P. M., Stallard, B. R., Wagner, G. C., Gunsalus, I. C., **1982**, "Resonance Raman detection of the Fe-S bond in cytochrome P450$_{cam}$", *J. Am. Chem. Soc.* **104**, 5469–5472.

Clarke, R. J., **2001**, "The dipole potential of phospholipid membranes and methods for its detection", *Adv. Colloid Int. Sci.* **89**, 263–281.

Cotton, F. A., **1990**, *"Chemical Applications of Group Theory"*, Wiley, New York.

Cotton, T. M., Schultz, S. G., Vanduyne, R. P., **1980**, "Surface-enhanced resonance Raman scattering from cytochrome *c* and myoglobin adsorbed on a silver electrode", *J. Am. Chem. Soc.* **102**, 7960–7962.

Dasgupta, S., Spiro, T. G., **1986**, "Resonance Raman characterization of the 7-ns photoproduct of (carbonmonoxy)hemoglobin: Implications for hemoglobin dynamics", *Biochemistry*, **25**, 5941–5948.

Desbois, A., Mazza, G., Stetzkowski, F., Lutz, M., **1984**, "Resonance Raman spectroscopy of protoheme-protein interactions in oxygen-carrying hemoproteins and in peroxidases", *Biochim. Biophys. Acta* **785**, 161–176.

Döpner, S., Hildebrandt, P., Rosell, F. I., Mauk, A. G., **1998**, "The alkaline conformational transitions of ferricytochrome c studied by resonance Raman spectroscopy", *J. Am. Chem. Soc.* **120**, 11246–11255.

Döpner, S., Hildebrandt, P., Rosell, F. I., Mauk, A. G., von Walter, M., Soulimane, T., Buse, G., **1999**, "The structural and functional role of lysine residues in the binding domain of cytochrome *c* for the redox process with cytochrome *c* oxidase", *Eur. J. Biochem.* **261**, 379–391.

Droghetti, E., Oellerich, S., Hildebrandt, P, Smulevich, G., **2006**, "Heme coordination states of unfolded ferrous cytochrome *c*", *Biophys. J.* **109**, 3022–3031.

Dyer, R. B., Peterson, K. A., Stoutland, P. O., Woodruff, W. H., **1994**, "Picosecond infrared study of the photodynamics of carbonmonoxy-cytochrome *c* oxidase", *Biochemistry* **33**, 500–507.

Frauenfelder, H., Parak, F., Young, R. D., **1986**, "Conformational substates in proteins", *Annu. Rev. Biophys. Biochem.* **17**, 451–479.

Friedman, J. M., **1994**, "Time-resolved resonance Raman spectroscopy as probe of structure, dynamics, and reactivity in hemoglobin", *Methods Enzymol.* **232**, 205–231.

Gorbikova, E. A., Vuorilehto, K., Wikström, Verkhovsky, M. I., **2006**, "Redox titration of all electron carriers of cytochrome *c* oxidase by Fourier transform infrared spectroscopy", *Biochemistry* **45**, 5641–5649.

Gouterman, M., **1979**, "Optical spectra and electronic structure of porphyrins and related rings", in *"The Porphyrins"*, Dolphin, D. (Ed.), Vol. III, part A, Academic Press, New York, pp. 1–156.

Han, S., Takahashi, S., Rousseau, D. L., **2000**, "Time-dependence of the catalytic intermediates in cytochrome *c* oxidase", *J. Biol. Chem.* **275**, 1910–1919.

Hanson, L. K., Sligar, S. G., Gunsalus, I. C., **1977**, "Electronic structure of cytochrome P450", *Croat. Chem. Acta* **49**, 237–250.

Heibel, G., Hildebrandt, P., Ludwig, B., Steinrücke, P., Soulimane, T., Buse, G., **1993**, "A comparative resonance Raman study of cytochrome *c* oxidase from beef heart and *Paracoccus denitrificans*", *Biochemistry* **32**, 10866–10877.

Heimburg, T., Marsh, D., **1993**, "Investigation of secondary and tertiary structural changes of cytochrome *c* in complexes with anionic lipids using amide hydrogen exchange measurements: An FTIR study", *Biophys. J.* **65**, 2408–2417.

Heitbrink, D., Sigurdson, H., Bolwien, C., Brezinski, P., Heberle, P., **2002**, "Transient binding of CO to Cu_B in cytochrome *c* oxidase is dynamically linked to structural changes around a carboxyl group: A time-resolved step-scan Fourier transform infrared investigation", *Biophys. J.* **82**, 1–10.

Hellwig, P., Behr, J., Ostermeier, C., Richter, O.-M., H., Pfitzner, U., Odenwald, A., Ludwig, B., Michel, H., Mäntele, W., **1998**, "Involvement of glutamic acid 278 in the redox reaction of the cytochrome *c* oxidase from *Paracoccus denitrificans*. Investigated by FTIR spectroscopy", *Biochemistry* **37**, 7390–7399.

Hellwig, P., Grzybek, S., Behr, J., Ludwig, B., Michel, H., Mäntele, W., **1999**, "Electrochemical and ultraviolet/visible/infrared spectroscopic analysis of heme *a* and heme *a3* contributions and assignment of vibrational modes", *Biochemistry* **38**, 1685–1694.

Hildebrandt, P., **1992**, "Resonance Raman spectroscopy of cytochrome P-450", in *Frontiers in Biotransformation*, Vol. 7, Ruckpaul, K., Rein, H. (Eds.), Akademie-Verlag/VCH, Berlin/Weinheim, pp. 166–215.

Howes, B. D., Schiodt, C., Welinder, K. G., Marzocchi, M. P., Ma, J.-G., Zhang, J., Shelnutt, J. A., Smulevich, G., **1999**, "The quantum mixed-spin heme state of barley peroxidase: A paradigm for class III peroxidases", *Biophys. J.* **77**, 478–492.

Hrabakova, J., Ataka, K., Heberle, J., Hildebrandt, P., Murgida, D. H., **2006**, "Long distance electron transfer in cytochrome *c* oxidase immobilised on electrodes. A surface enhanced resonance Raman spectroscopic study", *Phys. Chem. Chem. Phys.* **8**, 759–766.

Hu, S., Smith, K. M., Spiro, T. G., **1996**, "Assignment of protoheme resonance Raman spectrum by heme labeling in myoglobin", *J. Am. Chem. Soc.* **118**, 12638–12646.

Hu, S., Morris, I. K., Singh, J. P., Smith, K. M., Spiro, T. G., **1993**, "Complete assignment of cytochrome *c* resonance Raman spectra via enzymatic reconstitution with isotopically labeled hemes", *J. Am. Chem. Soc.* **115**, 12446–12458.

Ignarro, L. J., **2002**, "Nitric oxide as a unique signalling molecule in the vascular system: A historical overview", *J. Physiol. Pharmacol.* **53**, 503–514.

Iwaki, M., Puustinen, A., Wikström, M., Rich, P. R., **2003**, "ATR-FTIR spectroscopy of the P_M and F intermediates of bovine and *Paracoccus denitrificans* cytochrome c oxidase", *Biochemistry* **42**, 8809–8817.

Jayaraman, V., Rodgers, K. R., Mukerji, I., Spiro, T. G., **1995**, "Hemoglobin allostery: Resonance Raman spectroscopy of kinetic intermediates", *Science* **269**, 1843–1848.

Jiang, X., Wang, X., **2004**, "Cytochrome c-mediated apoptosis", *Annu. Rev. Biochem.* **73**, 87–106.

Kitagawa, T., Ozaki, Y., **1987**, "Infrared and Raman spectra of metalloporphyrins", *Struct. Bonding*, **64**, 72–113.

Kitagawa, T., **2000**, "Structures of reaction intermediates of bovine cytochrome c oxidase probed by time-resolved vibrational spectroscopy", *J. Inorg. Biochem.* **82**, 9–18.

Kozlowski, P. M., Rush III, T. S., Jarzecki, A. A., Zgierski, M. Z., Chase, B., Piffat, C., Ye, B.-H., Li, X.-Y., Pulay, P., Spiro, T. G., **1999**, "DFT-SQM force field for Ni porphine: Intrinsic ruffling", *J. Phys. Chem.* **103**, 1357–1366.

Li, X.-Y., Czernuszewicz, R. S., Kincaid, J. R., Spiro, T. G., **1989**, "Consistent porphyrin force field. 3. Out-of-plane modes in the resonance Raman spectra of planar and ruffled nickel octaethylporphyrin", *J. Am. Chem. Soc.* **111**, 7012–7023.

Li, X.-Y., Czernuszewicz, R. S., Kincaid, J. R., Su, Y. O., Spiro, T. G., **1990**, "Consistent porphyrin force field. 1. Normal mode analysis for nickel porphine and nickel tetraphenylporphine from resonance Raman and infrared spectra and isotope shifts", *J. Phys. Chem.* **94**, 31–47.

Matsukawa, S., Mawatari, K., Yoneyama, Y., Kitagawa, T., **1985**, "Correlation between the iron-histidine stretching frequencies and oxygen affinity of hemoglobins. A continuous strain Model", *J. Am. Chem. Soc.* **107**, 1108–1113.

Messerschmidt, A., Huber, R., Wieghardt, K., Poulos, T., **2001**, *"Handbook of Metalloproteins"*, Wiley, New York.

Michel, H., Behr, J., Harrenga, A., Kannt, A., **1998**, "Cytochrome c oxidase: Structure and spectroscopy", *Annu. Rev. Biophys. Biomol. Struct.* **27**, 329–356.

Murgida, D. H., Hildebrandt, P., **2001a**, "The heterogeneous electron transfer of cytochrome c adsorbed on coated silver electrodes. Electric field effects on structure and redox potential", *J. Phys. Chem. B* **105**, 1578–1586.

Murgida, D. H., Hildebrandt, P., **2001b**, "Proton coupled electron transfer in cytochrome c", *J. Am. Chem. Soc.* **123**, 4062–4068.

Murgida, D., Hildebrandt, P., **2002**, "Electrostatic-field dependent activation energies control biological electron transfer", *J. Phys. Chem. B* **106**, 12814–12819.

Murgida, D. H., Hildebrandt, P., **2004**, "Electron transfer processes of cytochrome c at interfaces. New insights by surface-enhanced resonance Raman spectroscopy", *Acc. Chem. Res.* **37**, 854–861.

Murgida, D. H., Hildebrandt, P., **2005**, "Redox and redox-coupled processes of heme proteins and enzymes at electrochemical interfaces", *Phys. Chem. Chem. Phys.* **7**, 3773–3784.

Murgida, D. H., Hildebrandt, P., **2006**, "Surface-enhanced vibrational spectroelectrochemistry: Electric field effects on redox and redox-coupled processes of heme proteins", *Top. Appl. Phys.* **103**, 313–334.

Myer, Y. P., Srivastava, R. B., Kumar, S., Raghavendra, K., **1983**, "States of heme in heme c model systems: Cytochrome c and heme c models", *J. Prot. Chem.* **2**, 13–43.

Nagai, M., Wajcman, H., Lahary, A., Nakatsukasa, T., Nagatomo, S., Kitagawa, T., **1999**, "Quaternary structure sensitive tyrosine residues in human hemoglobin: UV resonance Raman studies of mutants at $\alpha 140$, $\beta 35$, and $\beta 145$ tyrosine", *Biochemistry*, **38**, 1243–1251.

Nyquist, R. M., Heitbrink, K., Bolwien, C., Wells, T. A., Gennis, R. B., Heberle, J., **2001**, "Perfusion-induced redox differences in cytochrome c oxidase: ATR/FT-IR spectroscopy", *FEBS Lett.* **505**, 63–67.

Oellerich, S., Wackerbarth, H., Hildebrandt, P., **2002**, "Spectroscopic characterization of non-native states of cytochrome c", *J. Phys. Chem. B* **106**, 6566–6580.

Oellerich, S., Wackerbarth, H., Hildebrandt, P., **2003**, "Conformational equilibria and dynamics of cytochrome *c* induced by binding of SDS monomers and micelles", *Eur. Biophys. J.* **32**, 599–613.

Oellerich, S., Lecomte, S., Paternostre, M., Heimburg, T., Hildebrandt, P., **2004**, "Peripheral and integral binding of cytochrome *c* to phospholipid vesicles", *J. Phys. Chem. B* **108**, 3871–3878.

Ogura, T., Hirota, S., Proshlyakov, D. A., Shinzawa-Itoh, K., Yoshikawa, S., Kitagawa, T., **1996**, "Time-resolved resonance Raman evidence for tight coupling between electron transfer and proton pumping of cytochrome *c* oxidase upon the change from the Fe^V oxidation level to the Fe^{IV} oxidation level", *J. Am. Chem. Soc.* **118**, 5443–5449.

Oklejas, V., Sjostrom, C., Harris, J. M., **2003**, "Surface-enhanced Raman scattering based vibrational Stark effect as a spatial probe of interfacial electric fields in the diffuse double layer", *J. Phys. Chem. B* **107**, 7788–7794.

Parthasarathi, N., Hansen, C., Yamaguchi, S., Spiro, T. G., **1987**, "Metalloporphyrin core size resonance Raman marker bands revisited: Implications for the interpretation of hemoglobin photoproduct Raman frequencies", *J. Am. Chem. Soc.* **109**, 3865–3871.

Rich, P. R., Breton, J., **2002**, "Attentuated total reflection Fourier transform infrared studies of redox changes in bovine cytochrome *c* oxidase: Resolution of the redox Fourier transform infrared difference spectrum of heme *a3*", *Biochemistry* **41**, 967–973.

Richter, O.-M. H., Dürr, K. L., Kannt, A., Ludwig, B., Scandurra, F. M., Giuffrè, A., Sarti, P., Hellwig, P., **2005**, "Probing the access of protons to the K pathway in the *Paracoccus denitrificans* cytochrome *c* oxidase", *FEBS J.* **272**, 404–412.

Rivas, L., Murgida, D. H., Hildebrandt, P., **2002**, "Conformational and redox equilibria and dynamics of cytochrome *c* immobilised on electrodes via hydrophobic interactions", *J. Phys. Chem. B*, **106**, 4823–4830.

Rodgers, K. R., Su, C., Subramaniam, S., Spiro, T. G., **1992**, "Hemoglobin R → T structural dynamics from simultaneous monitoring of tyrosine and tryptophan time-resolved UV resonance Raman signals", *J. Am. Chem. Soc.*, **114**, 3697–3709.

Rousseau, D. L., Ondrias, M. R., **1983**, "Resonance Raman scattering studies of the quaternary structure transition in haemoglobin", *Annu. Rev. Biophys. Bioeng.* **12**, 357–380.

Schlereth, D. D., Mäntele, W., **1993**, "Electrochemically induced conformational changes in cytochrome *c* monitored by Fourier transform infrared difference spectroscopy: Influence of temperature, pH, and electrode surface", *Biochemistry* **32**, 1118–1126.

Scott, R. A., Mauk, A. G. (Eds.), **1995**, "Cytochrome c. A Multidisciplinary Approach", University Science, Mill Valley.

Shelnutt, J. A., Rousseau, D. L., Dethmers, J. K., Margoliash, E., **1981**, "Protein influences on porphyrin structure in cytochrome *c*: Evidence from Raman difference spectroscopy", *Biochemistry* **20**, 6485–6497.

Simpson, N. C., Millet, F., Pan, L. P., Larsen, R. W., Hobbs, J. D., Fan, B., Ondrias, M. R., **1996**, "Transient and time-resolved resonance Raman investigation of photoinduced electron transfer in ruthenated cytochromes *c*", *Biochemistry* **35**, 10019–10030.

Smulevich, G., Mauro, J. M., Fishel, L. A., English, A. M., Kraut, J., Spiro, T. G., **1988**, "Heme pocket interactions in cytochrome *c* peroxidase studied by site-directed mutagenesis and resonance Raman spectroscopy", *Biochemistry* **27**, 5477–5485.

Smulevich, G., Hu, S., Rodgers, K. R., Goodin, D. B., Smith, K. M., Spiro, T. G., **1996**, "Heme-protein interactions in cytochrome c peroxidase revealed by site-directed mutagenesis and resonance Raman spectra of isotopically labeled hemes", *Biospectroscopy* **2**, 365–376.

Spiro, T. G., **1983**, "The resonance Raman spectroscopy of metalloporphyrins and heme proteins", in *Iron Porphyrins*, Lever, A. B. P., Gray, H. B. (Eds.), Vol. 2, Addison-Wesley, Reading, chap. 3.

Spiro, T. G., **1985**, "Resonance Raman spectroscopy as a probe of heme protein structure and dynamics", *Adv. Protein Chem.* **37**, 111–159.

Spiro, T. G., Strekas, T. C., **1974**, "Resonance Raman spectra of heme proteins. Effects of oxidation and spin state", *J. Am. Chem. Soc.* **96**, 338–345.

Spiro, T. G., Zgierski, M. Z., Kozlowski, P. M., **2001**, "Stereoelectronic factors in CO, NO and O_2 binding to heme from vibrational spectroscopy and DFT analysis", *Coord. Chem. Rev.* **219–221**, 923–936.

Stein, P., Mitchell, M., Spiro, T. G., **1980**, "H-bond and deprotonation effects on the resonance Raman iron-imidazole mode in deoxyhemoglobin. Implications for hemoglobin cooperativity", *J. Am. Chem. Soc.* **102**, 7795–7797.

Varotsis, C., Zhang, Y., Appelman, E. H., Babcock, G. T., **1993**, "Resolution of the reaction sequence during the reduction of O_2 by cytochrome oxidase", *Proc. Natl. Acad. Sci. USA* **90**, 237–241.

Wackerbarth, H., Hildebrandt, P., **2003**, "Redox and conformational equilibria and dynamics of cytochrome c at high electric fields", *ChemPhysChem.* **4**, 714–724.

Wang, D., Zhao, X., Spiro, T. G., **2000**, "Chain selectivity of tyrosine contributions to hemoglobin static and time-resolved UVRR spectra in ^{13}C isotopic hybrids", *J. Phys. Chem. A*, **104**, 4149–4154.

Wilson, E. B., Decius, J. C., Cross, P. C., **1955**, *"Molecular Vibrations: The Theory of Infrared and Raman Vibrational Spectra"*, McGraw-Hill, New York.

Winkler, J. R., Gray, H. B., **1992**, "Electron-transfer in ruthenium-modified proteins", *Chem. Rev.* **92**, 369–379.

Yeh, S.-R., Han, S., Rousseau, D. L., **1998**, "Cytochrome c folding and unfolding: A biphasic mechanism", *Acc. Chem. Res.* **31**, 727–736.

Xavier, A. V., **2004**, "Thermodynamic and choreographic constraints for energy transduction by cytochrome c oxidase", *Biochim. Biophys. Acta* **1658**, 23–30.

8
Non-heme Metalloproteins

In metalloproteins, nature exploits the properties of transition metal ions in a large variety of structural scaffolds. The coordination sphere of the metal ion may be either provided by organic ligands such as porphyrins (see Chapter 7) or, as discussed in this chapter, by amino acid residues or small inorganic ligands including sulfur, oxygen, water, carbon monoxide or even cyanide (Messerschmidt et al. 2001). These non-heme metalloproteins function, for instance, as electron transfer (ET) proteins or enzymes. Among them, only a few classes, i.e., copper proteins, iron–sulfur proteins, none-heme iron proteins, and hydrogenases, will be treated in detail in the following sections to illustrate the potential of vibrational spectroscopy in elucidation structure–function relationships.

The main targets of the vibrational spectroscopic studies are the metal–ligand and intra-ligand vibrational modes. Metal–ligand stretching and bending vibrations appear in the low-frequency region between 200 and 800 cm^{-1}, and thus they are more readily accessible for RR rather than for IR spectroscopy. The electronic transitions that are required for the resonance enhancement are generally much weaker than those provided by organic prosthetic groups and typically originate from charge-transfer (CT) transitions either from the ligand to the metal (LMCT) or from the metal to the ligand (MLCT). At best, extinction coefficients are ca. 5000 LM^{-1} cm^{-1}, but in many instances they are even lower. Thus, high quality RR spectra require concentrations (>100 μM) that are significantly higher than those for heme proteins (<10 μM).

A substantial improvement of the spectral quality can be obtained in low-temperature experiments of frozen samples as the background scattering is strongly reduced and the bands are sharper. An additional constraint for the application of RR spectroscopy results from the nature of the CT transitions, which depends sensitively on the occupancy of the metal ion orbitals. Thus, frequently only one oxidation state of the metal exhibits a CT transition that is sufficiently separated from the UV absorption bands of the protein and, hence, is suitable for RR spectroscopy, whereas in other oxidation states the metalloprotein is RR-silent.

For hydrogenases, none of the oxidation states exhibit a CT in the visible or near-UV region. Hence, vibrational studies focus on the intra-ligand stretching modes of the bound CO and CN. These modes appear in a spectral region that

Vibrational Spectroscopy in Life Science. Friedrich Siebert and Peter Hildebrandt
Copyright © 2008 WILEY-VCH Verlag GmbH & Co. KGaA, Weinheim
ISBN: 978-3-527-40506-0

is free from any interference with vibrational bands of the protein and thus these ligand modes can even be detected in the absolute IR spectra. IR spectroscopy is, therefore, the method of choice for studying hydrogenases.

The interpretation of vibrational spectra of metalloproteins is currently largely restricted to empirical approaches. In most instances, the only promising approach is the comparison with vibrational spectra of metalloproteins, for which the three-dimensional structure is known. Normal mode analyses that are based on empirical force fields require isotopic labelling of the cofactor–protein complexes. These isotopomers can only be obtained by growing microorganisms, which produce the target proteins, on isotopically enriched media. Comparative vibrational analyses of synthetic models may further guide the interpretation of the spectra. However, this approach represents an enormous challenge as the cofactor structure strongly depends on the specific protein environment, which can rarely be mimicked by synthetic compounds. For the same reason, quantum chemical spectra calculations for *isolated* cofactors are problematic. These calculations require the optimised geometries of the cofactors, which typically correspond to very different structures in the absence of the surrounding protein matrices than in cofactor–protein complexes. This drawback is by far not as severe for organic prosthetic groups, such as in retinal or heme proteins, as these cofactors exhibit more rigid structures that, in most instances, are similar in the protein-bound and in the isolated states (e.g., see Section 7.1.4).

8.1
Copper Proteins

Copper proteins are divided into three major classes according to their UV–vis and EPR spectroscopic properties, denoted as type-1, type-2, and Cu_A sites (Fig. 8.1). The copper ion can exist in the oxidation states Cu(I) (reduced) or Cu(II) (oxidised) and thus may serve as redox sites for catalytic or ET processes (Messerschmidt et al. 2001). In type-1 copper proteins three ligands, including one Cys and one or two His, span a plane with the Cu ion located either in the centre (trigonal) or slightly above the plane (distorted tetrahedral). The latter geometry is stabilised by an additional weak ligand which may be, for instance, a Met residue. In the oxidised state, these type-1 proteins exhibit a relatively strong electronic transition at ca. 600 nm, which gives rise to the blue colour of these proteins. This transition, which results from a CT from the p-orbital of the Cys–sulfur to the $d_{x^2-y^2}$-orbital of Cu(II), is suitable for RR spectroscopy. A second CT at higher energies (<500 nm) exhibits much weaker oscillator strength in type-1 copper proteins. In the reduced state, the d-orbitals of the Cu ion are fully occupied, such that these CT transitions do not exist and RR spectroscopy of copper proteins is restricted to the oxidised forms.

Typical representatives of blue copper proteins are azurin and plastocyanin (Pc), which have been studied extensively using genetically engineered protein variants and isotopically labelled proteins. In both proteins the Cu ion is coordi-

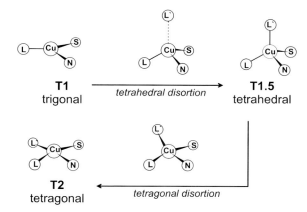

Fig. 8.1 Geometries of the various copper binding sites in proteins. "N" and "S" denote the histidine nitrogen and cystein sulfur, respectively. "L" and "L'" may be either other amino acid side chains, carbonyl functions of the peptide backbone, or small exogenous ligands such as water (illustration adapted from Andrew and Sanders-Loehr 1996).

nated by a Cys and two His residues in the trigonal positions and a Met in the "axial" position. The RR spectra excited in resonance with the LMCT transition show ca. 10 vibrational modes in the region between 200 and 500 cm^{-1} (Andrew and Sanders-Loehr 1996). This rather complex vibrational band pattern is surprising at first sight, as the coupling to the LMCT is expected to provide Franck–Condon-type resonance enhancement mainly for the modes that include the Cu–S stretching coordinate. In fact, this coordinate is distributed among a large number of modes in this region, as the comparison with spectra of isotopically labelled proteins demonstrated. These proteins have been produced by organisms grown on media including ^{34}S-Cys, ^{65}Cu, or ^{15}N-labelled NH$_4$Cl (Andrew and Sanders-Loehr 1996; Dong and Spiro 1998; Qiu et al. 1995). Most of the RR modes exhibit small isotopic shifts, indicating the involvement of the Cu–S stretching coordinate in the most intense bands in this region (Fig. 8.2). Moreover, the magnitude of the shifts, albeit small in each instance, correlates with the intensity of the corresponding bands. This finding confirms the intuitive expectation that the excited state displacement of the Cu–S stretching coordinate provides the main contributions to the RR spectrum upon excitation in resonance with the CT transition. One may, therefore, examine the RR intensities of the modes according to Eq. (2.72).

Taking into account that $\partial V/\partial q_k$ represents the classical "restoring" force that is related to the product of the force constant and the displacement coordinate (Blair et al. 1985), one can obtain an expression for the local oscillator frequency of the Cu–S entity, ν_{Cu-S}, that is the frequency of a hypothetical mode containing exclusively the Cu–S stretching (100% PED). Accordingly, one obtains

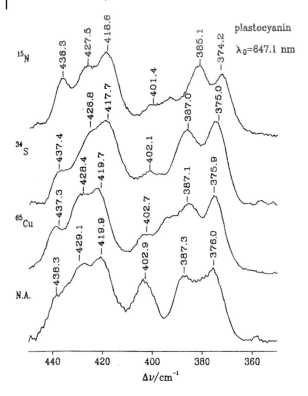

Fig. 8.2 RR spectra of Pc (plastocyanin) measured with 647-nm excitation. The bottom trace ("N.A.", natural abundance) refers to the non-isotopically labelled protein, whereas the upper traces display the RR spectra of ^{15}N-, ^{34}S-, and ^{65}Cu-labelled Pc. The figure is reproduced from Qiu et al. (1995) with permission.

$$\nu_{Cu-S} = \frac{\sum_i [I_{RR}(i) \cdot \nu_i^2]}{\sum_i [I_{RR}(i) \cdot \nu_i]} \tag{8.1}$$

Applying Badger's rule, the frequency ν_{Cu-S} can then be used to calculate the Cu–S bond distance (Fig. 8.3). The Cu–S bond lengths derived from the local oscillator frequency provide further insight into the overall geometry of the copper complex (Andrew et al. 1994; van Pouderoyen et al. 1996). The shortest bond lengths are expected for the trigonal ligation geometry with the Cu ion lying in the plane. This is in fact so for the wild-type azurin with a ν_{Cu-S} frequency of 414 cm^{-1} corresponding to a Cu–S bond length of ca. 2.13 Å. In pseudoazurin, the frequency is lower by ca. 30 cm^{-1} and the increased bond length is indicative of a distorted tetrahedral site with the Cu ion lying above the trigonal plane (Fig.

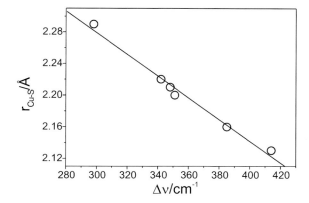

Fig. 8.3 Plot of the "diatomic oscillator" Cu–S stretching frequencies with the Cu–S bond distances of various copper proteins. The frequencies, which were obtained from the weighted-average of the experimental RR frequencies, and the bond distances were taken from Andrew et al. (1994).

8.1). A fully tetrahedral geometry is obtained when the axial ligand is of comparable strength to the ligands in the trigonal plane. In azurin, such a coordination shell can be obtained by genetically substituting the weak (axial) Met121 ligand by Glu, which leads to a further downshift of the ν_{Cu-S} frequency, concomitant with an increase in the bond length. The largest bond lengths refer to a tetragonal structure, which is established for the azurin mutant H117G after addition of exogenous His.

The trigonal (type-1) and tetragonal (type-2) copper sites constitute the limiting cases of a rich structural manifold with different Cu–S bond lengths that may be estimated on the basis of the RR spectra. In addition to this vibrational spectroscopic marker, the EPR hyperfine coupling constants and the energy and oscillator strengths of the CT transitions also represent characteristic parameters that allow classification of the copper proteins. The ratio of the oscillator strengths of the CT transitions and their energies vary with the Cu–S bond length, such that for type-2 proteins the strongest absorption band is observed at ca. 400 nm in contrast to type-1 copper proteins in which the 600-nm absorption band exhibits the strongest intensity.

The intensity-weighted analysis of the RR spectra of copper proteins not only provides key structural data but may also be applicable to predict thermodynamic parameters. It has been shown that the local oscillator frequency of the Cu–S stretching correlates with the enthalpy of reduction and the redox potentials (Blair et al. 1985).

The strategies for interpreting the RR spectra of copper proteins discussed so far do not require a comprehensive vibrational assignment. The underlying assumption is that all RR-bands include, at least to a small extent, internal coordinates from the Cu–Cys entity. More detailed surveys of the RR spectra have been

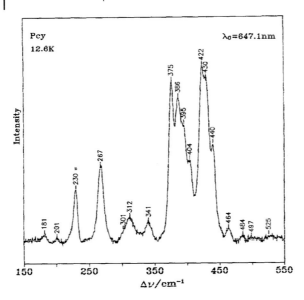

Fig. 8.4 Highly-resolved RR spectrum of Pc obtained with 647-nm excitation at 12 K. The spectrum is reproduced from Dong and Spiro (1998) with permission.

carried out for the type-1 copper protein plastocyanin (Pc) (Dong and Spiro 1998; Qiu et al. 1995). Well-resolved RR spectra have been obtained for Pc and isotopically labelled derivatives with ^{65}Cu- and ^{15}N-labelling and with Cys isotopomers, including ^{34}S/^{32}S substitutions and deuteration at the β-carbon atom. The quality of the data also allowed the spectrally cogent region between 360 and 450 cm^{-1} to be disentangled, such that more than 15 bands could be identified in the region of the fundamentals between 150 and 500 cm^{-1} (Fig. 8.4). Most surprisingly, most of these bands are not only affected by ^{34}S/^{32}S substitution (Fig. 8.2) but also show considerable downshifts (1–3 cm^{-1}) for Pc with the Cys deuterated at the β-carbon and ^{15}N-labelled at the amino group. Also, the band at 267 cm^{-1}, which is assigned to a mode dominated by the Cu–N(His) stretching, responds to Cys side chain labelling. Moreover, the overall ^{15}N/^{14}N labelling (all nitrogen atoms in the protein) causes more pronounced shifts as compared with the protein in which only the Cys is ^{15}N labelled. The results indicate a considerable mixing of many internal coordinates of the ligating amino acids. The mixing is not restricted to the Cys ligand but also includes the His residues. Moreover, even internal coordinates of the amide functions of the ligands appear to contribute to the RR-active modes.

Most of the fundamental modes in the region below 400 cm^{-1} give rise to combination modes and overtones that exhibit a considerable RR activity (Blair et al. 1985; Dong and Spiro 1998). The enhancement of these modes relative to those of the fundamentals is much higher than, for instance, in metalloporphyrins.

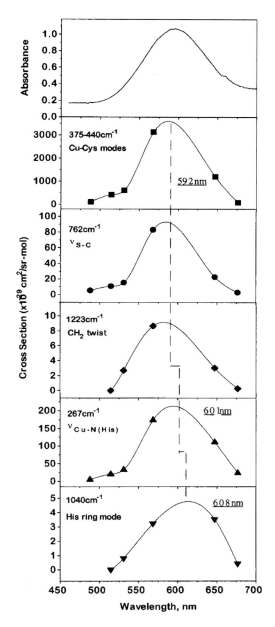

Fig. 8.5 Excitation profiles of various RR-bands of Pc (Cu–S stretch, C–S stretch, CH$_2$ twist, Cu–N stretch, His ring) in the region of the 600-nm absorption band (top). The figure is reproduced from Dong and Spiro (1998) with permission.

This effect suggests a large geometry change of the copper site in the excited state.

All of these results indicate that the excited state displacement of the Cu–S stretching coordinate is not the only quantity that governs the resonance enhancement. Furthermore, a careful analysis of the RR spectra of Pc allows identification of the bands between 1000 and 1400 cm^{-1} that originate from the internal modes of the Cys and His ligands (Dong and Spiro 1998). Whereas with Cys the resonance enhancement can be attributed to small but non-negligible excited-state displacements of the internal coordinates involved, such an explanation can hardly account for RR activity of the His modes. A likely explanation is based on the overlap of the π-orbitals of the imidazole ring of His87 with the π-orbital of the Cys sulfur and the half-filled d$_{x^2-y^2}$ orbital of Cu(II). This overlap is possible due to the nearly perpendicular orientation of the imidazole ring of His87 with respect to the trigonal N$_2$SCu plane, whereas the ring of His37 adopts a coplanar orientation. This conclusion implies that the electronic transition at 600 nm is not exclusively due to a S $\rightarrow$ Cu LMCT but may also include a CT from the His87 nitrogen. In fact, the excitation profile of the various RR-bands of the copper complex and the ligand modes support this interpretation as the intensities of the Cu–S stretching modes display a maximum at 592 nm, whereas that of the His ring mode is observed at 608 nm (Fig. 8.5). This conclusion suggests that the intensity-weighted analysis of the RR spectra in terms of the Cu–S stretching coordinate represents just a first approximation and may not be justified for all copper proteins.

Whereas type-1 and type-2 copper proteins include only one copper ion, the Cu$_A$ site, which is found in cytochrome c oxidase (CcO) (see Section 7.4), is constituted by two Cu ions connected via two sulfur (Cys) bridges. The RR spectrum of the Cu$_A$ site cannot be obtained from the intact enzyme because its vibrational modes are obscured by the much stronger RR-bands of the heme groups. However, the Cu$_A$ domain of bacterial CcO can be expressed separately and thus allows a thorough RR spectroscopic characterisation (Andrew et al. 1996). Moreover, as for azurin and Pc, isotopic labelling is possible using appropriate growth media for the microorganisms. Thus, two bands at 260 and 339 cm^{-1} could be assigned to modes dominated by the Cu–S stretching coordinates in view of the large ^{34}S/^{32}S isotopic shifts. The normal mode frequencies and the isotopic shifts could be reproduced well by an empirical normal mode analysis. Here, such analyses are more successful than for type-1 copper proteins, for which the force-field calculations cannot be restricted to the immediate coordination sphere but must include the atoms in a larger shell around of the copper ion (Qiu et al. 1998).

8.2
Iron–Sulfur Proteins

Iron–sulfur (FeS) proteins include one or more irons complexed by inorganic sulfide and/or cysteine residues. A large variety of structurally different FeS proteins

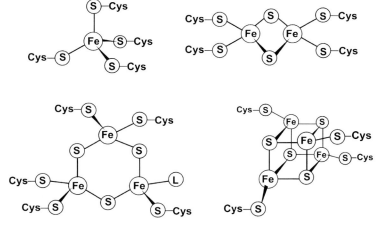

Fig. 8.6 Structures of various Fe–S clusters: 1-Fe (left top), 2-Fe (right top); 3-Fe (left bottom), 4-Fe (right bottom). Figure redrawn according to Spiro et al. (1988).

are known that may be classified in terms of the number of iron ions involved, i.e., one-iron (1-Fe) centre, two-iron (2-Fe) centre, etc. Most of the FeS proteins function as redox proteins in electron transfer chains through the reversible change of the Fe oxidation state from III to II. The direct involvement of FeS centres in enzymatic processes is rare.

Several structural motifs of FeS centres have been identified on the basis of crystal structure data (Fig. 8.6). To characterise FeS proteins, for which no three-dimensional structure data are available, EPR, Mößbauer, and RR spectroscopy are indispensable techniques. RR spectroscopy utilises the LMCT transitions, which in the oxidised forms, give rise to absorption bands in the visible region with typical extinctions coefficients of less than 5000 $M^{-1}cm^{-1}$ (Spiro et al. 1988). Only in very few instances have such transitions that are suitable for RR spectroscopy been identified for the (semi-)reduced states of FeS proteins (*vide infra*).

The simplest FeS site is constituted by four Cys residues ligating a central Fe in a tetrahedral coordination sphere. This structural motif holds for rubredoxins, which have been widely studied by RR spectroscopy (Fig. 8.7) (Czernuszewicz et al. 1986). The absorption spectrum exhibits a peak at ca. 495 nm with a shoulder at ca. 565 nm. Excitation in resonance with either of these transitions leads to RR spectra that are dominated by a strong band at 314 cm^{-1} attributed to the symmetric Fe–S stretching involving all four sulfur ligands. This band, therefore, shows no isotopic shift upon ^{54}Fe labelling. Instead, small downshifts (2.5–1.1 cm^{-1}) are observed for the considerably weaker bands at 348, 363, and 376 cm^{-1} and thus they are assigned to asymmetric Fe–S stretching modes. In general, the spectra excited at 497 and 568 nm display a similar vibrational band

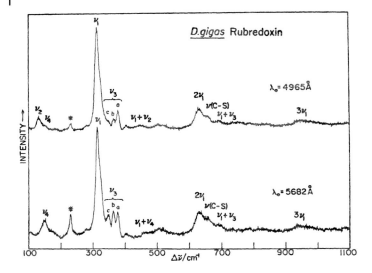

Fig. 8.7 RR spectra of Rubredoxin from *Desulfovibrio gigas* measured with 497- and 568-nm excitation. The figure is reproduced from Czernuszewicz et al. (1986) with permission.

pattern except for the bands at 130 and 150 cm^{-1} (S–Fe–S bending), which are preferentially enhanced in resonance with the short and long wavelength absorption band, respectively. This different enhancement pattern is related to the symmetry properties of the electronic transitions and the vibrational modes involved (see Section 2.2.3).

In 2-Fe proteins, the metal ions are linked via two sulfide bridges and the remaining two coordination sites of the Fe ions are occupied by Cys residues. These sites are found, for instance, in Rieske-type proteins or ferredoxins. In early RR spectroscopic studies, two strong bands at ca. 290 and 390 cm^{-1} were considered to be diagnostic for this type of FeS protein. However, upon accumulating further RR spectra of 2-Fe proteins, it was found that the region between 250 and 400 cm^{-1} exhibits a considerable variability in band positions and intensities (Kuila et al. 1992; Zhelyaskov et al. 1995). Moreover, the RR spectra of symmetrically substituted model compounds do not show the 290-cm^{-1} band. On the basis of empirical normal mode analyses, which have been fairly successful in reproducing the RR spectra of 1-Fe centres, this band is attributed to a mode that is IR-active and Raman-forbidden under C_{2h} symmetry (Spiro et al. 1988). The strong intensity observed in the RR spectra of most of the 2-Fe proteins reflects the lowering of the C_{2h} symmetry, which may be caused by hydrogen bond interactions with the thiolate ligands and different conformations of the Cys residues. Furthermore, different dihedral angles of the Cys side chains in the various 2-Fe proteins sensitively control the coupling between the S–C–C bending and the Fe–S stretching coordinates, which in turn is responsible for the variations in the

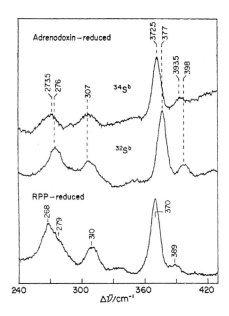

Fig. 8.8 RR spectra of reduced adrenodoxin and red paramagnetic protein (RPP) obtained with 530- and 568-nm excitation, respectively. The ^{34}S-labelling in adrenodoxin refers to the bridging S-atoms. The figure is reproduced from Han et al. (1989) with permission.

band positions and intensities between 300 and 400 cm^{-1} (Spiro et al. 1988; Loehr 1992). Bands in the region between 130 and 160 cm^{-1} most probably originate from the S–Fe–S bending, whereas the Fe–S–Fe bending is usually assigned to a very weak band at ca. 200 cm^{-1}.

The 2-Fe site of bovine adrenodoxin is one of the few FeS proteins for which RR spectra have been obtained in both the oxidised and the reduced state (Fig. 8.8) (Han et al. 1989). The RR bands of the reduced form appear to be broader than in the oxidised state, thereby leading to some poorly resolved broad features in the spectrum. Instructive information for the vibrational assignment was obtained by selective ^{34}S-labelling of the bridging sulfur atoms. As a result, the bridging modes at 398 and 377 cm^{-1} could be identified. The corresponding modes in the oxidised form are expected to be at higher frequencies and are, therefore, attributed to the bands at 421 and 393 cm^{-1}.

The cofactors in 4-Fe proteins adopt a cubane-type structure with four bridging sulfides and four terminal thiolate ligands. These centres are found, for instance, in ferredoxins and in high-potential iron proteins (HiPIP) for which detailed vibrational analyses have been carried out (Czernuszewicz et al. 1987). Most strikingly, there is good agreement between the RR spectra of the proteins and that of a model compound in which the four Fe ions are coordinated by $^-$SCH$_2$–phenyl ligands. For this "benzyl cubane" and its isotopomers an empirical normal mode

Table 8.1 Vibrational modes including the Fe–S stretching coordinates of ferredoxin, HiPIP, and the "benzyl-cube" model compound.[a] numbers in parenthesis are frequency downshifts upon ^{34}S substitution of the bridging atoms.

$D_{2d}(T_d)$ assignment	Benzyl cube, solid	Ferredoxin	HiPIP	Benzyl cube, solution	T_d assignment
Mainly terminal ν(Fe–S)					
A_1	391 (1)	395 (3.9)	397	384 (1)	A_1
$B_2(T_2)$	367 (1)	351 (0.7)	362	358 (1)	T_2
$E(T_2)$	359 (2)	363 (2.0)			
Mainly bridging ν(Fe–S)					
$B_2(T_2)$	385 (6)	380 (5.6)	390	384 (1)	T_2
$E(T_2)$					
A_1	335 (8)	338 (7.0)	337	333 (7)	A_1
$A_1(E)$	298 (5)	298 (4.9)	293	268 (3)	E
$B_1(E)$	283 (4)	276 (4.5)	273		
$E(T_1)$	283 (4)	276 (4.5)	273	268 (3)	T_1
$A_2(T_1)$	270 (3)	266 (4.0)			
$E(T_2)$	243 (5)				
$B_2(T_2)$	249 (6)	251 (6.2)	249	241 (6)	T_2
$E(T_2)$	243 (5)				

[a] Data taken from Czernuszewicz et al. (1987).

analysis has been performed to reproduce the IR- and RR-band frequencies and the isotopic shifts produced by ^{54}Fe/^{56}Fe substitution and ^{34}S-labelling at the bridging positions and the terminal ligands. The assignments can be readily extended to the 4-Fe centres of HiPIP and ferredoxin and even allow conclusions to be made about a symmetry lowering from T_d to D_{2d} (Table 8.1; Fig. 8.9). It appears to be that, unlike for the 2-Fe proteins, there are far-reaching similarities in the vibrational band patterns of 4-Fe proteins with one prominent at ca. 340 cm^{-1} and between 2 and 4 bands of much lower intensity between 240 and 300 cm^{-1} and between 350 and 400 cm^{-1} (Fu et al. 1992, 1993). All these modes are dominated by the Fe–S stretching coordinates. Those of the terminal Fe–S stretching vibrations provide the main contributions to the modes above 350 cm^{-1}. The strongest band at 340 cm^{-1} and those bands at lower frequencies are largely governed by the Fe–S bridge stretching coordinates.

Besides the ubiquitous 2-Fe and 4-Fe proteins, there are FeS proteins with more unusual structures and compositions. One example is the enzyme aconitase in which the FeS centre is involved in substrate binding. In the active form, it includes a cubane-type structure of four Fe atoms bridged by four sulfides. How-

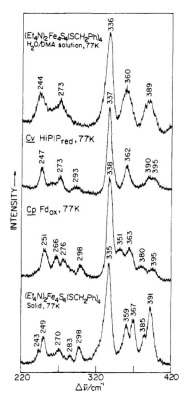

Fig. 8.9 RR spectra of reduced HiPIP (*Chromatium vinosum*, 488 nm) and oxidised ferredoxin (*Chlostridium pasteurianum*, 488 nm) compared with synthetic model compounds in solution (458 nm) and in the solid state (488 nm), all measured at 77 K. The figure is reproduced from Czernuszewicz et al. (1987) with permission.

ever, only three of the terminal ligands are coordinated by Cys residues whereas the fourth iron binds a hydroxyl group. This Fe also serves as the binding site for the substrate, a citrate, which is converted into isocitrate during the enzymatic process. In the inactive form of the enzyme, this Fe is released from the cluster leaving a 3-Fe (4-S) centre. RR spectra have been measured from the active, the substrate-bound, and the inactive forms of the enzyme (Kilpatrick et al. 1994). The active form displays the characteristic vibrational signature of the 4-Fe cluster with the most intense RR band at 339 cm^{-1}. The impact of the hydroxyl ligand on the spectrum is small but detectable. Although a mode dominated by the Fe–OH stretching is not enhanced, non-negligible contributions of this coordinate are involved in some of the modes between 360 and 400 cm^{-1}, leading to small isotopic shifts (ca. 1 cm^{-1}) in the RR spectra measured in $H_2^{18}O$.

Binding of the inhibitor nitroisocitrate has a substantial effect on the RR spectrum. In particular, there are pronounced changes in the relative intensities. The RR spectra measured after addition of substrates (citrate) correspond to a superposition of the RR spectra of the substrate-free and the inhibitor-bound enzyme, supporting the view of a direct interaction of the inhibitor/substrate with the 4-Fe cluster. However, the removal of the labile Fe atom from the cluster in the inactive form has a surprisingly small influence on the RR spectrum as only small upshifts (ca. 3 cm^{-1}) of some of the bands are noted while the overall appearance of the spectrum is still characteristic of a 4-Fe cluster.

Further examples of unusual FeS clusters are those containing six Fe and six sulfides, most probably arranged in a prismane-type structure. An RR spectroscopic study of one of these proteins, isolated from sulfate-reducing bacteria, has revealed an additional peculiarity (de Vocht et al. 1996). A relatively intense band at 801 cm^{-1} was observed that could not be attributed to an overtone or a combination mode. Instead, this band reveals an 18-cm^{-1} downshift in the protein dissolved in H$_2$^{18}O, such that it was assigned to the stretching mode of an Fe=O bond or an Fe–O–Fe bridge. The RR bands in the region of the Fe–S stretchings display only small but detectable shifts (ca. 1 cm^{-1}) implying that the iron-oxo entity is part of the 6-Fe cluster.

8.3
Di-iron Proteins

Owing to the high affinity towards basic ligands, Fe(III) and Fe(II) may form stable complexes with oxygen- and nitrogen-containing ligands. In proteins, these ligands can be provided by the carboxylate and imidazole residues of Glu/Asp and His, respectively. In many instances, these side chains stabilise dinuclear iron complexes, which typically also possess additional binding sites for small diatomic ligands such as oxygen. These ligation patterns constitute the structural basis for the functioning as redox enzymes or as transport or storage proteins (Que and True 1990).

A widely studied representative is hemerythrin (Hr), which serves as an oxygen carrier in several marine invertebrates – analogous to hemoglobin or myoglobin in mammals. It includes a mixed-valence dinuclear iron centre in which the two metal ions are bridged by two carboxylate side chains of Glu and Asp residues and an oxo ligand (μ-oxo bridge). Five of the remaining six coordination sites are occupied by His residues, whereas the sixth site allows for binding of an exogenous ligand (Fig. 8.10). The electronic transitions of the μ-oxo bridged di-iron centre in Hr and model compounds originate from LMCTs of the bridging oxygen to the ferric iron, typically observed in the near-UV region (320–360 nm with extinction coefficients between 7000 and 3000 LM^{-1} cm^{-1}) and from an LMCT involving the exogenous ligand and observed in the visible region (450–500 nm, $\varepsilon < 1000$ LM^{-1} cm^{-1}). These transitions may be used for probing the RR spectra.

With near-UV excitation (363 nm), the preferential enhancement of the modes involving the bridging oxygen is expected. In fact, two RR-bands are observed that are assigned to the symmetric (v_s) and asymmetric (v_{as}) Fe–O–Fe stretching, respectively (Sanders-Loehr et al. 1989; Shiemke et al. 1986). The actual frequencies found in various Hr complexes and model compounds span a considerable frequency range of 450 ± 80 and 800 ± 80 cm^{-1} for v_s and v_{as}, respectively. These variations are directly related to the angle of the Fe–O–Fe entity as shown for

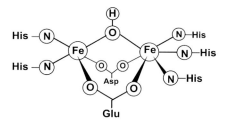

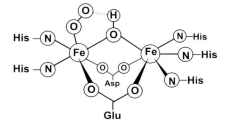

Fig. 8.10 Structural model of the di-iron site of deoxy-Hr (top) and oxy-Hr (bottom).

those species for which X-ray crystallographic data are available. Upon taking into account $^{18}O/^{16}O$ shifts, these frequencies also allow determination of the Fe–O–Fe angle within the framework of a simple three-atom oscillator model.

In the absence of any exogenous ligand, i.e., deoxy-Hr, the vacant coordination site can be occupied by a hydroxyl ligand in a pH-dependent association process. The presence of the hydroxyl ligand in deoxy-HrOH complicates the RR spectrum. Firstly, the Fe–OH stretching gives rise to a band at 562 cm^{-1}, whereas two bands are observed in the region of the symmetric Fe–O–Fe stretching (Fig. 8.11). The individual components can be better resolved and assigned in low-temperature experiments using isotopically labelled samples. The two closely spaced bands are attributed to Fe–O–Fe stretchings in the presence (492 cm^{-1}) and in the absence (506 cm^{-1}) of a hydrogen bond between the hydroxyl group and the bridging oxygen.

In the oxygen-bound form, oxy-Hr, excitation in resonance with long-wavelength transition leads to the enhancement of the modes involving the modes of the bound oxygen. These are the O–O stretching at ca. 845 cm^{-1} and the Fe–O$_2$ stretching at ca. 505 cm^{-1} (Kaminaka et al. 1992). In ^{2}H$_2$O, the former band shifts to higher and the latter band to lower frequencies, indicating a protonation of the bound oxygen.

The structural motif of μ-oxo bridges (i.e., oxygens linking two metal ions) are found in a variety of di-iron proteins among which the enzyme ribonucleotide reductase (RNR) is the most prominent one. This enzyme catalyses the conversion of ribonucleotides into deoxy-ribonucleotides (Que and True 1990). The activation of the apo-form, a Fe(II)Fe(II) centre bridged through carboxylate side chains, is initiated by oxygen binding and eventually leads to a high-valent dinuclear com-

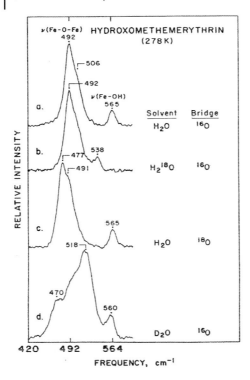

Fig. 8.11 RR spectra of (hydroxy)metHr measured with 364-nm excitation. The spectra were obtained after isotope labelling via solvent exchange. The figure is reproduced from Shiemke et al. (1986) with permission.

plex, which is capable of abstracting an electron and a proton from Tyr122 to form a tyrosyl radical as the catalytically active agent. In the resultant active state, the two ferric Fe are linked via a μ-oxo bridge. This state of the di-iron centre gives rise to a symmetric Fe–O–Fe stretching at 500 cm^{-1}. Using $^{18}O_2$, this band shifts down to 487 cm^{-1}, supporting the view that the bridging oxygen stems from molecular oxygen (Ling et al. 1994). The detailed mechanism of the activation process is not yet fully understood. To identify and to characterise intermediate states, freeze–quench RR spectroscopy has been employed to a genetically engineered protein variant (Moenne-Loccoz et al. 1998). The RR spectrum of the first species that could be trapped displays three bands at 458, 499, and 870 cm^{-1}, which shift down by 16, 22, and 46 cm^{-1} when $^{18}O_2$ is used instead of $^{16}O_2$ (Fig. 8.12). These bands could only be detected upon red excitation, i.e., in resonance with the broad 700-nm absorption band, but are not visible when using blue/violet excitation lines that meet the resonance conditions for the LMCT of the μ-oxo bridged di-iron centre. Comparison with model compounds supports the assignment of the two low-frequency bands to the asymmetric and

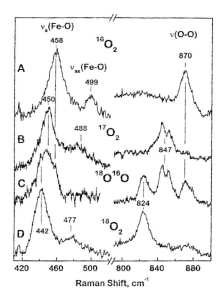

Fig. 8.12 RR spectra of the freeze-trapped peroxo-intermediate of RNR generated with $^{16}O_2$, $^{17}O_2$, $^{18}O^{16}O$, and $^{18}O_2$ and measured with 647-nm excitation. The figure is reproduced from Moënne-Loccoz et al. (1998) with permission.

symmetric Fe–O–O–Fe stretchings. For these two bands, the somewhat smaller isotopic shifts brought about by $^{17}O_2$ can readily be understood.

However, the O–O stretching at ca. 870 cm^{-1} reveals an unexpected splitting. On the basis of a two-atom oscillator model (cf. Section 2.1) the isotopic shifts for the O–O stretching can be calculated according to

$$v(^xO_2) = v(^{16}O_2) \cdot \sqrt{\frac{\mu(^{16}O_2)}{\mu(^xO_2)}} \tag{8.2}$$

where $v(^{16}O_2)$ and $v(^xO_2)$ are O–O stretching frequencies of the ^{16}O and ^xO ($x = 17, 18$) isotopes, respectively. The corresponding reduced masses are denoted by $\mu(^{16}O_2)$ and $\mu(^xO_2)$. For the ^{18}O isotope, the calculated frequency is 820 cm^{-1}, which is close to the experimental value of 824 cm^{-1}. For the ^{17}O isotope, the calculated frequency of 844 cm^{-1} nearly coincides with the low-frequency component of the doublet. Using $^{18}O^{16}O$, three bands are expected at 870, 845, and 820 cm^{-1} with an intensity ratio of 1:2:1. However, the observed band pattern does not fit the expectations. Whereas the lowest and the highest frequency band are in fact of similar intensity, the central band is split into two components, such that it is closely related to the doublet in the $^{17}O_2$ labelled sample. This feature has been attributed to a Fermi resonance effect, thus the low frequency component at ca. 845 cm^{-1} (^{18}O–^{16}O or ^{17}O–^{17}O stretching) provides intensity for a yet unidentified but energetically close-by mode at ca. 852 cm^{-1}, which is localised in the immediate vicinity of the Fe–O–O–Fe entity.

8.4
Hydrogenases

Hydrogenases are bacterial enzymes that metabolise molecular hydrogen either by producing H_2 from protons, or by oxidising H_2 to protons. In most representatives of these enzymes, which are of particular interest in view of their potential biotechnological importance in energy conversion, a dinuclear metal centre with an Ni and an Fe ([NiFe] hydrogenases) or with two Fe atoms ([FeFe] hydrogenases) constitutes the catalytic site (Frey 2002). For [NiFe] hydrogenases (e.g., from *Desulfovibrio gigas*), the two metal ions are bridged by two Cys sulfurs. The Ni ion carries two additional Cys ligands, whereas two CN ligands and one CO are bound to the Fe. In the oxidised (catalytically inactive) form, the two metal ions are linked via an additional bridging and an as yet unidentified ligand (oxygen or sulfur), which is removed upon transformation into the activated state via treatment with H_2. The activated form is ready for the H_2 metabolism and the catalytic cycle runs through a series of intermediates.

The [NiFe] cofactor is RR-silent as it does not give rise to an electronic transition in the visible or near-UV region that is sufficiently separated from the absorption bands of the protein. It is only the additional cofactors, i.e., usually FeS clusters that are part of the electron transfer chain, that can be probed by RR spectroscopy. For the catalytic centre, however, IR spectroscopy is the method of choice. The CO and CN ligands that contribute to a fairly unusual coordination pattern lead to strong IR bands in the region between 1900 and 1970 cm^{-1} and between 2060 and 2100 cm^{-1}, which originate from the CO and CN stretching vibrations, respectively. These bands can be readily detected in the absolute spectra and do not require the IR difference technique (Albracht 1994). Moreover, the exact frequencies of these modes are very sensitive to the electron density distribution in the metal complex and thus respond to changes of the redox state and binding or dissociation of other ligands. In this way, IR spectroscopy is very important for monitoring the catalytic process and to identify the various states involved.

A particularly elegant approach is based on IR-spectroelectrochemistry (see Section 4.2.1.3), which allows for the *in situ* generation and characterisation of the various states of the catalytic centre (de Lacey et al. 1997; Best 2005). As an example, Fig. 8.13 shows a selection of IR spectra measured from the activated enzyme at different potentials. At the most positive potential (−0.05 V), we note a prominent band at 1946 cm^{-1} attributable to the stretching of the CO ligand of the Fe atom. The two weaker bands at 2090 and 2079 cm^{-1} originate from the stretching vibrations of the CN ligands of the Fe. Upon decreasing the potential the three bands of this so-called Ni–B state decrease in intensity concomitant with the growing-in of the four and two bands in the CN and CO stretching region,

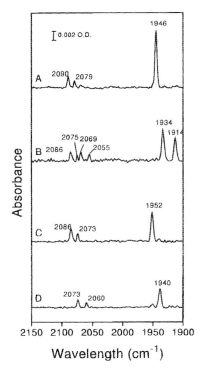
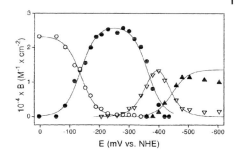

Fig. 8.13 IR spectra of the activated [NiFe] hydrogenase from *Desulfovibrio gigas* measured in a spectroelectrochemical cell at various potentials, i.e., −0.05 V (A), −0.225 V (B), −0.395 V (C), and −0.475 V (D) (left side). The potential-dependence of the intensities of the various CO stretching modes follow the Nernstian equation (right side; solid lines). The open circles, filled circles, open triangles, and filled triangles denote the intensities of the IR intensities at 1946, the sum of 1914 and 1934, 1952, and 1940 cm^{-1}, respectively. The figure is reproduced from de Lacey et al. (1997) with permission.

respectively. These findings suggest the formation of a state (denoted as SI) that includes two species, each of them characterised by a set of two CN stretching and one CO stretching modes. This view is confirmed by pH-dependent studies indicating that the species involved are coupled in an acid–base equilibrium. At −0.395 V, both species are quantitatively converted into the so-called Ni–C state, which is characterised solely by a set of three bands at 2086, 2073, and 1952 cm^{-1}. The fact that only the Ni–B and Ni–C forms are associated with EPR signals is in agreement with the assumption that each transition (i.e., Ni–B → SI → Ni–C) is associated with a one-electron reduction. This view is indeed confirmed by fitting the Nernstian equation to the intensities of the CO stretchings at 1946 (Ni–B) and 1952 cm^{-1} (Ni–C) and to the sum of the intensities of the 1934 and 1914 cm^{-1} bands (SI) (Fig. 8.13). The final transition then leads to the reduced state of the activated enzyme, the so-called R-state, which can be identified on the basis of the CN and CO stretching modes at 2073, 2060, and 1940 cm^{-1}. In a first approximation, one may assume that the absorption cross sections of the CO stretchings are the same for each species. Then, the Nernstian plots readily allow the determination of the redox potentials for the individual transitions. A structural characterisation of the various states of the active site based solely on the CO and CN stretching modes is not really possible, even though the frequencies respond to changes in the electron density at the metal sites or to structural alteration of the dinuclear centre. It is therefore necessary to extend the IR spectroscopic studies to lower frequencies to probe the vibrational modes involving the metal ions (50–800 cm^{-1}) and the modes of the polypeptide backbone and amino acid side chains (500–1800 cm^{-1}).

References

Albracht, S. P. J., **1994**, "Ni hydrogenases – in search of the active site", *Biochim. Biophys. Acta* **1188**, 167–204.

Andrew, C. R., Yeom, H., Valentine, J. S., Karlsson, B. G., Bonander, N., van Pouderoyen, G., Canters, G. W., Loehr, T. M., Sanders-Loehr, J., **1994**, "Raman spectroscopy as an indicator of Cu-S bond length in type 1 and type 2 copper cysteinate proteins", *J. Am. Chem. Soc.* **116**, 11489–11498.

Andrew, C. R., Sanders-Loehr, J., **1996**, "Copper-sulfur proteins: Using Raman spectroscopy to predict coordination geometry", *Acc. Chem. Res.* **29**, 365–372.

Andrew, C. R., Fraczkiewicz, R., Czernuszewicz, R. S., Lappalainen, P., Saraste, M., Sanders-Loehr, J., **1996**, "Identification and description of copper-thiolate vibrations in the dinuclear Cu$_A$ site of cytochrome *c* oxidase", *J. Am. Chem. Soc.* **118**, 10436–10445.

Best, S. P., **2005**, "Spectroelectrochemistry of hydrogenase enzymes and related compounds", *Coord. Chem. Rev.* **249**, 1536–1554.

Blair, D. F., Campbell, G. W., Schoonover, J. R., Chan, S. I., Gray, H. B., Malmstrom, B. G., Pecht, I., Swanson, B. I., Woodruff, W. H., Ch, W. K., English, A. M., Fry, H. A., Lunn, V., Norton, K. A., **1985**, "Resonance Raman studies of blue copper proteins: Effect of temperature and isotope substitutions. Structural and thermodynamic implications.", *J. Am. Chem. Soc.* **107**, 5755–5766.

Czernuszewicz, R. S., LeGall, J., Moura, I., Spiro, T. G., **1986**, "Resonance Raman-spectra of rubredoxin – new assignments and vibrational coupling mechanism from ^{54}Fe/^{56}Fe isotope shifts and variable-wavelength excitation", *Inorg. Chem.* **25**, 696–700.

Czernuszewicz, R. S., Macor, K. A., Johnson, M. K., Gewirth, M., Spiro, T. G., **1987**, "Vibrational mode structure and symmetry in proteins and analogues containing Fe$_4$S$_4$ clusters: Resonance Raman evidence for different degrees of distortion in HiPIP and ferredoxin", *J. Am. Chem. Soc.* **109**, 7178–7187.

Dong, S., Spiro, T. G., **1998**, "Ground- and excited-state mapping of plastocyanin from resonance Raman spectra of isotope-labelled proteins", *J. Am. Chem. Soc.* **120**, 10434–10440.

Frey, M., **2002**, "Hydrogenases: Hydrogen-activating enzymes", *ChemBioChem.* **3**, 153–160.

Fu, W., Drozdzewski, P.-M., Morgan, T. V., Mortenson, L. E., Juszczak, A., Adams, M. W. W., He, S.-H., Peck Jr., H. D., DerVartanian, D. V., LeGall, J., Johnson, M. K., **1993**, "Resonance Raman studies of iron-only hydrogenases", *Biochemistry* **32**, 4813–4819.

Fu, W., O'Handley, S., Cunningham, R. P., Johnson, M. K., **1992**, "The role of the iron-sulfur cluster in *Escherichia coli* endonuclease III", *J. Biol. Chem.* **267**, 16135–16137.

Han, S., Czernuszewicz, R. S., Kimura, T., Adams, M. M. W., Spiro, T. G., **1989**, "Fe_2S_2 Protein resonance Raman-spectra revisited – structural variations among adrenodoxin, ferredoxin, and red paramagnetic protein", *J. Am. Chem. Soc.* **111**, 3505–3511.

Kaminaka, S., Takizawa, H., Handa, T., Kihara, H., Kitagawa, T., **1992**, "Resonance Raman study on the active-site structure of a cooperative hemerythrin", *Biochemistry* **31**, 6996–7002.

Kilpatrick, L. K., Kennedy, M. C., Beinert, H., Czernuszewicz, R. S., Qui, D., Spiro, T. G., **1994**, "Cluster structures and H bonding in native, substrate-bound and 3Fe forms of aconitase as determined by resonance Raman spectroscopy", *J. Am. Chem. Soc.* **116**, 4053–4061.

Kuila, D., Schoonover, J., Dyer, R. B., Batie, C. J., Ballou, D. P., Fee, J. A., Woodruff, W. H., **1992**, "Resonance Raman studies of Rieske-type proteins", *Biochim. Biophys. Acta* **1140**, 175–183.

de Lacey, A. L., Hatchikian, E. C., Volbeda, A., Frey, M., Fontecilla-Camps, J. C., Fernandez, V. M., **1997**, "Infrared-spectroelectrochemical characterisation of the [NiFe] hydrogenase of *Desulfovibrio gigas*", *J. Am. Chem. Soc.* **119**, 7181–7189.

Ling, J., Sahlin, M., Sjöberg, B.-M., Loehr, T. M., Sanders-Loehr, J., **1994**, "Dioxygen is the source of the μ-oxo bridge in iron ribonucleotide reductase", *J. Biol. Chem.* **269**, 5595–5601.

Loehr, T. M., **1992**, "Detection of conserved structural elements in ferredoxins and cupredoxins by resonance Raman spectroscopy", *J. Raman Spectrosc.* **23**, 531–537.

Messerschmidt, A., Huber, R., Wieghardt, K., Poulos, T., **2001**, "*Handbook of Metalloproteins*", Wiley, New York.

Moenne-Loccoz, P., Baldwin, J., Ley, B. A., Loehr, T. M., Bollinger Jr., J. M., **1998**, "O_2 activation by non-heme diiron proteins: Identification of symmetric μ-1,2-peroxide in a mutant of ribonucleotide reductase", *Biochemistry* **37**, 14659–14663.

van Pouderoyen, G., Andrew, C. R., Loehr, T. M., Sanders-Loehr, J., Mazumdar, S.,
Hill, H. A. O., Canters, G. W., **1996**, "Spectroscopic and mechanistic studies of type-1 and type-2 copper sites in *Pseudomonas aeruginosa* azurin as obtained by addition of external ligands to mutant His46Gly", *Biochemistry* **35**, 1397–1407.

Qiu, D., Dong, S., Ybe, J. A., Hecht, M. A., Spiro, T. G., **1995**, "Variations in the type I copper protein coordination group: Resonance Raman spectra of ^{34}S-, ^{65}Cu-, and ^{15}N-labeled plastocyanin", *J. Am. Chem. Soc.* **117**, 6443–6446.

Qiu, D., Dasgupta, S., Kozlowski, P. M., Goddard III, W. A., Spiro, T. G., **1998**, "Chromophore-in-protein modelling of the structures and resonance Raman spectra for type 1 copper proteins", *J. Am. Chem. Soc.* **120**, 12791–12797.

Que Jr., L., True, A. E., **1990**, "Dinuclear iron- and manganese-oxo sites in biology", *Progr. Inorg. Chem.* **38**, 97–200.

Sanders-Loehr, J., Wheeler, W. D., Shiemke, A. K., Averill, B. A., Loehr, T. M., **1989**, "Electronic and Raman spectroscopic properties of oxo-bridged dinuclear iron centers in proteins and model compounds", *J. Am. Chem. Soc.* **111**, 8084–8093.

Shiemke, A. K., Loehr, T. M., Sanders-Loehr, J., **1986**, "Resonance Raman study of oxyhemerythrin and hydroxyhemerythrin. Evidence for hydrogen bonding of ligands to the Fe-O-Fe center", *J. Am. Chem. Soc.* **108**, 2437–2443.

Spiro, T. G., Czernuszewicz, R. S., Han, S., **1988**, "Iron-sulfur proteins and analog complexes", in *Biological Applications of Raman Spectroscopy*, Vol. III, Spiro, T. G. (Ed.), chap. 12, Wiley, New York.

de Vocht, M. L., Kooter, I. M., Bulsink, Y. B. M., Hagen, W. R., Johnson, M. K., **1996**, "Resonance Raman evidence for non-heme Fe-O species in the [6Fe-6S]-containing iron-sulfur proteins from sulfate-reducing bacteria", *J. Am. Chem. Soc.* **118**, 2766–2767.

Zhelyaskov, V., Yue, K. T., LeGall, J., Barata, B. A. S., Moura, J. J. G., **1995**, "Resonance Raman study on the iron-sulfur centers of *Desulfovibrio gigas* aldehyde oxidoreductase", *Biochim. Biophys. Acta* **1252**, 300–304.

Subject Index

a

Ac-part 66, 91
Adrenodoxin 293
Adsorption 118, 124, 126 ff.
Ag/AgCl electrode 15, 121
Amide A 156 ff., 216
Amide I 156 ff., 159, 161 ff., 169, 199, 200, 213 ff., 216
Amide I' 157, 159, 160, 195, 212, 216
Amide II 156 ff., 195, 198, 212, 216
Amide III 156 ff., 171, 176
Amyloid fibrils 164
Anharmonic oscillator 24
Anharmonicity 168
Anion binding 216
Anion pump 214
Anion uptake 214 ff., 217
Anti-stokes 28 ff.
Adipodisation 92 ff.
ATR 108 ff., 119, 122 ff., 161, 214 ff., 216, 276
– effective thickness 110 ff.
– multiple reflection 111
– penetration depth 109 ff.
– single reflection 123
ATR-SEIRA 122 ff., 258
Attenuated total reflection, *see* ATR
Azurin 284 ff.

b

Backbone 155
Background subtraction 103, 149
Backscattering 88, 116 ff., 131
Bacteriorhodopsin 5 ff., 133, 181, 215
– BR570 5
– IR 206 ff.
– M410 5
Badger's rule 286
Band fitting 150, 160, 173
Band widths, *see* line widths

Bathorhodopsin 185, 188 f., 195, 197
Beam collimation 96
Beam splitter 65 ff.
Blue-shifted intermediate, *see* BSI
Born–Oppenheimer approximation 33
BSI 194

c

C=N stretching vibration 192, 199, 200, 208
^{13}C-labelling 163 ff.
Caged compound 141, 181, 217 ff.
Caged GTP 218 ff.
Capillary flow system 105, 135 ff., 195, 207
Carboxyl group 202 ff., 255, 273
– protonated 201, 208
Cartesian displacement coordinates 54
CCD 86 ff., 147, 149
CD, *see* circular dichroism
Charge–transfer 283, 296
– transition 36, 233
Charged coupled device, *see* CCD
Circular dichroism 160, 161
– IR 167
– UV 160 f., 167
– vibrational 167
Confocal spectrometer 87 f., 117
Copper center, vibrational analysis 285 ff.
Copper proteins 284 ff.
– Cu_A 290
– isotopic labelling 285 ff.
– RR 284 ff.
– type-1 284 ff., 287, 288, 290
– type-2 284 ff., 287, 290
Core-size frequency relationship 234 ff., 237, 270
Counter electrode 114 ff., 121
Counterion 194, 204, 206, 209, 215
Cryotrapping 207
Cylic voltammetry 114

Cytochrome c 126 ff., 230, 247
 – conformational dynamics 248 ff., 267
 – conformational equilibria 248 ff.
 – electron transfer 260 ff.
 – electrostatic immobilisation 260 ff.
 – immobilisation 255 ff.
 – IR 247 ff., 251
 – isotopic labelling 245
 – photoreduction 247
 – redox equilibria 246 ff., 256, 258 ff.
 – SEIRA 258 ff.
 – SERR 254 ff.
 – reorganisation energy 264
 – RR 245, 246, 249 ff., 253, 260
 – unfolding 249, 253, 255
Cytochrome c oxidase 244, 268 ff.
 – catalytic cycle 274 ff.
 – CO complex 275
 – Cu_A 290
 – electron transfer 274
 – IR 271, 277
 – IR spectroelectrochemistry 273
 – redox transitions 271 ff.
 – $^{18}O/^{16}O$ isotopic shifts 175
 – RR 268, 275

d
2D-IR 167 ff.
 – cross peak 169
Database 160
Dc-part 66
Delay time 76, 104, 133 ff., 144, 146 ff., 240 ff., 261, 275
Depolarisation ratio 30 ff., 230 ff.
DFT 23 ff., 36, 192, 234
Diamond ATR cell 113
Difference frequency generation 78
Diffunction 91
Dihedral angle 171
Di-iron center, vibrational analysis 296
Di-iron proteins 296 ff.
Dipole moment 25, 26, 55 ff.
Disc membranes 186, 195
Double difference spectrum 201

e
Elastic scattering 29, 85
Electric field effect 127, 129, 257 ff., 266
 – conformational equilibrium 257 ff.
Electrical double layer 123, 127 ff.
Electron transfer 118, 244, 260 ff., 268
Electron tunnelling 263
Electronic transition 31 ff., 34, 37, 101 ff.

Ethylenic stretching mode 287, 290, 292, 302, 208, 216
Excited state displacement 34
Excitonic sytem 162

f
Factor analysis 160
Faraday rotator 82 ff.
Felgett's advantage 70, 89
Ferredoxin 292, 294, 295
FG matrix 19 ff., 21, 23 ff., 53 ff.
 – F-matrix 21, 22, 23
 – G-matrix 20, 22, 23, 53
Fluorescence 101 ff.
 – quenching 104
Force constant 13, 16, 23 ff., 55 ff., 59 ff.
Fourier transform 64 ff., 91, 94, 95
Fourier transform Raman 89 ff.
Fourier-transform infrared, see FTIR
Franck–Condon 33 ff., 37 ff., 58 ff.
Fresh sample condition 136 ff., 185
FTIR 90

g
Gated cw-excitation 137, 138, 147
Global fit 174, 210
 – amplitude spectra 174, 210
Globular proteins 164
Gouy–Chapman theory 124
G-protein coupled receptor 204
Grating 84 ff.
Group theory 22
GTP hydrolysis 221
GTPase 218

h
$^1H/^2H$ exchange 157, 159 ff., 173
$H/^2H_2$ 108
H_2O, IR spectrum 100, 157
^2H_2O, IR spectrum 100, 157
Halorhodopsin 182, 214
Harmonic approximation 24, 26, 42
Harmonic oscillator 16, 44 ff., 55
Harmonic vibration 12
α-helix 156 ff., 160, 163, 173
Helmholtz layer 124, 126
Heme 227
 – coordination 228, 235 ff., 249
 – formyl modes 232, 270
 – high spin 235 ff.
 – low spin 235 ff.
 – Q-band 228 ff., 231
 – Soret band 228 ff., 231
 – type-a 228, 231, 269

– type-*b* 228, 231, 237
– type-*c* 228, 233
– vibrational analysis 237 ff., 245, 269
– vinyl modes 232, 238 ff., 270
Heme protein 227 ff., 235
Hemerythrin 296 ff.
– $^{18}O/^{16}O$ isotopic shifts 297 ff.
– RR 236 ff.
Hemoglobin 236 ff.
– CO complex 240 ff.
– deoxy 237, 238, 239 ff.
– IR 237, 240 ff.
– met 237, 238
– oxy 238
– R state 237, 239, 240 ff.
– RR 239, 240 ff.
– T state 237, 239, 240 ff.
– UV RR 241 ff.
Herzberg–Teller 59
HiPIP 293, 294, 295
HOOP 190, 194, 198, 208, 211, 216
Hückel 58
Hydrated film 107, 195, 198
Hydrogen bond 24, 157, 158, 164, 171, 193 ff., 202, 253, 271
Hydrogenase 300 ff.
– IR 300 ff.
– IR-spectroelectrochemistry 300 ff.
Hydrogen-out-of-plane, *see* HOOP
Hydroxylamine 184

i
Inelastic scattering 29
Infrared detector 69, 149
– array 77, 78 ff.
– photon 69
– pyroelectric 69
– semiconductor 69
– thermal 69
Infrared spectroscopy
– aqueous solution 99
– electrochemical cell for 113 ff., 247
– instrumentation 63 ff.
– reaction induced difference spectroscopy 108, 195
– spectroelectrochemistry 114 f., 247
– *see also* IR and FTIR
Interference filter 82 ff.
Interferogram 67 ff., 72, 76, 91, 95
– non-symmetric 94
Interferometer 67 ff., 72, 74, 90

Internal coordinate 18 ff., 22 ff., 23, 25, 34, 35 ff., 55 ff.
– bending 19, 52 ff.
– out-of-plane deformation 19
– stretching 18, 35 ff., 52 ff.
– torsional 19
Ion pump, light-driven 181
IR-active 26, 57, 163, 228, 229
IR dichroism 27
IR intensity 26, 42
IR pump/IR probe 167
Iron oxygen modes
– cytochrome *c* oxidase 276
– hemerythrin 297
– hemoglobin 239
– ribonucleotide reductase 298 ff.
Iron sulfur center, vibrational analysis 291 ff.
Iron–sulfur proteins 290 ff.
– isotopic labelling 291, 293, 294
– RR 291 ff., 296
Isorhodopsin 184, 189, 192
Isotopic labelling 7, 22, 55, 163, 166, 167, 169, 184, 188, 200, 208
Isotopic shift 164

j
Jaquinot's advantage 77, 96

k
Kerr cell 104
K-state 199, 211

l
Laser
– Ar ion 80, 186, 188
– cw 81, 105, 133 ff.
– diode 176
– dye 186, 188
– frequency multiplying 80
– Kr ion 80, 170
– lead salt laser diodes 74, 176
– Nd:YAG 80, 89, 138, 170, 176, 208
– pulsed 81 ff., 105, 138 ff.
– Ti:sapphire 76, 78, 80, 139, 170
Ligand modes 239 ff., 283
Light-induced reaction, *see* photoinduced processes
Line width 85, 93
– homogeneous 169
– inhomogeneous 169
Lorentzian band shape 104, 150
Low temperature measurements 117 ff., 145, 184, 185, 195, 197 ff., 240

L-state   211
Lumirhodopsin   194, 197, 200

m

MI   183 ff., 195 ff., 201
MI/MII equilibrium   201
MII   183 ff., 195 ff., 201
 – decay   204
Mass-weight Cartesian coordinates   15, 17 ff.
Membrane protein   161, 181
Metal–ligand modes   232, 239 ff., 283
Metalloporphyrin   102, 228 ff., 231, 233
 – core size   234
 – RR   229, 232
 – marker bands   234, 237 ff., 245
Metarhodopsin I, see MI
Metarhodopsin II, see MII
Michelson interferometer   64 ff.
Microfabricated flow cell   194, 201
Monochromator   84 ff.
M-state   209
Multiple reflection   111 ff.
Myoglobin   236 ff.
 – CO complex   240 ff.
 – deoxy   237 ff.
 – IR   163
 – met   237 ff.
 – oxy   238
 – RR   237 ff., 240 ff.

n

Natronobacterium pharaonis   214
Noise   69, 70
Non-heme metalloproteins   283 ff.
Normal coordinates   17, 20, 26, 33 ff.
Normal hydrogen electrode   115
Normal mode   15 ff., 28, 34, 44, 50, 55 ff.
 – analysis   15, 33, 55, 162, 233 ff., 245 ff., 269
 – totally symmetric   51, 55 ff., 58 ff.
Notch filter   87
NpHR, see halorhodopsin and natronobacterium pharaonis
Nyquist criteria   73, 75

o

^{18}O-labelling   219, 221
Opsin state   204
Optical path difference   67, 90, 93
O-state   214 ff.

p

PED, see potential energy distribution
Peptide   168, 176
 – backbone   155
 – chain   162
 – DFT force fields   164, 167
 – empirical force field   162
 – helical   166, 170
 – two-strands   164
Peptide bond   159
 – dihedral angle   155, 156
 – electronic transition   169
 – trans   155
α-phosphate   219 f.
β-phosphate   219 f.
γ-phosphate   219 f.
Photocycle   214
Photoinduced process   104 ff., 118, 132 ff., 139, 141, 181 ff.
Photomultiplyer   85, 117
Photon counting   85, 117
Photon flux   81 ff., 105
Photoreceptor   181 ff.
Photorhodopsin   195
Plastocyanin   284 ff.
Plasmons resonance   131
Plastocyanin   284 ff.
 – RR   286, 288, 290
Point group   22, 46 ff.
 – C_{2h} 292
 – C_{2v} 46, 52, 54 ff., 57
 – C_{4v} 232
 – D_{2d} 232, 294
 – D_{3h} 48, 51, 52
 – D_{4h} 228 ff., 231, 233 ff., 245
 – D_{6h} 58
 – irreducible representation   50, 55, 229
 – T_d 294
Polariser   31, 82 ff.
Polarisibility   28 ff., 31, 55 ff., 82 ff.
Polyalanine   162
Polyglycine   162
Polylysine   158
Potential energy   16, 20, 23, 44 ff., 53
Potential energy distribution   21
Potential jump   146 ff., 261
Potential of zero charge   124, 126, 128, 254
Potentiostat   114, 121
Protein folding   171 ff., 249 ff.
 – kinetics   173, 174
 – two-state model   173, 175, 176
Protein immobilisation
 – covalent   129, 160
 – electrostatic   129, 260
 – hydrophobic   129, 260
 – protein-tethered membrane   129, 174

Protein, unfolding   128, 147, 171 ff.
   – kinetics   173
Proteorhodopsin   182
Proton gradient   206
Proton pump   5, 271, 273
   – light-driven   183, 206 ff.
Proton transfer   206, 207, 211, 273
Protonated Schiff base   215

q
Q band   231
Quantum chemical calculation   23 ff.

r
Raman
   – cross section   29
   – intensity   40
   – microscopy   88, 148
   – scattering, tip-enhanced   130 ff.
   – shifting   176
Raman spectroscopy   12 ff.
   – instrumentation   86 ff.
Raman-active   31, 57, 228, 229
Random coil   156 ff., 160, 166, 170, 175
Rapid flow   144 ff., 185, 189, 194
Rapid freeze–quench   145 ff., 252, 298
Rapid mixing techniques   141 ff., 173, 249
Ras protein   218
Ras-GAP protein   221
Rayleigh
   – limit   39 ff., 148
   – scattering   28 ff., 87, 102
Red paramagnetic protein   293
Reference electrode   114 ff., 121
Relaxation methods   146 ff., 262
Renhancement   160
Resonance Raman, see RR
Retinal
   – 11-*cis*   186 f.
   – ^{13}C isotopic labelling   192
   – chromophore   198
   – 13-*cis*   186, 207
   – 13-demethyl   199
   – 9-*cis*   186, 188
   – all-*trans*   186, 204
   – deuterated Schiff base   192
   – deuteration   190
   – electronic excited state   183
   – isomer   181
   – isomerisation   184, 195, 208, 214
   – oxime   189
   – proteins   181 ff., 184
   – protonated Schiff base   181, 186, 199, 201, 206
   – Schiff base   184, 207
Retinylidene Schiff base, see retinal Schiff base
Rhodopsin   183, 185
   – intermediates   185 ff.
   – IR   195 ff.
   – photoreaction   183
   – RR   185 ff.
Ribonuclease A   173, 174
Ribonuclease T$_1$ 174
Ribonucleotide reductase   297
   – ^{18}O/^{17}O/^{16}O isotopic shifts   298 ff.
   – RR   298 ff.
Right-angle scattering, see 90° scattering
Rotating cell, see rotating cuvette
Rotating cuvette   105, 116, 135, 138
Rotating electrode   122
RR   32, 132 ff.
   – A-term   33 ff., 58 ff., 229, 234
   – B-term   33 ff., 59, 229, 234
   – intensity   33 ff., 191, 192
Rubredoxin   291 ff.

s
SAM   128 ff., 255, 261
Sandwich cuvette   106 ff., 144, 195, 198, 201
Scaling factor   24 ff.
Scanning optical near-field microscopy   148
90° scattering   82, 83, 116 ff.
Scattering tensor   30 ff.
Scrambler   82 ff.
Secondary structure   155 ff., 160, 161 ff., 170, 176
Second-derivative   160
SEIRA   38, 118 ff., 125, 129, 267
Self-assembled monolayer, see SAM
Self-deconvolution   159, 172
SER, see SERS
SERR   129, 267
SERS   38, 104, 118 ff., 125, 129
   – colloids   40, 111 ff.
   – electrochemical cell   120
   – electrode   40, 119 ff.
β-sheet   156 ff., 161 ff., 166, 173, 175
Single-reflection element   109
Singular value decomposition   149, 150 ff.
   – basis set   160
Spectral resolution   71, 85 ff., 93, 96
Spectrograph   86 ff.
Spin-label   214
Stokes   28 ff.

Surface
- enhanced infrared absorption, *see* SEIRA
- enhanced Raman, *see* SERS
- enhanced resonance Raman, *see* SERR
- enhancement factor 40 ff., 44
- plasmon 39 ff.
- roughness 120

Symmetry 46, 55 ff.
- coordinates 30 ff.
- lowering 231 ff.
- normal coordinate 51
- normal mode 22
- wave-function 59

Symmetry operation 22, 46 ff.
- character 49
- class 50

t

Temperature jump 147, 174, 176
Time resolution 72, 76
Time-resolved 79 ff., 131 ff., 146
- IR 79 ff., 142, 144 ff.
- pump-probe 74, 76 ff., 83, 132, 133, 138, 140, 141, 241 ff.
- rapid–scan FTIR 72 f., 218
- RR 132 ff., 139, 142, 144 ff., 241 ff., 260, 275
- SEIRA 147, 265
- SERR 146 ff., 261
- step-scan ATR 216
- step-scan IR 74 ff.
- step scan FTIR 207, 212
- studies 74
- ultra-short 76, 140

Tip-enhanced Raman 149
Transform limit 81
Transform theory 35
Transition
- dipole moment 26, 31, 33 ff., 55, 58 ff., 171
- dipole moment coupling 162, 163 ff.
- probability 25 ff., 31 ff.

Transmembrane helix 169
β-turn 160, 163
γ-turn 160, 163
Two-dimensional IR, *see* 2D IR

u

Ultraviolet RR, *see* UV RR
Unfolding 157, 159, 166, 249
- step 176
- temperature-induced 172

UV RR 169 ff., 176

v

Vibrational analysis, *see* normal mode analysis
Vibrational coherence 167
Vibronic state 31, 34 ff., 101 ff.
Vibronic transition 31 ff.
Visual pigment 181, 183

w

Wave packet 36
Working electrode 114 ff., 121